T0127993

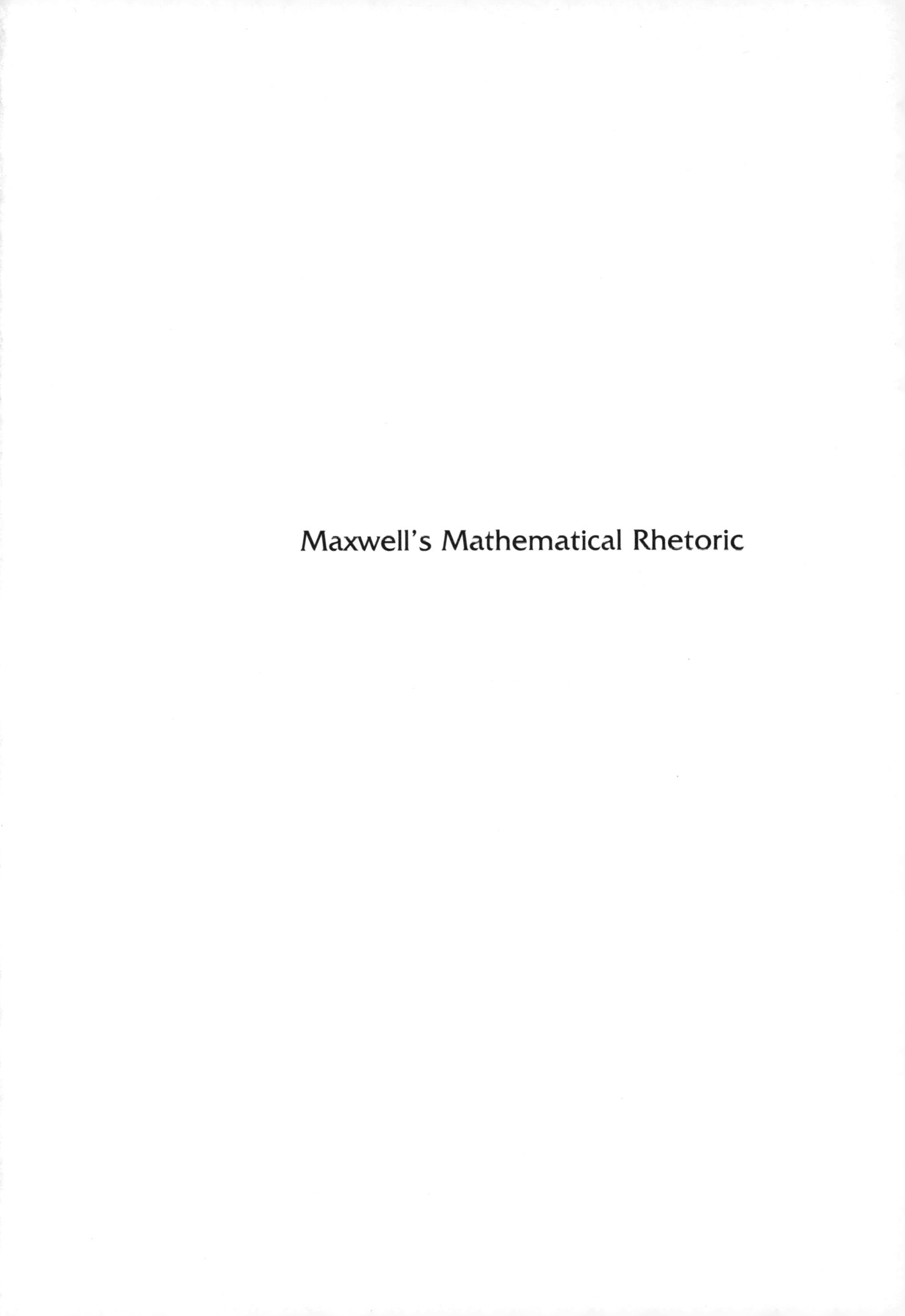

Maxwell's Mathematical Rhetoric

Maxwell's Mathematical Rhetoric

Rethinking the *Treatise on Electricity and Magnetism*

Thomas K. Simpson

Green Lion Press

Santa Fe, New Mexico

© Copyright 2010 by Green Lion Press

All rights reserved. No part of this publication may be reproduced, stored in a retrieval system, or transmitted, in any form or by any means, electronic, mechanical, photocopying, recording, or otherwise, without prior written permission of the publisher.

Manufactured in the United States of America

Published by Green Lion Press,
Santa Fe, New Mexico, USA

www.greenlion.com

Set in 11-point New Baskerville. Printed and bound by
Sheridan Books, Inc., Chelsea, Michigan.

Cover design by Eric Simpson. The cover illustration combines two images: that of the field, so central to Faraday's view of the physical world, and the portal of the Cavendish Laboratory. The conception and construction of this laboratory was the culminating achievement of Maxwell's career at Cambridge University. These images bracket the trajectory of the *Treatise* itself, which from the beginning directs the reader to the style of Faraday, and in the end brings us to the doorway of a new way of conceiving the natural world through carefully-shaped experiments like Faraday's own. They signify a new program in experimental natural philosophy, one that comes to birth within the pages of Maxwell's *Treatise.*

Cataloguing-in-Publication data:
Simpson, Thomas K.
Maxwell's mathematical rhetoric: Rethinking the treatise on electricity and magnetism / by Thomas K. Simpson

Includes index, introductions, bibliography, biographical notes.

ISBN 978-1-888009-36-1 (sewn softcover binding)

1. Maxwell, Treatise on Electricity and Magnetism. 2. History of Science. 3. Physics. 4. Electricity. 5. Magnetism.

I. James Clerk Maxwell (1831–1879). Simpson, Thomas K.. (1924–). III. Title

Library of Congress Control Number 2010925404

To the memory of

Richard T. Cox

1898–1991

Professor of Physics, The Johns Hopkins University, 1945–1965

Table of Contents

List of Illustrations

The Green Lion's Preface

Green Lion Press is pleased indeed to offer a new edition of *Maxwell's Mathematical Rhetoric*. After many years of circulating in typescript form, this remarkable study by Thomas K. Simpson has become something of an underground classic among a small but appreciative cadre of students and scholars. With the present (newly-titled) edition, we hope it will reach the wider audience it so deserves.

Yet, one might ask, can a forty-year-old book be at all relevant to the world of Maxwell studies today? Remarkably, in the decades that have elapsed since this study was first undertaken, little attention has been given to Simpson's principal focus, which is the deep literary structure of Maxwell's *Treatise*. While historians of science have begun to show increased interest in scientific rhetoric, their efforts have for the most part excluded mathematics. This practice may stem in part from an excessively narrow conception of rhetoric as being primarily forensic in character—ordained to persuasion. Simpson offers a distinctively contrasting voice, arguing that Maxwell's rhetoric is directed to illumination and intelligibility, and that it does not exist as an addition to mathematical reasoning in physics but as intrinsic to it.

The heart of the study is Simpson's careful reading of a crucial set of chapters in Part IV of the *Treatise*, chapters in which Maxwell develops his "dynamical theory of the electromagnetic field." By means of this investigation, the author proposes to inquire whether mathematical physics has a rhetoric, and if so, whether attention to the rhetoric will ultimately advance our understanding of the physics. We believe Simpson's effort will lead many readers to answer "yes" to both questions.

Simpson worked directly with Maxwell manuscripts at the University Library, Cambridge in 1968. He also had access to numerous microfilmed Maxwell materials. Today's researchers enjoy the inestimable benefit of P. M. Harmon's three-volume *The Scientific Letters and Papers of James Clerk Maxwell* (Cambridge University Press 1990, 1995, 2002). Accordingly, Simpson's references to the Cambridge materials have been revised to refer to the Harmon volumes. Although a few additions have been made to Simpson's bibliography, there has been no attempt to bring it fully up to date.

Readers of this volume will also be interested in Simpson's *Figures of Thought, A Literary Appreciation of Maxwell's Treatise on Electricity and*

Magnetism. Originally intended as the author's preface to the Green Lion edition of *Maxwell's Mathematical Rhetoric*, it grew to achieve book length in its own right and was published by Green Lion in 2005. *Figures of Thought* is an excellent companion to *Maxwell's Mathematical Rhetoric*.

Dana Densmore
William H. Donahue
for Green Lion Press

Editor's Note

In an attempt to locate the microfilmed Maxwell materials which Thomas Simpson worked with in the late 1960's, I received generous assistance from Adam J. Perkins, Curator of Scientific Manuscripts for the University Library, Cambridge. Regrettably, the films no longer exist. In keeping with University Library policy at the time, they were made available on a temporary basis, then destroyed after use. My sincere thanks go to Mr. Perkins, whose sad duty it was to confirm the fate of those films.

It is a pleasure to acknowledge the assistance of Katie Kelley, who worked with Simpson's manuscript at an early stage of production, and whose keen eye for usage and sentence construction contributed greatly to the editing process.

Howard J. Fisher
Editor for the Green Lion edition

Author's Preface to the Green Lion Edition (2010)

Perhaps the chief aspect of *Maxwell's Mathematical Rhetoric*, and the source of the importance of its other aspects, is its conception as a *literary* study of Maxwell's *Treatise*, that is, its assumption that the *Treatise* must be approached as a work of literature, whose contribution is not merely to "science" narrowly defined, but to the unfolding stream of human thought. In short, *Maxwell's Mathematical Rhetoric* represents an attempt at reading Maxwell's *Treatise* as a great book.

This literary reading reveals that Maxwell himself wrote with an expressed interest in questions of rhetoric—particularly of mathematical rhetoric. Maxwell writes with an evident intent to shape new mathematical forms, designed to lead the reader to new conceptual approaches to the natural world. What results is a strong reaffirmation in practice of the concept of natural philosophy, as distinct from merely technical scientific advance. This yields a number of important features.

First among them must be Maxwell's dramatic turn to the Lagrangian form in which to present the concept of the electromagnetic field and, in the same mode, to reveal the electromagnetic theory of light. Narrow readings of the *Treatise* have regarded this as a mere matter of convenience, or even whim, on Maxwell's part. Critics have been slow to catch the fundamental difference between a Newtonian and a Lagrangian world view. It is true that under proper conditions, Lagrange's equations can be derived from Newton's, and vice versa. But from the standpoint of Maxwell and the *Treatise*, this by no means renders them equivalent! Lagrange addresses a world in which the cosmos is whole and connected; Newton, a world of isolated particles acting upon one another over distances. Lagrange's equations appear in the *Treatise* as the culmination of a long series of rhetorical moves, including (among others) Green's theorem, Gauss's potential theory, and Faraday's lines of force—all of which have prepared the reader for the Lagrangian vision of a natural world that is whole and connected: a veritable sea-change from Newton's vision.

One indication that the *Treatise* is truly a literary work is our increasing awareness as we read that we are witnessing an unfolding portrait of Faraday as thinker, researcher, and unassuming human being. Maxwell's devotion to Faraday is undisguised and consistent, even to the formulation of the equations of Maxwell's final theory: Faraday's "exploring wire"[1] persists as a paradigm for *envisioning* relationships which the equations express in analytic terms. The new world that Lagrange describes requires investigators like Faraday, not like the "high mathematicians" of the universities. Despite Maxwell's gentle style, his *Treatise* must be read as an act of quiet defiance—a call to transform at once the concept and content of science, as well as the character of its practice and practitioners.

Curiously, scholars have failed to notice that the *Treatise* thus constitutes a watershed event, a dialectical moment in the history of science with wide-ranging implications for society as a whole. Maxwell is—perhaps a bit slyly—aware of such implications as he writes the *Treatise*. Ostensibly, he is writing the text that will implement a new Tripos component on electricity and magnetism at Cambridge, and the *Treatise* does achieve that important goal. But Maxwell quite evidently has also something else very much in mind, for in focusing on Faraday he is celebrating a scientist of a very unacademic sort, alien to the halls of Cambridge and its mathematical ways, intimately in touch with nature by way of experiment, and deeply averse to the lofty theory of Newton's *Principia*. Equally unorthodox is Maxwell's turning to Bacon and adopting that rhetorician's notion of "experiment of illustration."

Other projects express a similar spirit of reform. Maxwell will soon be engaged in reducing (perhaps we should say "elevating") the *Treatise* to its true elements in a new *Elementary Treatise on Electricity and Magnetism*. This, though never completed, is meant to take its place beside other elementary treatises, such as *Matter and Motion*, which he has written over the years to implement courses he has taught for working people. And he will soon be conniving with the Duke of Devonshire to build a new teaching laboratory (the Cavendish) where, over the opposition of the University faculty, he will implement within the hallowed halls of Cambridge a new program in *experimental* natural philosophy. Like other watershed events in the history of the sciences, the *Treatise* carries social and political implications reaching far beyond the sciences themselves, at least as they are commonly understood.

A certain moment in the *Treatise* deserves special attention, both as representing Maxwell's quiet yet potent literary style, and his own characterization of the watershed the *Treatise* is traversing. This is the chapter in which Ampère and his elaborate mathematical theory—intricately contrived to preserve the form and spirit of Newton—is contrasted with the modest approach of Faraday, who, innocent of such mathematical formalities, listens to nature and painstakingly follows her ways. Maxwell, who magically turns Ampère's theory into Faraday's in the course of the chapter, clearly delights in this contrast between the way of the aristocrat of science, and

1. See, for example, Faraday's *Experimental Researches in Electricity*, Twenty-eighth Series.

the ways of this uncommon common man, unmathematical, but keenly observant, who thinks in a different, far more effective, mode. It might not be altogether out of order to see in this move an ironic Scottish rebellion against the pretensions of British aristocracy; but its implications are not so limited. Maxwell himself is curiously positioned as at once agent and reporter of his revolution, living, as he literally was, in both camps.

All this is interesting as pertaining to England of the 19th century, but its interest is multiplied by its application to pressing issues of our own time. The text of *Maxwell's Mathematical Rhetoric* may be forgiven for not having anticipated developments which ensued forty years after its composition. In any case, the question of "wholeness" in physical theory is very much before us today, and our inability to deal effectively with whole systems has become a matter of urgent importance.

The problem is evident in many arenas of a now hugely interconnected world, but it will be sufficient here to speak of our increasing awareness of the interconnectedness of our environment, from local ecosystems to the atmosphere and biosphere. As a result, theorists today are making the same move Maxwell made in the *Treatise*, away from the Newtonian paradigm in which systems are understood as force-coupled assemblages of separate parts—still the commonplace of physicists and biologists alike—to the Lagrangian alternative in which the wholeness of a system becomes the starting point for thought and for an investigation characterized in terms of energy exchanges and material flows. Today, Lagrange's equations are regularly invoked to characterize such processes, often in conjunction with a first-order term representing the second law of thermodynamics. Strong light is thus focusing on the very dialectical turn that is the central theme of the *Treatise*, though generally with insufficient sense of its true meaning or importance. Might this not be an excellent time for us to turn to Maxwell for considered advice about the significance of the path on which we are embarking, and the importance of the decision now in our hands?

Author's Preface (1968)

The purpose of this study can be very simply stated: to read Maxwell's own account of his electromagnetic theory in his *Treatise on Electricity and Magnetism*, and to understand this account as he intended it. It will be an historical investigation in the sense of the larger purpose of history—to preserve or recover our understanding of the thought of the past. It is even possible at a remove of a century that we might achieve a more comprehensive understanding than was attainable at the time the *Treatise* was first read. But it is not easy to read Maxwell's book today. Maxwell wrote with constant concern for questions that are no longer regarded as valid, and even in this text on electricity and magnetism, he sought to achieve philosophical goals which modern readers do not share, and which only seem to render his account obscure. To read him in his own terms, it is necessary—even if only as an exercise of scholarship—to re-establish an understanding of the concepts and questions which were in his mind as he wrote.

Science is conceived as an inherently progressive discipline, and it is perhaps more difficult in this field than in others to justify the time and effort of spirit required to study old and outdated works. I believe, however, that in another sense the alternatives for thought represented by the classic statements of earlier theories remain ever-present. We know where we are, intellectually, only insofar as we remain clear about the alternatives which our present position implicitly rejects. More than this, I believe that the philosophical questions which ultimately motivated and shaped Maxwell's scientific inquiry are permanent, even though we reject the form in which they were put. We, too, desire to know what entities and powers constitute the natural world, and we too are groping for any possible relation between such human questions and the mathematical symbols which are all that our theories and experiments are able to give us. As I hope this study will make clear, Maxwell was neither naive in his approach to natural philosophy, nor disillusioned. I think it is quite possible that we may yet be able to learn from him something about the right way to question nature, and the form in which the answers might come.

This study is based on the hypothesis, perhaps a reflection of the above concern, that a serious work of science is *ipso facto* a literary work, and that a major scientific statement such as Maxwell's *Treatise* must be read from a literary, as well as a logical point of view, if its meaning and worth are to be grasped. The present study is, then, in a sense an experiment: to discover whether such an approach will in fact prove rewarding in the case of the

Treatise. The study will consist of a careful reading of one crucial section of the *Treatise*, that in which Maxwell unfolds his "dynamical theory of the electromagnetic field"—namely, Chapters I through IX of Part iv. To a very modest extent, it will be a *literary* criticism, in the sense that it will be concerned with the kind of questions one would ordinarily expect to ask, for example, about a novel: questions of meaning, of structure, and of style.

To borrow a term from a tradition which was alive for Maxwell though not in our own time, these are questions of "rhetoric," that art which concerns itself with style—the way things are said insofar as this can be distinguished from the abstract content of the argument. In part, the inquiry will help to determine whether mathematical physics has a "rhetoric," and if so, whether attention to the rhetoric will ultimately advance our understanding of the physics.

The "rhetoric" which I have in view does not concern mere questions of ornament or persuasiveness—it does not exist as an *addition* to the scientific statement but is intrinsic to it, as the very means by which the statement is made. One point of view would have it that in science, the matter alone counts, and that questions of manner are distractions, or at best concern a desirable but quite unnecessary elegance. This is, I think, what Hertz meant when he remarked, in desperation, in his attempt to understand Maxwell's meaning on the subject of electric charge: "Maxwell's theory is Maxwell's system of equations."[1]

I am encouraged in the alternative view both by the traditional concern of scientists for their own methods, and by Maxwell himself, who frequently revealed his preoccupation with questions of form and style. Such questions arise whenever we find that the same thing can be said in more than one way, and that the choice among these ways makes a difference. Consider, then, this statement from Maxwell's famous "Address to the Mathematical and Physical Sections of the British Association" (1870):

> As mathematicians, we perform certain mental operations on the symbols of number or of quantity, and, by proceeding step by step from more simple to more complex operations, we are enabled to *express the same thing in many different forms.* The equivalence of these different forms, though a necessary consequence of self-evident axioms, is not always, to our *minds*, self-evident; but the mathematician, who by long practice has acquired a familiarity with many of these forms, and has become expert in the processes which lead from one to another, can often transform a perplexing expression into another *which explains its meaning in more intelligible language.* (*SP* ii/216; emphasis added)[2]

1. Heinrich Hertz, *Electric Waves*, trans. D. B. Jones (London: 1893; reprinted New York: Dover, 1962), p. 21.

2. On this notation for references, see page xxiii, below.

In this passage Maxwell makes three points that suggest the role of rhetorical considerations—questions of form and style—in mathematical physics. First, mathematical expressions are subject to wide-ranging transformations. Second, though mathematically equal, these are not always equivalent to our minds—that is, they have different meanings for us. Third, mathematics is a language, and differences among mathematical expressions may affect their intelligibility to the scientist. In physics, it is not only the mathematical theory which can shift in this way: Maxwell continues, in the succeeding paragraph of the same lecture, to point out the transformations which a laboratory procedure can undergo in the hands of a skillful experimenter. Here, too, the form in which a phenomenon is presented can make a great difference in the meaning we draw from it. There is a rhetoric of experiment.

These are points about which Maxwell feels very strongly. Without embarking here on the discussion of a topic which will occupy us in Chapter 5, we may simply mention that Maxwell has a deep interest in questions of learning and meaning in science. He is alarmed to see, with the spectacular proliferation of mathematical physics in the middle of the nineteenth century, the imminent loss of meaning from the mathematical symbols. More than any one science, Maxwell wants to master the symbols themselves—to learn to use "physical analogy," "illustration," and the mathematical transformations themselves to wring meaning out of the seemingly abstract and empty symbols, and thus to preserve science for philosophy and the human understanding.

The *Treatise* is, I believe, the work in which he brings to bear all the arts be has developed for this purpose. The *Treatise* is frankly a work dedicated to a rhetorical purpose: Maxwell knows, and explains in his Preface, that there are two general ways, *mathematically equivalent*, in which electromagnetic theory can be presented. They predict the same phenomena; Maxwell acknowledges at the outset that even the most distinctive aspect of his theory, the electromagnetic theory of light, is predicted equally by the retarded potentials of Lorenz. But the difference between these "equivalent" theories is precisely the source of the personal energy with which Maxwell writes his *Treatise*; his purpose is to reveal the intellectual results of following one of these alternatives consistently:

> These physical hypotheses, however, are entirely alien from the way of looking at things which I adopt, and one object which I have in view is that some of those who wish to study electricity may, by reading this treatise, come to see that there is another way of treating the subject... (*Tr* i/*x*)[3]

3. On this notation for references, see page xxiii, below.

Throughout the *Treatise*, Maxwell throws his equations into that form which will, as he often says, "suggest the ideas" he wishes to present to the reader's mind. These are Faraday's ideas, and the *Treatise* can in my opinion best be understood as a strategic effort to transform the equations of electromagnetic theory into a new structure which conforms to Faraday's geometrical and physical vision. The principal thesis which I shall develop in this study is the assertion that the *Treatise* is a translation of Faraday's thought into the form of mathematical physics.

Maxwell distinguishes his purpose from that of one who treats all equivalent forms as equals, by saying in the Preface:

> I have therefore taken the part of an advocate rather than that of a judge, and have rather exemplified one method than attempted to give an impartial description of both. (*Tr* i/*xi*)[4]

Archbishop Whately made the same distinction at the outset of his *Elements of Rhetoric*, where I suspect Maxwell found it. For Whately, this is precisely the distinction between the work of the rhetorician and that of the logician; he contrasts his work on rhetoric with an earlier treatise on logic:

> I remarked in treating of logic, that Reasoning may be considered as applicable to two purposes, which I ventured to designate respectively by the terms "Inferring," and "Proving"; i.e., the ascertainment of the truth by investigation, and the establishment of it to the satisfaction of another: and I there remarked, that Bacon, in his Organon, has laid down the rules for the conduct of the former of these processes, and that the latter belongs to the province of Rhetoric: and it was added, that to infer is to be regarded as the proper office of the philosopher, or the Judge;—to prove, of the Advocate. It is not however to be understood that Philosophical works are to be excluded from the class to which Rhetorical rules are applicable; for the Philosopher who undertakes, by writing or speaking, to convey his notions to others, assumes, for the time being, the character of Advocate of the doctrines he maintains. The process of investigation must be supposed completed, and certain conclusions arrived at by that process, before he begins to impart his ideas to others in a treatise...[5]

In Whately's sense, then, I propose to study the *Treatise* as a work of rhetoric: both to learn the conclusions Maxwell had arrived at in his own investigations, and to understand better the methods he devised for conveying those ideas to his readers, both through physical illustrations, and through the equations of a new form of mathematical physics. To do this, I think, will be simply to attempt to read the *Treatise* responsibly as a work of scientific literature.

4. On this notation for references, see page xxiii, below.

5. Richard Whately, *Elements of Rhetoric: Comprising the substance of the Article in the Encyclopaedia Metropolitana* (Boston: 1844), pp. 5–6.

A few words should be added about the method and plan of this study. A critical study of the sort here proposed must first of all be a commentary—it must first enunciate the critic's understanding of the text itself. The very term "commentary" seems a bit dusty and suggests a plodding medieval piling of texts upon texts; and I am afraid I can offer very little to dispel any suspicion that this will be slow going. I must ask the reader in effect to join me in this study of the *Treatise*; and in order to avoid excessive quotations, I shall assume that he has a copy of the *Treatise*, and sometimes Faraday's *Experimental Researches in Electricity* or William Thomson's *Reprint of Papers on Electrostatics and Magnetism* open before him. Insofar as it is successful, the fruit of such commentary will be an unveiling of significance in the text, which would not otherwise be immediately apparent.

The study is divided into two parts. The first deals with those chapters of Part iv of the *Treatise* which form a kind of entrance-way to the dynamical theory; this *introit* is through Faraday's thoughts on electromagnetism, and it amounts to Maxwell's translation of Faraday's theories on a first level. Since I believe that the *Treatise* as a whole is best understood as a translation of Faraday, I devote sections of Part I successively to my own review of Faraday's thought, to the concept of "translation" in this context, and to the work of other translators who preceded Maxwell—especially William Thomson—in order to make clear the nature of the problem which Maxwell undertook to solve. Chapter 3 of the first part of this study is a commentary on Chapters I–IV of Part iv of the *Treatise*.

The second part of this study deals with those chapters of the *Treatise* in which the dynamical theory itself is developed. I regard this as the translation of Faraday at the second level, which Maxwell believes to be the true formulation of the ideas toward which Faraday was moving. This section requires an excursion, in Chapter 5, into questions of the nature of science as Maxwell understood it; and I find help here in the lectures of a teacher of Maxwell's, William Hamilton of Edinburgh.

What I think emerges is that the *Treatise*, at least in the section studied here, is a work in which questions of structure and manner are of the highest importance, and which relate immediately to the very purpose of the book, and its meaning. Maxwell is not interested in presenting a set of equations which merely predict the appearances economically. Rather, he has very carefully shaped the equations and ordered the argument into a form which he believes is an appropriate image of the truth as Faraday had already glimpsed it.

In writing a study in the history of mathematical physics, one is immediately confronted with problems of symbols and notation. In this case, I think very little is lost, and a very great deal is gained, if equations are thrown into the notation of the vector calculus. This is by no means out of the spirit of Maxwell's work, as he enthusiastically contributed to the development and popularization of this notation. He used it in the *Treatise* to reformulate his results in order to make them as clear as possible, although he did not feel that he or his readers were ready to put the entire argument in this form.[6]

For consistency and for the convenience of the modern reader, I have at some points substituted symbols which are conventional today for those used by Maxwell or others. I employ the notation

$$\varphi \leftrightarrow \Phi$$

to mean that I have substituted the symbol φ for Φ in the original text; and similarly

$$\mathbf{B} \leftrightarrow a, b, c$$

to denote the substitution of the vector **B** where the original text wrote the components a, b, c separately. I hope that this device will make it easy to move back and forth between the commentary and the original text.

In places in which Maxwell does use quaternion notation, it is necessary for the modern reader to keep in mind that the scalar part of the quaternion product was the *negative* of the modern scalar product of two vectors. Thus, if **a** and **b** are quaternions, Maxwell writes S·(ab) to denote the scalar part of their product; and if **a** and **b** in turn denote the vector parts of the same two quaternions, we have the following relationship between our notation and Maxwell's:

$$-(\mathbf{a} \cdot \mathbf{b}) \leftrightarrow \mathrm{S} \cdot (\mathrm{ab}).$$

Maxwell defines the nabla operator ∇ as we do, but because of the difference in the scalar product, the sign of the Laplacian operator is reversed. Maxwell therefore writes (*Tr* i/31):

$$\nabla^2 = -\left(\frac{\partial^2}{\partial x^2} + \frac{\partial^2}{\partial y^2} + \frac{\partial^2}{\partial z^2}\right).$$

He himself disliked the fact that the scalar product ("the square") in quaternions was negative, but the system was not of his own inventing, and he tried to follow Tait's prescriptions.[7]

6. Maxwell wrote to his friend Peter Guthrie Tait, who was promoting the use of quaternions, "The like of you may write everything ... in pure 4ions, but in the transition period the bilingual method may help to introduce and explain the more perfect." Maxwell regarded his text on electricity as a "bilingual treatise" and asked, "May one plough with an ox and an ass together?" (Cargill G. Knott, *Life and Scientific Work of Peter Guthrie Tait* (Cambridge: Cambridge University Press, 1911, p.151.) See the interesting fragment, "On the Application of the Ideas of the Calculus of Quaternions to Electromagnetic Phenomena" (H.347).

7. *Loc. cit.* Earlier, Maxwell pointed out that William Thomson was getting the sign of ∇^2 wrong (*ibid.*, p. 150).

For a few works which are cited very often, I have used the following short-form notations:[8]

(*Tr* i/100): James Clerk Maxwell, *Treatise on Electricity and Magnetism* (3rd ed., 2 vols; Oxford: 1892), volume i, page 100.

(*SP* i/100): W. D. Niven, ed., *The Scientific Papers of James Clerk Maxwell* (2 vols.; Cambridge: 1890), volume i, page 100.

(H.100): P. M. Harman, ed., *The Scientific Letters and Papers of James Clerk Maxwell* (3 vols.; Cambridge: Cambridge University Press 1990, 1995, 2002), item 100.

(*XR* i/100): Michael Faraday, *Experimental Researches in Electricity* (3 vols.; London: 1839–1855), volume i, page 100.

(*L*: 100): Lewis Campbell and William Garnett, *Life of James Clerk Maxwell* (revised edition; London: 1884), page 100.

(*EM*: 100): William Thomson, *Reprint of Papers on Electrostatics and Magnetism* (2nd ed.; London: 1884), page 100.

Maxwell's *Treatise* and *Scientific Papers* are currently (2008) available in Dover Publications reprints, which preserve the original editorial material and pagination. Faraday's *Experimental Researches in Electricity* is available in a 3-volume facsimile reprint from Green Lion Press.

In general, I have tried to refer to editions which are in print in preference to earlier editions, unless the prior version is important to the substance of my remarks; and I have referred to collected works rather than to the scattered original papers. I have frequently referred the reader to translations, as being in general more accessible rather than original works, although I have felt free to leave quotations in the text in the original French or German when I felt the original wording was of interest.

I wish to express my gratitude to Dr. Robert Kargon and to the late Dr. Harry Woolf of the Johns Hopkins History of Science Department, both of whom most kindly encouraged my work on this project. I am grateful too to Mr. A. E. B. Owen, then Keeper of Manuscripts in the University Library, Cambridge, for making the Maxwell manuscript collection available on film to the Johns Hopkins University in 1967 and 1968, where it was available to me for this study.

Except for the figures reproduced from Maxwell's *Treatise*, illustrations are based on original drawings generously executed by Marcia Simpson, who thereby added a touch of interest to an otherwise unbroken expanse of text. Finally, I cannot sufficiently express my indebtedness to the late Dr. Richard T. Cox, who helped immensely at every stage of preparation of

8. For more complete citations of these works, see the Bibliography.

this work, and with whom I enjoyed a long and delightful series of conversations on problems and possibilities suggested by its argument.

Thomas K. Simpson

PART ONE

The Account of Faraday's Theory in the *Treatise*

Chapter 1.
Faraday's Thought on Electromagnetism

In an extraordinary way, Maxwell's *Treatise on Electricity and Magnetism* presupposes that the reader is already familiar with another book, Faraday's *Experimental Researches in Electricity*. One can indeed master the *Treatise* without having read Faraday, but such an approach cannot lead to the understanding which Maxwell deeply desires to share with the reader. Many works refer repeatedly to a single principal source, but there must be few which *command* their readers as Maxwell does in a footnote to Chapter III of Part iv: "Read Faraday's Experimental Researches, Series i and ii" (*Tr* ii/178). As I have asserted earlier, Maxwell has designed his *Treatise* as a faithful translation of Faraday's *Researches*; and throughout, Faraday is the mainspring of the work and the guide to its development. But Maxwell not only draws upon Faraday as his source, in a curious way he makes this source the object of his work as well. We are directed back to Faraday, to read with renewed appreciation and understanding:

> If by anything I have here written I may assist any student in understanding Faraday's modes of thought and expression, I shall regard it as the accomplishment of one of my principal aims—to communicate to others the same delight which I have found myself in reading Faraday's *Researches*. (*Tr* i/xi)

It is therefore essential that the present study of Maxwell's *Treatise* should begin with some consideration of Faraday's *Experimental Researches*.

No summary can convey either the content or the character of Faraday's writings. As perhaps the ultimate example of their genre, the literature of scientific inquiry organized episodically in linked series, they are a texture of speculation and intelligent experiment so interwoven that it is virtually impossible to isolate significant sections from their context of detail. The point of the work is in effect this very mass of theme and variation. What I hope to do in this chapter, rather, is first to discuss the nature of science as Faraday conceives and practices it, and then to identify certain principal insights which Faraday tended to bring to bear on electromagnetism in his later work. These last constitute the primary material with which Maxwell works in Part iv of the *Treatise*. Since it appears to me that these insights, however powerful, by no means constitute a theory, I have titled the present chapter simply "Faraday's Thought on Electromagnetism," and not "Faraday's Theory." As we shall see, one way of looking at the "translation" of Faraday's ideas which Maxwell accomplishes is exactly as the transformation from the mode of empirical inquiry in which the *Researches* are cast, to that of theory, demonstrated in a systematic treatise.

THE NATURE OF SCIENCE FOR FARADAY

Faraday developed two concepts which became increasingly significant in his thought about magnetic and electromagnetic effects. One was of course his concept of "lines of force." He was impressed by this notion early when he spoke of the lines merely as representations, and called them, as others did, the "magnetic curves."[1] Toward the end of his work, they bore nearly the whole burden of his thought on these topics, and he speculated increasingly about their possible "physical existence." The other fundamental concept was that of the *electrotonic state.* This is now less familiar to students of science, but for Faraday it was both significant and deeply troubling, and he returned to it, as Maxwell points out, again and again from its first introduction in connection with the induction experiments of Series I in 1831.[2] Essentially, Faraday was convinced that the surge of current which he observed in the secondary circuit in electromagnetic induction could not occur unless the secondary conductor had been initially in a state of electric tension, which was relaxed when the primary circuit was broken. The term "electrotonic" was manufactured to express this electric tension.[3] As we shall see, Maxwell's translation of Faraday's thought on electromagnetism turns about these two notions, the "lines of force" and the "electrotonic state."

For Maxwell, then, these will become fundamental concepts in a connected theory. But what sort of use does Faraday himself make of them? I think it becomes clear on a reading of the *Experimental Researches* that Faraday does not use them as elements of a scientific theory in any formal sense. Even if we were not to ask for a connected or complete theory, doubt arises whether concepts in Faraday's hands even tend toward inclusion in a provisional theoretical structure. The question, indeed, becomes more fundamental: does science, for Faraday, grow toward a theoretical shape, or does it develop toward something else altogether, not theory, but an account of a different kind?

We might attempt to answer the question by putting Faraday's account to certain tests, asking it to do the kinds of things theories are expected to do: to solve "problems," to "predict" phenomena on the basis of certain given

1. See pages 53 ff., below.

2. See page 14 below.

3. H. Bence Jones, *The Life and Letters of Faraday* (2 vols.; Philadelphia: 1870), *2*, p. 9.

conditions, or to yield general "theories" through logical argument from assumed premises. But these functions, though they are surely among the proper tasks of an effective scientific theory, are precisely those which Faraday does not undertake.

Faraday was almost totally uneducated in mathematics.[4] It is difficult to grasp the significance of this fact for science. However brilliantly he succeeded in self-education in other areas, he apparently never felt it necessary to acquire the mathematics he had missed. The result is that he does not have before him with any vividness that universal paradigm of a reasoned deductive system, toward which virtually all competent physics over the ages has tried to shape itself. In particular, Faraday apparently never studied Euclid; he has no working notion of a system of axioms and postulates, or of reasoning leading with logical rigor to universal theorems.

There is at least the possibility that this innocence of mathematics represents not merely accidental ignorance, but a deliberate rejection. We are blessed with the record of a remarkable exchange of correspondence between Faraday and André-Marie Ampère, the French philosopher and mathematical physicist, whose work on electrodynamics Duhem once called "a theory which dispenses with the Frenchman's need to envy the Englishman's pride in the glory of Newton."[5] Here Faraday confronts his opposite, a brilliant mathematical theorist, and in the course of the correspondence Faraday finds occasion to describe his view of his own role. We shall return to this in greater detail below, but note this reaction of Faraday's when confronted with a mathematical argument of Ampère's:

> I regret that my deficiency in mathematical knowledge makes me dull in comprehending these subjects. I am *naturally skeptical in the matter of theories* and therefore you must not be angry with me for not admitting the one you have advanced immediately...[6]

This skepticism of theory, I believe, turns him away from mathematics almost on principle. He tends to regard mathematicians as operating on a height, while his own work, as experimentalist, is close to nature, and to fact. It would be a mistake, I believe, to overlook the element of pride which mixes with humility in his descriptions of his more modest work. In a moment of triumph following upon the discovery of electromagnetic induction and the explanation it afforded of the Arago disk phenomenon, he wrote to his friend Richard Phillips:

4. L. Pearce Williams, Michael Faraday (New York: Basic Books, 1965), p. 7.

5. Pierre Duhem, *The Aim and Structure of Physical Theory*, trans. Philip Wiener (New York: Atheneum, 1962), p. 127.

6. Faraday to Ampère, Sept. 3, 1822. L. de Launay, *Correspondance du Grand Ampère* (3 vols; Paris: Gauthier-Villiars, 1936–1943), 3, p. 911. Emphasis added.

> It is quite comfortable to me to find that experiment need not quail before mathematics, but is quite competent to rival it in discovery; and I am amazed to find that what the high mathematicians have announced as the *essential condition* to the rotation ... has so little foundation...[7]

There is a suggestion of a moral note in Faraday's rejection of theory, as will become clearer when we look at his characterizations of the true form which science should take. In relation to mathematics, this passage from the work which Faraday early adopted as his pocket guide, Isaac Watts' *Improvement of the Mind*, may be significant:

> But a penetration into the abstruse difficulties and depths of modern algebra and fluxions, the various methods of quadratures ... and twenty other things that some modern mathematicians deal in, are not worth the labour of those who design either of the three learned professions... This is the sentence of a considerable man ... who was a very good proficient writer on these subjects; he affirms, that they are but barren and airy studies for a man entirely to live upon... He adds further, concerning the launching into the depth of these studies, that they are apt to beget a secret and refined pride, and over-weening and over-bearing vanity, the most opposite temper to the true spirit of the gospel. This tempts them to presume on a kind of omniscience in respect to their fellow-creatures, who have not risen to their elevation...[8]

Faraday remained, in a phrase he used without, I think, any hint of apology, an "unmathematical philosopher."[9]

Maxwell grew up solving problems; as a child, he carried them with him for spare moments, calling them his "props"; as a man, he filled his *Treatise* with them. By contrast, for all we know, no teacher may ever have set Faraday a problem, ever assigned him certain "givens" and required him to find the corresponding "unknowns." Naturally enough, then, he does not think of his electromagnetic notions as serving this purpose and a reader cannot make much of the *Experimental Researches* if he comes to them with this criterion in mind. It is not only that Faraday cannot perform the operations of algebra or geometry, that he can make only a first approximation to a quantitative argument of any complexity. It is not only that he uses spatial images and models where others would work with number and symbol. It is not that he uses brief arguments where others would construct theorems. To a degree, these may all be true of Faraday, but as I express them

7. Bence Jones, *op. cit.*, *2*, p. 10. Emphasis added.

8. Isaac Watts, *The Improvement of the Mind* (Washington, DC: 1813), p. 213.

9. (*XR* iii/528). He is speaking of the reader for whom Euler composed his *Letters on Different Subjects* (see page 52, below).

they suggest defects; in Faraday's case, they must be construed as aspects of his special strength. His steps are not weak and faltering; his works are not those of a man who hesitantly approaches a language he has not properly learned. Faraday (for all the humility which has been attributed to him) steps forth proudly, and while he is generally cautious and of course always modest in his public claims, he speaks with a voice as confident and firm as that of any scientist who knows the strengths of his own method. He is *speaking a different language* from that of the theoretical scientist. He is not stepping hesitantly toward a mathematical physics; he is marching confidently along a different road. He has indeed no notion of theory as a goal of science.

In retrospect, this is a fantastic situation. In the land of Newton, at a time when mathematical physics was again flourishing, some of the most creative scientific work of the century in a new and more difficult area of physics was done over a long period of years by a man who had no notion of the *Principia* and did not share either Newton's methods or his goals. For Faraday, Newton's classic triumph—the triumph which had polarized the intellectual life of Europe for a century—was quite meaningless.

How can the historian of science understand the success of such a deep and total disruption of any reasonable continuity in the scientific enterprise? Perhaps the ultimate challenge for the present study is to lay the groundwork for finding the significance of the Faraday-Maxwell episode in the development of mathematical physics. Physics seems indeed to have profited immensely from this instance of amnesia in its work, but it is well to keep in mind that the fact that this one course was followed does not mean that it was in principle a necessary "stage" of development, or that it was ultimately for the "best." There is a sense in which the serious historian must always weigh alternative histories. On the Continent there were of course a number of highly competent scientists at work on the topics which concerned Faraday, both mathematicians and experimentalists, and I think one may reasonably suppose that most of Faraday's discoveries would before long have been made by others whose thought was in the tradition stemming from Newton. The great Newtonian counterpart of Faraday was Ampère, but there were many others: Weber, the Neumanns (father and son), Kirchoff, Lorenz, Helmholtz, Boltzmann, Hertz. These were at once capable mathematicians and skillful researchers; had electromagnetic theory developed in the hands of such workers without the peculiar turn of thought given it by Faraday and Maxwell, it is quite conceivable that physics might have passed into the modern era, more or less on schedule, as a more purely mathematical discipline, and without the digression marked by the "field" concept. According to a number of critics, the field concept was never systematically necessary, and has been the source of much confusion, beginning with the *Treatise* itself.[10] Faraday's unmathematical physics, therefore, developed and interpreted as it was by Maxwell, may have diverted science from a more rational course of development which would have made little or no use of the concept of the "field," but would have left none of the phenomena unaccounted for.

What then is Faraday's concept of science? It is evident that he is not *merely* an experimentalist, cleverly providing data for theoreticians to work up into theories. On the contrary, Faraday had one of the most fertile and insistent of speculative minds; in a certain sense, he was constantly producing new hypotheses and his mind was constantly reasoning from them. The result of this, reported in the thousands of paragraphs of the *Experimental Researches* and the still more numerous paragraphs of the *Diary* is not theory, but a vast weaving and unweaving of powers, a process of discovery and identification, a great, highly unified formulary for the production and classification of effects. Faraday, as Tyndall proclaimed and all the world agreed, is the great "discoverer"; the paradigm for Faraday is Odysseus rather than Euclid: he travels from land to land, reporting wonders, guided by legend and myth, rumor or divine love. For Odysseus, the dominant desire is to see men's cities and to know their minds, and to gather all this together in the return to Ithaca. For Faraday, it is to investigate all the powers of nature, and to unveil them as essentially one, in the lecture hall on Albemarle Street. If Faraday did not learn this from Homer, he was moved to it by Dr. Watts:

> Let the hope of new discoveries, as well as the satisfaction and pleasure of known truths, animate your daily industry. Do not think learning in general is arrived at its perfection, or that the knowledge of any particular subject in any science cannot be improved, merely because it has lain five hundred or a thousand years without improvement. The present age, by the blessing of God on the ingenuity and diligence of men, has brought to light such truths in natural philosophy, and such discoveries in the heavens and the earth, as seemed to be beyond the reach of man. ... Do not hover always on the surface of things, nor take up suddenly with mere appearances; but penetrate into the depth of matters.[11]

10. The classic criticism of Maxwell's theory is Pierre Duhem, *Les Théories Electriques de J. Clerk Maxwell* (Paris: Hermann, 1902). Outstanding among the more recent critics is O'Rahilly, who recommends a direct particle-particle interaction equation in the tradition of Wilhelm Weber. Alfred O'Rahilly, *Electromagnetics: a Discussion of Fundamentals* (1935), reprinted as *Electromagnetic Fundamentals* (2 vols.; New York: Dover Publications, 1965), especially pp. 645 ff. See also J. A. Wheeler and R. P. Feynman, "Classical Electrodynamics in Terms of Direct Interparticle Action," *Reviews of Modern Physics, 21* (1949) pp. 425 ff., and articles by Parry Moon and D. E. Spencer, among them "A New Electrodynamics," *Journal of the Franklin Institute, 257* (1954) pp. 369 ff., and "Electromagnetism without Magnetism," *Amer. J. Physics, 22* (1954), pp. 120 ff.

11. Watts, *op. cit.*, p. 24.

Here is Faraday's own characterization of his hope for his chosen science of electricity, in a passage which opens the famous Series XI on induction:

> The science of electricity is in that state in which every part of it requires experimental investigation; not merely for the discovery of new effects, but what is just now of far more importance, the development of the means by which the old effects are produced, and the consequent more accurate determination of the first principles of action of the most extraordinary and universal power in nature:—and to those philosophers who pursue the inquiry zealously yet cautiously, combining experiment with analogy, suspicious of their preconceived notions, paying more respect to a fact than a theory, not too hasty to generalize, and above all things, willing at every step to cross-examine their own opinions, both by reasoning and experiment, no branch of knowledge can afford so fine and ready a field for discovery as this. Such is most abundantly shown to be the case by the progress which electricity has made in the last thirty years: Chemistry and Magnetism have successively acknowledged its overruling influence; and it is probable that every effect depending upon the powers of inorganic matter, and perhaps most of those related to vegetable and animal life, will ultimately be found subordinate to it. (*XR* i/360)

I think I do not need to apologize for including this lengthy quotation entire; in one paragraph Faraday has given both his vision of nature, and the method of science proportioned to it. He describes an impasse in the science of electricity—one in which phenomena are known in plenty, but first principles are lacking. To another man, this would look like a situation which called for a theory; but for Faraday, this is precisely the situation which calls for "experimental investigation"; it is a "fine and ready ... *field for discovery.*" What is to be sought? Not axioms or laws, but the principles of action of a *power*—the most universal in nature, which moves all of the inorganic world and perhaps for the most part the plants and animals as well. This is not something to be resolved with paper and pencil, but by facts: we see that for Faraday, the causes, the "first principles," are not laws but themselves *facts*, to be unveiled to observation in the laboratory. We should remember that it was given to Odysseus to talk with Athena face to face.[12]

This is not theoretical physics; as has been suggested, it is essentially chemistry, not in the modern sense of Lavoisier, but as a science of powers, in the tradition of von Helmont and Stahl.[13] We might best understand

12. It is impossible not to point out the striking appropriateness of the epithet which Homer regularly assigns to Odysseus: in his laboratory, Faraday is indeed the man "of many devices," πολυμήχανος.

13. "In a very real sense, all his discoveries were chemical if chemistry be defined (as it was by Faraday) as the science of the powers of matter." L. P. Williams, in Lancelot Whyte, ed., *Roger Joseph Boscovich* (London: Allen & Unwin, 1961), p. 163.

Faraday as the disciple of Davy, and the *Researches* as the evolution of a coherent *chemistry* of electromagnetism. The translation of Faraday which Maxwell is to accomplish in the *Treatise* must be, among other things, the transformation of this chemistry into the form of a mathematical theory worthy of admission to the halls of Trinity. In this, it must be like the historic transformation of Greek thought from the mode of Homer to that of Euclid.

The *Experimental Researches* is dense with questions—Faraday's method is that of unremitting inquiry. The very notion of a "series" of researches is, in a sense, that of a chain of linked questions and answers. Before we bring our own questions to bear on Faraday's work, let us listen briefly to the questions Faraday asks of himself.

The underlying question for Faraday is always the same: what really exists in nature? The practical form which this takes is that of the *test*: what will happen if I do this? Can I produce the phenomenon, the visible or tangible evidence, which will be the sure symptom of the existence of this or that suspected power or state? Think, for example, of the discovery of the diamagnetic force.[14] Faraday had first sought to reveal the state of strain in a dielectric, to which the curved lines of electric action were already a clue; using polarized light, he hoped to make manifest the existence of the hypothesized lines of action of contiguous particles. This failing, he asked the analogous question for the curved lines of action of the magnetic force: do lines of strain really exist in a diamagnetic medium? When this succeeded, and the plane of polarized light was rotated on passage through his "heavy glass" in a strong magnetic field, he announced that he had

> at last succeeded in *magnetizing and electrifying a ray of light, and in illuminating a magnetic line of force.* (*XR* iii/2, emphasis added)

By this, he says in a note, he

> intended to express that the line of magnetic force was illuminated as the earth is illuminated by the sun, or the spider's web illuminated by the astronomer's lamp. Employing a ray of light, we can tell *by the eye*, the direction of the magnetic lines through a body. (*loc. cit.*)

This I believe is a paradigm of Faraday's concept of science: to make manifest to the eye what is suspected to exist in nature. In one of his last writings about the magnetic lines of force, he carefully drew a distinction between the limited powers of mathematical physics as a mode of *representing* the forces of nature, and his own search for "the one true physical signification" of the phenomena:

14. *XR*, Series XIX and XX. Faraday's work on this topic is reviewed extensively in Williams, *Michael Faraday*, pp. 381 ff.

> Indeed, what we really want, is not a variety of different methods of representing the forces, but the one true physical signification of that which is rendered apparent to us by the phænomena, and the laws governing them. ... [S]upposing that ... mathematical considerations cannot at present decide which of the three views [of magnetism] is either above or inferior to its co-rivals; it surely becomes necessary that physical reasoning should be brought to bear upon the subject as largely as possible. For if there be such physical lines of magnetic force as correspond (in having a real existence) to the rays of light, it does not seem so very impossible for experiment to touch them; and it must be very important to obtain an answer to the inquiry respecting their existence, especially as the answer is likely enough to be in the affirmative. (*XR* iii/531)

An hypothesis or a theory is nothing more than an unresolved suspicion; a part, as Tyndall suggested, of the scaffolding, not of the edifice of science itself, which moves on to deal with existences.

He then demands of himself, what it is which has been revealed by the new phenomena. It is a condition of tension, and therefore of force, because it relaxes as soon as the magnetic induction is removed; but is it the same as the magnetic force, or different? He tests, by determining whether the diamagnetic body responds to a magnet; not observing any motion, he concludes that:

> [T]he molecular condition of these bodies, when in the state described, must be specifically distinct from that of magnetized iron, or other such matter, and must be *a new magnetic condition*; ... the force which the matter in this state possesses and its mode of action, must be to us a *new magnetic force* or *mode of action* of matter. (*XR* iii/21)

In other words, having made a force manifest, Faraday proceeds to identify it, by asking whether it is the same as, or specifically different from, previously known forces.

The same line of inquiry, of course, in time revealed that indeed the diamagnetic material does move under the action of a magnet, but moves in a way specifically distinct from the notion of a magnetic material. The existence of the diamagnetic force, first revealed only optically, is now revealed by a second token, a specific type of motion. This motion is summarized in a law:

> All the phænomena resolve themselves into this, that a portion of such matter, when under magnetic action, tends to move from stronger to weaker places or points of force. (*XR* iii/69)

There follows a brilliant experimental inquiry as to the universality of the new effect, concluding with the generalization:

> All matter appears to be subject to the magnetic force as universally as it is to the gravitating, the electric and the chemical or cohesive

> forces; for that which is not affected in the manner of ordinary magnetic action, is affected in the manner I have now described... (*XR* iii/70)

To reveal, to identify, and to generalize—the greatest part of the stream of Faraday's working questions serves these three ends. He frequently reasons by analogy, but for Faraday an analogy functions most often as a tentative identity, drawing him on (as in the case of the diamagnetic and magneto-optic forces) to decide whether the two analogous powers are finally the same or different.

Admittedly, Faraday is never content to rest with an unexplained phenomenon, and he moves on from a law such as that of diamagnetic action above to ask why the action occurs. Again, this might seem to be the step into theory, which I have denied Faraday takes. Indeed, he begins a paragraph shortly after with the words, "Theoretically, an explanation of the diamagnetic bodies ... might be offered...," and proposes an account of diamagnetism in terms of induced polarity, a theory which was, as we shall see, beautifully successful in the hands of Weber, though Faraday himself soon disowned it. But a theory for Faraday is merely a makeshift explanation, a temporary and unsatisfactory stage of science, which is dispelled as the science progresses. The explanation of diamagnetism which proved more fruitful in Faraday's hands, and which accounted as well for magnetism and magnecrystallic action (the pointing of crystals in the magnetic field), is a good example. This is the theory of the *conduction* of lines of force:

> I cannot resist throwing forth another view of these phænomena which may possibly be the true one. The lines of magnetic force may perhaps be assumed as in some degree resembling the rays of light, heat, &c.; and may find difficulty in passing through bodies, and so be affected by them, as light is affected. ... [T]he position which the crystal takes ... may be the position of no, or of least resistance; and therefore position of rest and stable equilibrium. (*XR* iii/122–23)

I submit that the decisive words in this paragraph, for Faraday, are these: "which may possibly be the true one." Even if this explanation accounted for all the known phenomena, I do not believe Faraday would have rested for a moment until he had found out the lines of force, which here take on new importance, and had given tangible and visible evidence of their presumed passage through bodies. Only thus would the "truth" of the view be exhibited to his satisfaction.

His renewed efforts to reveal the "physical lines" of force are well known. They culminate in papers of 1852 and 1855, published, significantly I think, in the *Philosophical Magazine* rather than the *Philosophic Transactions*.[15] Here the concern is openly for the real ("physical") existence of the lines,

15. *XR* iii/437, 528.

and these papers are therefore "speculative" in a way in which the disciplined researches were not; but I think it is apparent that the whole thrust of the *Experimental Researches* has been toward this end: to discover the powers of nature, to find out their true characters, and, finally, to produce the guarantees of their "physical" existence. In an apology for speculation, which prefaces the 1852 paper but which really speaks for the role speculation has played throughout the *Researches*, Faraday says:

> It is not to be supposed for a moment that speculations of this kind are useless, or necessarily hurtful, in natural philosophy. They should ever be held as doubtful, and liable to error and to change; but they are wonderful aids in the hands of the experimentalist and mathematician. For not only are they useful in rendering the vague idea more clear for the time, giving it something like a definite shape, that it may be submitted to experiment and calculation; but they *lead on*, by deduction and correction, to the discovery of new phænomena, and so cause an increase and advance of *real physical truth*, which, unlike the hypothesis that led to it, becomes *fundamental knowledge not subject to change.* (*XR* iii/408; emphasis added)

Speculation is a thread which "leads on" toward the goal of science; the path is through "discovery of new phenomena," and the terminus is not a completed theory, but "real *physical* truth." What then is "fundamental knowledge not subject to change"? As *immune to change,* it must be manifest in the phenomena; as *fundamental,* it must consist in those select phenomena which directly reveal the primary, universal powers. Such would be phenomena which made manifest the "physical lines" of magnetic force. In his late, unpublished researches on "Time in Magnetism," Faraday was hard at the effort to capture such primary evidence.[16]

In his valuable biography of Faraday, L. Pearce Williams makes a striking observation about the law known as "Faraday's," relating the current induced in a circuit to the change of flux through it. Faraday certainly states the law:

> They also prove, generally, that the quantity of electricity thrown into a current is directly as the amount of curves intersected. (*XR* iii/346)

Williams points out, however, that in its context in a search for sure evidence of the lines of force, this law "was not directed at electricity at all":

> Faraday was trying to prove that there was a certain specific amount of 'power' associated with every magnet; the induced currents merely detected this power.[17]

16. *Faraday's Diary,* ed. Thomas Martin (7 vols.; London: G. Bell & Sons, 1932–1936), *6,* pp. 434–444; *7,* pp. 255–333. See T. K. Simpson, *Isis,* 57, (1966) pp. 423–425.

17. Williams, *op. cit.,* p. 463 n. 51.

Williams adds that Faraday was interested in the induced electricity in Series I, though in the passages cited there Faraday was not stating the quantitative "law." I think this distinction is valid, and significant for an understanding of Faraday's purposes. Whereas the world has taken from Faraday a quantitative law relating motion in a magnetic field to induced current, Faraday himself had his eye on the problem of detecting the "sphondyloid of power" about a magnet; he was seeking out an entity in nature, and was using the moving wire with its law of action merely as a highly prized instrument in the search.

This is the physics of the explorer, the discoverer. Throughout the *Researches* Faraday sought what he called "contiguity" in nature; understandably, he seeks the same contiguity in the *account* of nature. A work of science should record a completed exploration, a detailed mapping, without gaps, of contiguous substances and powers. Maxwell, I believe, sees this about Faraday, so that it is not merely Faraday's clarity of view and inventiveness which attract Maxwell, but an image of the form physics might take, a physics of contiguity. In the transformation which Maxwell made of Faraday's thought, I believe his objective was not only to find a mathematical expression appropriate to Faraday's electromagnetic concepts, but to bring into analytic form Faraday's insight about the nature of physics itself as a connected system. This calls for a new kind of mathematical physics, field physics.

I mentioned earlier that two concepts are particularly important in Faraday's thinking about electromagnetism, namely the "lines of force" and the "electrotonic state." Thus far, the discussion of Faraday's inquiry into diamagnetism has emphasized only the former, but curiously it was the "electrotonic state" which seemed to Faraday himself the more fundamental idea, and it is this latter which Maxwell takes as the key in his translation of Faraday. For anyone who holds as I do that Faraday did not work with theories, the electrotonic state presents a special problem, since it is a concept which he never abandoned, and yet was never able to support empirically. He held it with tenacity, as we shall see, even though it remained a pure speculation.

Without attempting to trace the history of this elusive notion through the *Experimental Researches,* I should like to try to indicate its role in Faraday's thought. In Series I, it appears in effect as the vehicle for Faraday's perplexity at the unexpected finding that an electric current is induced only by *variation* of the current in a primary circuit, or by *motion* of a permanent magnet. Like Fresnel, Faraday had expected a magnet to produce an electric effect in a conductor; and even after he had discovered the pulse of current on *make* or *break* of the primary circuit, he continued to look for the anticipated effect during steady flow of the primary current. Taking the inductive pulse as evidence of a change of this supposed state, he names it, "after advising with several learned friends," the *electro-tonic state,* signifying that it is a tension in the direction of current flow in the conductor; and Series I is conceived more as the announcement of the discovery of *a new state of matter*, than as the discovery of the phenomenon of induced currents:

> Whilst the wire is subject to either volta-electric or magneto-electric induction, it appears to be in a peculiar state. ... This electrical condition of matter has not hitherto been recognized, but it probably exerts a very important influence in many if not most of the phænomena produced by currents of electricity. (*XR* i/16)

He has to confess, however, that he had found no evidence whatever for the existence of the newly named and announced state:

> This peculiar condition shows no known electrical effects whilst it continues; nor have I yet been able to discover any peculiar powers exerted, or properties possessed, by matter whilst retained in this state. (*XR* i/16–17)

The *Experimental Researches* thus opens with a somewhat embarrassing blunder—for he never was able to find any evidence of the "state," and in Series II he formally withdrew the claim that it exists, though at the same time he reasserted his own conviction that it must:

> Thus the reasons ... have disappeared, and though it still seems to me unlikely that a wire at rest in the neighbourhood of another carrying a powerful electric current is entirely indifferent to it, yet I am not aware of any distinct *facts* which authorize the conclusion that it is in a particular state. (*XR* i/69)

Before making this reluctant retraction, he had made great efforts to exhibit the existence of some such static state in conductors in a magnetic field; his efforts to produce an effect in a conductor due to the mere presence of a strong magnetic field did not stop with the discovery of electromagnetic induction but were, if anything, accelerated by it. The electrotonic state was the surrogate for the effect which had for years been expected, and which he still felt must exist. The experiments which he performed then, on copper bars and leaves in a magnetic field, were the equivalent of a search for diamagnetism or diamagnetic polarity; and the actual discovery of diamagnetism seven years later, which we have discussed above, was the outcome of essentially the same search for a state of tension due to a steady magnetic field. In effect, the persistent search for the electrotonic state yielded the diamagnetic state.

Faraday himself speaks eloquently of his unremitting dedication to the search for this missing "state." Three years after the retraction, he writes:

> Notwithstanding that the effects appear only at the making and breaking of contact, (the current remaining unaffected, seemingly, in

> the interval) I cannot resist the impression that there is some connected and correspondent effect produced by this lateral action of the elements of the electric stream [that is, the magnetic action of the current] during the time of its continuance ... [T]here appears to be a link in the chain of effects, a wheel in the physical mechanism of the action, as yet unrecognized. If we endeavour to consider electricity and magnetism as the results of two forces of a physical agent, or a peculiar condition of matter, exerted in determinate directions perpendicular to each other... (*XR* i/342)

He is seeking in the electrotonic state a "physical agent" of which electricity and magnetism are merely two manifestations. In 1852, some thirty years after Series I, he reasserts his faith, now linking the search with that for the physical lines of magnetic force:

> Again and again the idea of an *electro-tonic* state ... has been forced on my mind; such a state would coincide and become identified with that which would then constitute the physical lines of magnetic force. Another consideration tends in the same direction. I formerly remarked that the magnetic equivalent to *static* electricity was not known... (*XR* iii/420–421)

He then sketches what amounts to a complete scheme of nature, in which he takes the magnetic line as dynamic by analogy to electric currents, and then inserts the electrotonic state as a static state of magnetism, a magnetic tension analogous to the electrostatic tension which he has asserted precedes all conduction.

The conviction expressed in Series I is reiterated in the last pages of the *Experimental Researches*; speaking of the wire which experiences a current when it is moved in a magnetic field, he demands:

> Now, how is it possible to conceive that the copper or mercury could have this power in the moving state, if it had no relation at all to the magnetic force in the fixed state? ... The mere addition of motion could do nothing, unless there were a prior static dependence of the magnet and the metal upon each other... (*XR* iii/551)

Even the complex, never-completed experiment on "Time in Magnetism" with which the *Diary* closes is a search, strictly speaking, for the electrotonic state; he writes to Maxwell in 1857:

> I hope this summer to make some experiments on the *time* of the magnetic action, or rather on the time required for the assumption of the electrotonic state... (*L*: 200)

It would hardly be going too far to say that the *Experimental Researches* begins, and the *Diary* ends, with abortive efforts to find the one thing Faraday most wanted to discover, yet for which he was never able to adduce a single definite fact. The momentum of this search carries over, however, into Maxwell's reformulation of Faraday's views, where, as we shall see, the electrotonic state holds the central place.

We see here, as others have emphasized, that the "experimental" researches are shaped and motivated by great speculative forces. Does this mean that science is, for Faraday, ultimately a theoretical enterprise? I think it is only so to the extent that there is a gap which has not yet been filled, as it should be, by something other than such speculative concepts. His own terms in a quotation above are revealing: there is "a link in the chain of effects, a wheel in the physical mechanism of the action, as yet unrecognized." As the gap is filled, the need for theory will disappear. As science takes its completed form, hypothesis and speculation (which is the only sense of "theory" for Faraday) drop out, and a completed "physical mechanism" with no gears missing takes their place. True science, for Faraday, is the machine revealed.

It is useful, finally, to note what Faraday does *not* ask himself. He does not ask questions about functional relationships: he does not ask, for example, the amount of the repelling force on a diamagnetic body, or the dependence of this on the strength of the field in which it is placed. He does not work with ratios and proportions. Not only does he almost never write an equation: he never asks the kind of question which has an equation as the natural form of its answer.

Faraday's discomfort with the notion of a functional relation in mathematics is revealed poignantly by a remark he made very late in his career, at a time when he had finally been brought into confrontation with the dreaded inverse-square law of gravity. He rebels at the formulation, "with a strength VARYING INVERSELY..." The capital letters are his, expressing his outrage at what he considers a blatant violation of the principle of conservation of the force: how can it then "vary"?[18] He understands, indeed, the algebraic relation as describing the effect, but the equation which for Newton and many generations of scientists after him had fully characterized the force, seems to Faraday utterly unjust to it. "Why, then, talk about the inverse square of the distance?" he says, commenting on the dismissal of his theory by the astronomer-royal, Sir George Airy; "I had to warn my audience against the sound of this law and its supposed opposition on my Friday evening..."[19]

18. Faraday, "On the Conservation of Force," *Proc. of the Royal Inst., 2* (1857) pp. 352 ff. This essay is collected in Faraday, *Experimental Researches in Chemistry and Physics* (London: 1859), pp. 443 ff., where the remark quoted is found on p. 463. The essay is of special interest to the present study, as Maxwell discusses it at length in a letter to Faraday (*L*: 202 ff.). Note that Williams' reproduction of the letter omits Maxwell's figure (Williams, *op. cit.*, pp. 511 ff.).

19. Bence Jones, *op. cit.*, 2, p. 354.

Reference was made above to Faraday's correspondence with Ampère. The following passage from a letter to Ampère, written early in Faraday's career, reveals the extent to which he was aware of the divergence of his concept of science from that of the mathematicians:

> I am unfortunate in a want of mathematical knowledge and the power of entering with facility into abstract reasoning; I am obliged to feel my way by facts closely placed together. ... On reading your papers and letters, I have no difficulty [in] following the reasoning, but still at last I seem to want something more on which to steady the conclusions. I fancy the habit I got into of attending too closely to experiment has somewhat fettered my power of reasoning and chains me down and I cannot help now and then, comparing myself to a timid navigator who, though he might boldly and safely steer across a bay or an ocean by the aid of a compass which in its action and principles is infallible, is afraid to leave sight of the shore because he understands not the power of the instrument that is to guide him.[20]

Could there be a more revealing contrast between Faraday's steady effort to satisfy his mind with a dense series of "facts closely placed together," and the elegant demonstrations of the mathematical physicist, embodied in the electrodynamics of Ampère? The latter had announced his program at the beginning of his *Théorie Mathématique*:

> Observer d'abord les faits, en varier les circonstances autant qu'il est possible, accompagner ce premier travail *de mesures précises pour en déduire des lois générales,* uniquement fondées sur l'expérience, et déduire de ces lois, indépendamment de toute hypothèse sur la nature des forces qui produisent les phénomènes, *la valeur mathématique de ces forces, c'est-à-dire la formulequi les représente,* telle est la marche qu'a suivie Newton ... c'est elle qui m'aservi de guide dans toutes mes recherches.[21]

One should not, I think, be put off by the modesty which Faraday assumes, however sincerely, in his letter to Ampère. It is clear that he really makes no apology for his physics. It is a science, not of mathematics, but of facts. It does not *lack* mathematics, for it does not *need* mathematics. As the letter reveals, Faraday sees mathematics as useful (for others) as a short-cut, in leaping over gaps. His own idea, however, is of a science without such gaps, a science essentially nonquantitative, not needing either mathematical equations or chains of argument, but only intelligent experiment and clarity of view. Faraday certainly reasons incessantly. But for him the

20. Launay, *op. cit.*, 3, p. 929.

21. André-Marie Ampère, *Théorie mathématique des phenomènes électro-dynamiques, uniquement déduite de l'expérience* (1827) (reprinted Paris: Librarie Scientifique Albert Blanchard, 1958), p. 2 (emphasis added).

motions of the mind are constantly checked by reference to fact, so that the result is a dense structure, a closely-spaced series, directed throughout by experiment:

> Let the imagination go, guiding it by judgment and principle, but holding it in and directing it by experiment.[22]

Throughout the *Experimental Researches*, it is true, speculation and imagination run ahead of experiment, but seldom by more than a paragraph; this is not theory-building in the mathematician's sense, but envisioning new, possible things, and they are no sooner envisioned, than they are sought.

This is of course a naive view, profoundly naive. Faraday built the world of the *Experimental Researches* according to this naivete, and if we are to view it by any other criteria, or use his work for other purposes, we must first to a certain extent destroy it, and then rebuild. It is, perhaps, the literary triumph of Maxwell's *Treatise* that it effects this translation of Faraday's thought into mathematical physics with such gentleness and understanding, preserving so much of Faraday's concept of nature and of science.

FARADAY'S ULTIMATE VIEWS ON ELECTROMAGNETISM

In general, three phases might be distinguished in Faraday's thinking about electromagnetism. The first is that of Faraday's early discovery of the continuous rotations of a magnetic pole about a current and of a current about a magnet.[23] At that time Faraday thought in Wollaston's terms of a *rotational* power about a wire carrying current, represented by the magnetic curves, and this was explicitly contrasted with Ampère's theory of forces acting in direct lines between current elements, according to which all of the observed rotational motions were merely compound effects.[24]

Later, in the second phase, Faraday's attention is directed away from the lines of force; he distinguishes two effects of electricity, the *direct effect* and the *lateral effect*, the second including both magnetic and inductive actions of currents. Probably because of his interest in the induction and self-induction of currents which he had discovered, Faraday concentrates here on the

22. Faraday, *Diary*, 7, p. 337. The remark in its context in the *Diary* pages is not a reflection, but a Dionysian outcry in the midst of the chase. It is surrounded by a cascade of ideas, as much speculative as experimental, about a wished relation of gravity and electricity.

23. A number of these papers, which precede Series I of the *Experimental Researches in Electricity* by some ten years, were collected at the end of Volume Two of the *Researches*. The principal of these are "On Some New Electro-Magnetical Motions" (1821) (*XR* ii/127 ff.); "Description of an Electro-Magnetical Apparatus" (1822) (*XR* ii/184 ff.); and "Historical Statement Respecting Electro-Magnetic Rotation" (1823) (*XR* ii/159).

24. (*XR* ii/136) As it happens, Ampère responded directly to this passage, so that the issue was clearly drawn: *Annales de chimie et de la physique, 18* (1827) p. 137.

interactions of currents, and the characterization of the effect as "transverse" takes priority over emphasis on circular or rotational aspects of magnetism.[25] He regards the current itself as a case of electric induction, a dynamic instance of the same phenomenon which appears at rest as static induction: the current flows just insofar as static insulation is constantly breaking down in the conducting medium, the relaxing static induction constituting electric current in this persistent speculation (*XR* i/418–420). He labors to bring "direct induction" in this enlarged sense and the lateral induction of currents into some kind of schema, wondering whether they are actions of the same or different types, and whether they are perhaps two manifestations of a single power in nature (*XR* i/530, 342).

During this period, his attitude toward Ampère's theory is curiously ambivalent. Formally, he admits the possibility that the lateral action may well be a "higher relation" (*XR* i/527), by which he means an action-at-a-distance of the kind Ampère supposes (*XR* i/530). At one point, at which the argument has swung around toward denying such action-at-a-distance in the case of the lateral discharge, he even maintains implausibly that he "would rather have proved the contrary" (*XR* i/550). But his every impulse, in truth, is evidently to understand the lateral effect as an action of the particles of an intervening medium, as he has already interpreted the direct electric induction in Series XI. When a number of experiments on magnetism and the lateral action fail to reveal any effect of interposed materials, he flatly declines to accept the negative evidence of his own experiments, and quite characteristically interprets them according to his own convictions, undeterred (*XR* i/549–551). In all of this discussion, however, the lines of force as such have relatively little role.

The third phase is on the whole coextensive with Volume Three of the *Experimental Researches*, but is most fully articulated in the later researches on lines of force, Series XXV–XXIX, and in several of the later papers, particularly those "On the Lines of Magnetic Force" (*XR* iii/402 ff), "On the Physical Character of the Lines of Magnetic Force" (*XR* iii/407 ff), and "On Some Points of Magnetic Philosophy" (*XR* iii/528 ff). Here the magnetic lines of force have once again come to the fore, and have indeed become the principal object of study, and the chief vehicle of explanation. Because this third phase constitutes Faraday's ultimate thought on electromagnetism, it demands more careful attention on our part before proceeding to the question of the translation of Faraday's views into the more conventional forms of mathematical physics.

When Faraday collected his *Experimental Researches in Electricity* for publication in book form, he chose not to alter the text or the sequence of their

25. This "second phase" corresponds roughly to Series IX through Series XIV of the *Experimental Researches* (1834–1838).

original composition. The result is that although the *Researches* are by no means disorderly, their order is that of a course of inquiry and speculation—"a faithful reprint or statement of the course and results of the whole investigation" (*XR* i/iii)—and is very different from that of a systematic treatise. The turns of Faraday's thought are often surprising and sometimes perplexing, following as they do the motions of an extraordinary and in a sense very private mind. Although some of the papers bound with the third volume of the *Researches* attempt to review aspects of the earlier work, even these reflect the special concerns of Faraday's growing effort to determine the physical character of the magnetic "lines of force," and hence do not really constitute a balanced report of the understanding of electromagnetism which Faraday had achieved.

It seems necessary, then, to attempt to extract from the later *Researches* some such integrated synopsis of Faraday's diverse insights concerning electromagnetism, before turning to the study of Maxwell's translation of them in his own *Treatise.* What point had Faraday in fact reached in his search for a coherent understanding of electricity and magnetism, and what problems remained unresolved? A reader of Faraday soon discovers that these questions are, however, not so easily answered. Faraday's numerous viewpoints are presented, in the spirit of inquiry, as tentative and partial, and any effort to collect them must appear somewhat kaleidoscopic, revealing the extent to which the sum of Faraday's notions falls short of constituting an orderly theory. There are points at which alternative accounts overlap, while other topics fail to attract Faraday's attention and are left as gaps in the overall view.

Nonetheless, having acknowledged the inherent difficulty of the enterprise, in the remainder of this section I shall make a systematic cross-cut through Volume Three of the *Experimental Researches in Electricity*, with the hope of presenting in a single account the final position which Faraday's thinking had reached with respect to electromagnetism. In doing so, I shall set aside questions of the chronological order in which the ideas were developed, and I shall on the whole disregard the distinctions of empirical or speculative status which Faraday assigned to them. Insofar as it is successful, this exercise should help to determine the starting point of the *Treatise*.

As we have already seen, the principal element in Faraday's later thinking about electromagnetism, as it developed in the researches following the discovery of diamagnetism with which Volume Three opens, is the "line of force." The concept of the "line of force," however, undergoes fundamental changes during the course of the *Experimental Researches*, from its introduction in Series I as a line traced by iron filings or a compass needle, to a "physical" line which, as part of a system of power surrounding a magnet, is one of the principal agents in nature. While the filings and the tracing needle remain of great value as auxiliaries, Faraday comes to prefer the current induced in a moving wire as a more "philosophical" indicator of the lines (*XR* iii/332, 563), and with this change of measuring instrument, there comes a new insight into the significance of the lines. The "moving wire" is able to trace them into the interior of a bar magnet, and it indicates their true strength in situations, such as recessed cavities in bar magnets, in which the indications of a compass needle are feeble and misleading (*XR* iii/435, 563). The result is that the line of force acquires a new definition, based on the moving wire (*XR* iii/328), and it becomes possible to think of the lines quantitatively in terms of the amount of current induced in a given conductor. In this new guise, the lines are no longer primarily representations of the direction of mechanical action on a body, but rather "axes" of a *force* or *power* (essentially equivalent terms for Faraday, often with the approximate sense of "energy"). This power may manifest itself in more than one way, and not necessarily in a motion tangential to the line itself. Such a quantitative understanding of the lines is important to Faraday, not in yielding a mathematical law, which as I have argued is a notion foreign to him, but in revealing the *conservation* of this power, and hence its permanence as a physical entity:

> [These lines] represent a determinate and unchanging amount of force. Though, therefore, their forms, as they exist between two or more sources of magnetic power, may vary very greatly, and also the space through which they may be traced, yet the sum of power contained in any one section of the lines is exactly equal to the sum of power in any other section of the same lines. (*XR* iii/329)

To reach this conclusion empirically as he did, Faraday had formulated the criterion for what has subsequently been termed the "ballistic" use of the galvanometer, in which a pulse of current is discharged during the first few degrees of deflection; and he distinguished this clearly from the ordinary use of the instrument, which in application to the study of magnetic fields would give a deflection dependent upon both the number of lines cut, and the velocity of motion (*XR* iii/375). Under the ballistic conditions Faraday specified, he concludes that

> ... in this mode of applying and measuring the magnetic powers, the number of degrees of swing deflection are for small arcs nearly *proportional to the magnetic force* which has been brought into action on the moving wire. (*XR* iii/377; emphasis added)

He quickly moves to speaking of the magnetic force in terms of the "amount of lines of magnetic force," or "the number of lines of force," intersected (*XR* iii/383, 406).

A crucial outcome of these investigations with the moving wire is the universal principle that magnetic lines of force form closed curves in all cases:

> That in fact every line of magnetic force is a closed curve, which in some part of its course passes through the magnet to which it belongs. (*XR* iii/405)

These closed curves represent a certain distribution of power, both inside and outside the magnet, and for Faraday this "atmosphere of power" (*XR* iii/422) becomes a highly interesting natural object with a definite spatial form. To designate this object, he invokes, as he so often does, "the advice of a kind friend" to coin an appropriate term (*XR* iii/422). Since in the case of the isolated bar magnet the lines suggest to Faraday the form of a beetle, the word made to his prescription is "sphondyloid," or beetle-like (σφονδύλη, beetle). He defines the *sphondyloid of power* in this way:

> If, in the case of a straight bar-magnet, any one of these lines, E, be considered as revolving round the axis of the magnet, it describe a surface; and as the line itself is a closed curve, the surface will form a tube [torus] round the axis and inclose a solid form. Another line of force, F, will produce a similar result. The sphondyloid body may be either that contained by the surface of revolution of E, or that between the two surfaces of E and F... (*XR* iii/422)

The "power" of the sphondyloid is then measured by the deflection of the ballistic galvanometer when the moving soil is introduced so as to cut all the lines within the sphondyloid, this having been shown to be the same for any intersection with the same sphondyloid. The quotation above suggests either a bounded torus (the sphondyloid within E), or a shell (the sphondyloid contained between E and F). He goes on to consider a sphondyloid which contains the entire power of a given magnet:

> I have no doubt ... that the sphondyloid representing the total power ... would have equal power upon the moving wire, with that infinite sphondyloid which would exist if the small magnet were in free space. (*XR* iii/424)

Faraday does not define the sphondyloid for cases which lack axial symmetry, though he has no hesitation in speaking of the sphondyloid in cases in which the natural symmetry has been distorted. It is clear that he thinks of the sphondyloid as an entity in nature subject to vicissitudes in interaction with other sphondyloids belonging to other magnets. Thus he says of one magnet approached by another:

> [A]s the magnet is approached, its external sphondyloid of power is compressed inwards ... and at last the magnet is self-contained. (*XR* iii/562)

The "power" of the sphondyloid is the number of lines it contains, the quantity subsequently termed the "flux," and since this is not a measure of energy, we see that Faraday's term "power" is related to energy in only a very general way. Faraday does not formally introduce the term "flux," though it occurs naturally, in passing, as an alternative designation of the flow of lines of force in connection with a fluid analogy which we shall discuss below; in that context, he speaks of "the idea of fluxes or currents" (*XR* iii/526). By "power" Faraday means an ability to act; the action in this case is primarily the surge of current registered by the ballistic galvanometer. The terms and quantities of conventional mechanics do not enter Faraday's discourse except under disguise; one of Maxwell's efforts will be to identify them behind their guise.

Faraday's association of a definite sphondyloid of power with a given magnet is much aided by his grasp of the limiting concepts of two ideal magnetic types: the perfectly "hard" and the perfectly "soft" material. A perfect magnet retains its power unaltered when other magnets, however strong, are brought to or removed from its vicinity:

> Unchangeable magnets are, therefore, required, and these are best obtained, as is well known, by selecting good steel for the bars, and then making them exceedingly hard... (*XR* iii/392; cf. 395)

In thinking about the interactions of magnets, it is a great help to Faraday to confine the discussion to such perfect magnets, whose sphondyloids are strictly unalterable in magnitude, however they may be distorted in form. Correspondingly, the concept of the magnetically soft material is equally useful in discussions of magnetic induction. Thus Faraday speaks of a soft iron sphere "so good in character" as to retain but very slight traces of magnetism when taken out of the magnetic field (*XR* iii/559). These two ideal substances play a role in Faraday's researches analogous, for example, to the theoretical concepts of the perfectly elastic and perfectly inelastic bodies of early discussions of mechanics.

More and more, Faraday comes to regard the system of lines of force, the sphondyloid of power, as of the essence of the magnet, very much as he had, in Series XI, been led to identify "charge" with the system of electric lines of force. With the discovery that magnetic lines of force make closed curves, the traditional magnetic "poles," or magnetic "fluids" distributed over the surfaces of magnets, tended to lose their significance altogether. These poles or surface fluids were alternative representations of the sources of a magnetic force which, in the mathematical theory of magnetism, was supposed to act directly at a distant point, by "attraction" or "repulsion." Thus with the rejection of such sources, and their replacement by the systems of lines of force, Faraday turns his back upon the traditional principles on which theories of magnetism had been based, the principles that "like poles repel" and "unlike poles attract," together with the force laws which express these distant actions quantitatively. "Such poles do not exist," Faraday concludes (*XR* iii/432). The system of lines of force becomes as essential as the magnet itself, and what appeared to be "attractions" and "repulsions" are revealed to be no more than consequences of events actually occurring in the space between magnets (*XR* iii/427). Faraday several times reformulates his new perception of the role of the "atmosphere of power" surrounding the magnet:

> In this view of a magnet, the medium or space around it is as essential as the magnet itself, being a part of the true and complete magnetic system. (*XR* iii/426)

> Contemplating a bar magnet by itself, I see in it a source of dual power. I believe its dualities are essentially related to each other, and cannot exist but by that relation. ... The relation externally appears to me to be through the space around the magnet; in which space a sphondyloid of power is present consisting of closed curves of magnetic force. ... The magnet could not exist without a surrounding medium or space, and would be extinguished, if deprived of it. (*XR* iii/564)

This "medium or space" he occasionally calls "the magnetic field":

> Any portion of space traversed by lines of magnetic power, may be taken as such a field, and there is probably no space without them. (*XR* iii/203)

This formulation, and others which are similar, suggests that Faraday refers to the "lines of force" or the "sphondyloid of power" as the magnetic entity, the atmosphere of power, and to the "field" as the containing place. If so, it would be more proper in his own terms to refer to Faraday's new view as suggesting a "sphondyloid theory" than a "field theory" of magnetism.

The sphondyloid, as an entity in the space about the magnet and essential to its existence, is prior to all the magnetic and electromagnetic effects we observe. The tendency of a magnet to point or move when inserted in the sphondyloid, and the tendency for current to flow in a wire moved across the same sphondyloid, are merely "like correlative and congruent effects of the magnetic force" (*XR* iii/563). However it manifests itself, or whatever transformations it undergoes, as a fundamental power in nature the sphondyloid is strictly conserved:

> As to the suppression of force, ... I conceive that the creation, annihilation, or suppression of force ... is as impossible as the like of matter. All that is permitted under the general laws of nature is to displace, remove, and otherwise employ it. (*XR* iii/546)

Note that since "force" is not simply equivalent to "energy," this intuitive conviction of Faraday's by its very imprecision is probably more far-reaching than the principle of conservation of energy. He considers, for example, the case of a north magnetic pole which is at first acting on the south pole of a neighboring magnet (acting "externally," in the language of the quotation below). What happens, Faraday wonders, to the power of that pole when the second magnet is removed?

> [T]here remains, therefore, in my mind, but two suppositions; either the N polar force when taken off from external compensating S polar force, is not exerted elsewhere as magnetic force at all; or else it is externally thrown upon and associated with the S polar force of the same magnet, and so sustained and disposed of, for the time, in its natural, equivalent, and essential state. (*XR* iii/549)

He certainly admits that it might be converted into a "new form of power," but since in this case he sees no evidence of this, he concludes that the whole magnetic force, quantitatively intact, ("equivalent"), swings around and associates with the south pole of its own magnet, leaving the sphondyloid of the isolated magnet in its "natural," undistorted state.

Faraday holds, without special concern, what would seem an untenable view of how the system of lines of force behaves when the magnet is put into motion. If the magnet moves in translation, he has no doubt that the sphondyloid advances with it; indeed, the essential binding of the sphondyloid of power to the ponderable body of the magnet is the basis of all of Faraday's account of the interactions of magnets. Yet if a bar magnet is rotated on its axis, in Faraday's view the sphondyloid is left unmoved. This is not a point for which he argues; he assumes it with as little question in analyses of phenomena in the later researches, as he did in Series I (*XR* i/53; iii/404).

Beginning with Series XXVI, Faraday regularly regards the lines of force as in some sense "conducted" by paramagnetic or diamagnetic materials, and ultimately, as we shall see, by free space itself. The notion was first suggested in Series XXII, by analogy to the passage of light, the relation having been suggested by the fact that the optical axis and the magne-crystallic axis are the same in diamagnetic crystals. In a passage we have already quoted, he speculates:

> I cannot resist throwing forth another view of these phænomena which may possibly be the true one. The lines of magnetic force may perhaps be assumed as in some degree resembling the rays of light, heat, &c.; and may find difficulty *in passing through bodies*, and so be affected by them, as light is affected. ... the position which the crystal takes in the magnetic field with its magne-crystallic axis parallel to the lines of magnetic force, may be the position of no, or of least resistance; and therefore the position of rest and stable equilibrium. (*XR* iii/122; emphasis added)

Analogies to light frequently occur in the discussion of lines of force in the *Researches*—thus Faraday speaks of their "refraction" (*XR* iii/424), and wonders in one notable speculation whether they may be "ray-vibrations" (*XR* iii/447 ff.). We get an insight, I think, into the very special significance for Faraday of the *curvature* of lines of force, when we find him speaking of them in relation to propagating rays of light; he refers to the lines of force as "subject to inflection *in their course*" (*XR* iii/410), or as "curved *beams*" (*XR* iii/531).

Whether he has in view the particular analogy to the passage of light, or some less specific ability to "transfer the power onwards" (*XR* iii/549), Faraday finds the notion of conduction extremely inviting. He traces the origin of the notion to Euler, whose theory he met in the *Letters to a German Princess*; but he avoids any such specific commitment as Euler's to an actual flow of a fluid (*XR* iii/529). Although we may catch Faraday in an unguarded moment speaking of a magnet "pouring forth lines of force" (*XR* iii/432), he usually is more careful to avoid any such definite image:

> When bodies (media) occupy the space around the magnet, they modify its capability of *transmitting* and *relating* the dual forces of the magnet, and as they increase or diminish that capability, are paramagnetic or diamagnetic in their nature; giving rise to the phænomena which come under the term of magnetic conduction. (*XR* iii/564; emphasis added)

> As yet … I only state the case hypothetically, and use the phrase *conducting power* as a general expression of the capability which bodies may possess of effecting the transmission of magnetic force; implying nothing as to how the process of conduction is carried on. (*XR* iii/200)

By interpreting diamagnetic materials as poor "conductors" in this very general sense, and paramagnetic materials (including hard magnets when favorably oriented) as good ones, he finds he is enabled to interpret the patterns of lines observed in all cases: the lines diverge about diamagnetics, as around obstacles, and converge to flow with ease through paramagnetics. He constructs diagrams to illustrate this (*XR* iii/212), and uses the principle to "read" configurations revealed by the iron filings:

> When there are several magnets in presence and in restrained conditions, the lines of force, which they present by filings, are most varied and beautiful … but all are easily read and understood by the principles I have set forth. As the power is definite in amount, its removability from place to place, according to the changing disposition of the magnets, or the introduction of better or worse conductors into the surrounding media, *becomes a perfectly simple result.* (*XR* iii/435; emphasis added)

Here, by combining the principles of conservation of the total power of the permanent magnets present ("the power is definite in amount"), and that of the "better or worse" conductivity of magnetic materials, the interpretation of a pattern of iron filings as a diagram of fluid flow becomes "perfectly simple." In his own view, this concept of the lines as transmitted or conducted entirely accounts for systems of lines of force which arise under any circumstances.

The "reading" of the pattern to which Faraday refers has, however, a second and more interesting aspect, to which we shall turn presently: what he wishes most of all to "understand" from the filings is the distribution of the power and the tendencies to motion of all the bodies present. He has principles of interpretation to accomplish this as well.

For Faraday, however he may disclaim a literal belief in the flow of a fluid in the field, the flow analogy is always vivid and compelling:

> If we take a large bar-magnet, and place a piece of soft iron, about half the width of the magnet ... about its own width from one pole, and ... then observe the forms of the lines of force by iron filings; it will be seen how beautifully those issuing from the magnet converge, by fine inflections, on to the iron, entering by a comparatively small surface, and how they pass out in far more diffuse streams [at the far end of the bar] ... I, at least, am satisfied that a section across the same lines of force in any part of their course, however or whichever way deflected, would yield the same amount of effect. (*XR* iii/434)

In thinking about the forms which the sphondyloids may be caused to take, Faraday is assisted by an analogy to conduction in an electric circuit, where the actual passage of a "fluid" is for him equally problematic. He imagines, for example, lines of current flow in an electrolyte, or in the water about an irritated electric eel:

> When, therefore, a magnet, in place of being a bar, is made into a horseshoe form, we see at once that the lines of force and the sphondyloids are greatly distorted... [T]he power gathers in ... just because the badly conducting medium, i.e. the space or air between the poles, is shortened. A bent voltaic battery in a surrounding medium ... or a gymnotus curved at the moment of its peculiar action ... present exactly the like results. (*XR* iii/428)

The electric-circuit analogy leads him in turn to the idea of a magnetic "circuit," and to the application to magnetism of the valuable distinction of electric *quantity* and *intensity* which he attributes to Cavendish (*XR* i/81). In effect, then, he employs at least an approximation to the later notions of flux and magnetomotive force, which have become important aids in the design of magnetic devices.

Turning now to the second aspect of Faraday's "reading" of the patterns of lines of force—the determination of the forces and motions to which they give rise—we find that Faraday employs principles of three different kinds, which may overlap in the discussion of any one case. The most universal and unequivocal law is very simply stated:

> All the phænomena resolve themselves into this, that a portion of such diamagnetic matter, when under magnetic action, tends to move from stronger to weaker places or points of force. (*XR* ii/69)

For a paramagnetic body, the effect is the reverse: it moves to places of greater "force," i.e., greater field intensity (*XR* iii/72). A striking corollary follows immediately, one which lies behind the superiority of the moving wire as an indicator of the magnetic lines:

> When the substance is surrounded by lines of magnetic force of equal power on all sides, it does not tend to move, and is then in marked contradistinction with a linear current of electricity under the same circumstances. (*XR* iii/69)

A "pole" would move in a uniform magnetic field, but Faraday is interested, not in hypothetical "poles," but in the actual magnetized bodies with which he experiments. A magnet, he has found, would not move in a uniform magnetic field, however strong: its motions are responses only to *variations* in the field. Faraday's thought adapts naturally to the phenomena as he encounters them; he is not, therefore, inclined to regard the whole magnetic bodies which he observes as composed of poles, but would rather begin with a rule for the motion of the whole body, and derive from it a completely revised notion of "polarity."

Faraday therefore introduces the term *conduction polarity* to distinguish his new understanding of polarity from that of the mathematicians, whose theory assumes poles as centers of force; he divides the "common" or "old notion" from the "true" in a way reminiscent of Newton's distinction of the "true" and "vulgar" concepts of space and time (*XR* iii/533–35). Since the lines of force are concentrated at the points of inflow or outflow at the ends of a magnet, a second magnet, by the new principle, will move bodily toward these regions. The *appearance of attraction*, which has for so long misled philosophers, is thus accounted for and dispelled:

> If a portion of still higher conducting power be brought into play, it will approach the axial line and displace that which had just gone there; so that a body having a certain amount of conducting power, will appear as if attracted in a medium of weaker power. (*XR* iii/201)

To make the law of action universally applicable, it is necessary to add a further refinement, which Faraday is in fact adding in the quotation above—for the same body may move either toward or from the north pole of a strong magnet, depending on the surrounding medium in which it is placed. The passage above concludes:

> ... a body having a certain amount of conducting power, will appear as if attracted in a medium of weaker power, and as if repelled in a medium of stronger power by this differential kind of action. (*loc. cit.*)

"Paramagnetism" and "diamagnetism" thus become *relative* terms, expressing the relation of any given body to the medium in which it is immersed. A body which is paramagnetic when studied in air will become diamagnetic when immersed in a fluid which is more paramagnetic than itself, as Faraday showed in a graded sequence of experiments with a solution of iron sulfate (*XR* iii/58 ff). The *medium* itself is acted on as well as the body, so the result is, as Faraday says, a *differential* action, the result of a kind of magnetic Archimedes' principle:

> Perhaps both magnetic attraction and repulsion, in all forms and cases, resolve themselves into the differential action … of the magnets and substances which occupy space, and modify its magnetic power. … So, then, a source of magnetic lines being present, and also magnets or other bodies affecting and varying the conducting power of space, those bodies which can convey on the most force, may tend, by differential actions, with the others present, to take up the position in which they can do so most freely, whether it is by pointing or approximation. (*XR* iii/437)

> It is the principle of the hydrometer or of Archimedes in respect of gravity applied in the case of the magnetic forces. (*XR* iii/500)

Faraday gradually recognized that the increasing role of the lines of force demanded a reconstruction of the concepts of the magnet, or polarity, and of attraction and repulsion. Now, in this extension to a relative or differential law of action, in which the medium has an active dynamic function, Faraday finds himself committed to a revision of the concept of space itself. In the statement above, space appears as a conductor of the lines of force, with a specific "conducting power" to be compared with that of other materials. For Faraday, this restructuring of the notion of the vacuum, or "mere space," is a fundamental and difficult step. He had long doubted whether a vacuum could sustain electric induction, a doubt which amounted, in connection with the concepts of Series XI, to the question whether a body could be electrified in a vacuum. At one point he arranged a crucial experiment by pointing a reflector toward the night sky, in which the lines of force might find no medium and no terminus, with the hope that it would prove impossible to place a charge upon the reflector (*XR* i/414–15). This was a relatively early experiment, but he is still subject to a similar concern some fifteen years later when, in 1852, he doubts whether the curved electric lines of force could arise *in vacuo* (*XR* iii/410). Curved lines always signify for Faraday the steering action of a deflecting medium, so that when he becomes convinced that magnetic lines must curve from pole to pole even in free space, he necessarily concludes that there exists in the vacuum and imponderable, but magnetically active medium: " … experimentally mere space is magnetic" (*XR* iii/443). What this may mean is one of the abiding speculative problems of the *Researches*, and remains unresolved at the close of the work.

In this view, then, "pure space" takes its place in a series of substances ranging from the most diamagnetic to the extreme paramagnetic; in this sequence, the vacuum appears as a zero or midpoint (*XR* iii/65–66, 71). Space, Faraday says, has "a magnetic relation of its own":

> Before determining the place of zero amongst magnetic and diamagnetic bodies, we have to consider the true character and relation of *space* free from any material substance. Though one cannot produce a space perfectly free from matter, one can make a close approximation to it in a carefully prepared Toricellian vacuum. Perhaps it is hardly necessary for me to state, that I find both iron and bismuth in such vacua perfectly obedient to the magnet. From such experiments ... it seems manifest that the lines of magnetic force ... can traverse pure space, just as gravitating force does, and as static electrical forces do ... and therefore space has a magnetic relation of its own. (*XR* iii/194)

Faraday continues in this passage to argue that the mode of action of "mere space" cannot be the same as that of matter: "Mere space cannot act as matter acts," and therefore the "magnetic relation" of space is unique in the sequence. The midpoint is then not arbitrary, but a "true zero" (*XR* iii/194–95)

Whatever the state of the vacuum under the action of a magnetic field, Faraday says that it must not be assumed "that the space is in a state of magnetic darkness" (*XR* iii/540). Here again, the analogy to light is convincing to Faraday:

> It is as if one should say, there is no light or form of light in the space between the sun and the earth because that space is invisible to the eye. Newton himself durst not make a like assumption even in the case of gravitation. (*loc. cit.*)

However it is to be visualized, then, the sphondyloid of power clearly exists in Faraday's view in the vacuum as fully as it does in other media. Space is acted upon dynamically, and the processes of magnetic action take place differentially with respect to pure space as with respect to ponderable media.

This principle of differential action is particularly applicable to paramagnetic and diamagnetic materials which are subject to induction under the action of a magnet. In addition to this principle, however, Faraday has other ways of thinking about mechanical actions in various magnetic arrangements, considering the actions in terms of the lines of sphondyloids themselves. One way of looking at this is to regard the sphondyloids of two magnets in separation, the actions arising from the interactions of the sphondyloids. Another way is to consider the single system of lines of force which arises from the two magnets together, and to regard the tendencies to motion as due to stresses within this common system of lines. Faraday does not distinguish these two viewpoints quite so formally; for him, the unity of all analyses in the natural world tends to blur the sharp edges of a logical distinction.

In either of these views, the forces arise in the magnetic field; the powers act primarily within the sphondyloids, and only secondarily and incidentally upon the magnets:

> As magnets may be looked upon as the habitations of bundles of lines of force, they probably show us the tendencies of the physical lines of force where they occur in the space around. (*XR* iii/435)

The sense of this passage is, I believe, that the magnets we handle and observe are *merely* the seats ("habitations") of powers; the powers are the object of our real interest, but we cannot directly observe them. Instead, we learn about the powers indirectly, by observing the bodies which respond to them; as we observe the magnets, we are learning about the real tendencies in nature, which in fact "occur in the space around." The powers in nature are always primary for Faraday; it is appropriate, then, that the laws of magnetism should be recast as rules of action of lines of forces, which are "axes of power," or sphondyloids, the "atmospheres of power."

When the two systems of lines corresponding to two magnets are viewed as separate, Faraday offers a rule for the interaction of the *lines* which is very much like the abandoned rule for the interaction of magnetic fluids or *poles*. Thinking of two bar magnets placed side-by-side with corresponding poles adjacent, a case in which we know the magnets repel one another, Faraday asserts that *parallel lines repel*. If one of the magnets is reversed end-for-end, so that the two bars remain parallel but now have unlike ends together, the magnets attract, while the lines run anti-parallel; Faraday concludes therefore that *anti-parallel lines attract* (*XR* iii/419). We may think equally of the interaction of parallel currents as due to the interaction of their magnetic lines of force: if one thinks of the meeting of the two circular fields between the two conductors, it is clear that parallel currents yield anti-parallel field lines at the point of tangency of the two sets of circles, and hence attract. Anti-parallel currents give rise to circular fields which run parallel at the point of meeting, and hence by the rule, repel (*loc. cit.*). Thus, not only the action-at-a-distance theory of magnetism is dispelled by these new rules for the mechanics of fields, but Ampère's theory of the interaction of currents as well.

If the two interacting magnets are placed end-to-end, we may think of the lines likewise encountering each other end-on as they issue from the juxtaposed magnets. Since these lines will be oppositely directed when like poles face one another, Faraday concludes as a rule for this case:

> Thus, unlike magnetic lines, when end on, repel each other, as when similar poles are face-to-face. (*XR* iii/420)

Behind these rules phrased in terms of the lines, is undoubtedly an image of the meeting of the two sphondyloids. Faraday describes vividly the approach of a large, "dominant" magnet toward a small spherical magnet, arranged with their senses opposing:

> How easily all these effects present themselves in a consistent form, if read by the principle of representative lines of force. ... [A]s the

> [spherical] magnet is approached [by the dominant], its external sphondyloid of power is compressed inwards ... and at last the magnet is self-contained ... so that it gives no induced currents. Within that distance the effect of the superior and overpowering force of the great magnet appears ... which, though it can take partial possession of the little magnet, still, when removed, suffers the force of the latter to develope itself again. (*XR* iii/562)

Here the interaction is described as if the two sphondyloids existed as separate entities, so that at the dramatic moment which particularly interests Faraday, the lesser sphondyloid has been driven back precisely to the boundaries of its own magnet. In a certain sense, Faraday undoubtedly insists on the independent existence of the sphondyloids belonging to two interacting magnets, for if they are hard magnets each one's power remains "unalterable" however much their powers may intermingle. But when the two magnets are placed close together and their pattern is taken with the iron filings, all that we see is a spatial configuration in which the two sphondyloids have merged. Perhaps Faraday's fullest intuition of the interaction of magnets is revealed in his interpretation of such a resultant pattern, a single diagram in which the tendencies of the entire system are implicit. Since this is the vision which the iron filings actually present, it is I think the approach to which Faraday is most naturally sympathetic.

By what principles are the forces to be read from a *combined* field pattern? Perhaps the best statement is this, from "Physical Lines of Magnetic Force" (1852):

> The association of magnet with magnet, and all the effects then produced ... are in harmony, as far as I can perceive, with the idea of a physical line of magnetic force. If the magnets are all free to move, they set to each other, and then tend to approach; the great result being, that the lines from all the sources tend to coalesce, to pass through the best conductors, and to contract in length. When there are several magnets in presence and in restrained conditions, the lines of force which they present by filings, are most varied and beautiful ... but all are easily read and understood by the principles I have set forth. (*XR* iii/435)

As in the case of the principle previously discussed, Faraday here first describes what we observe—that the magnets "set to" and tend to approach; the "great result" of this is a trio of principles concerning the lines. It is a "result" only in the sense that we infer it from the observed behavior; actually, of course, it is the other way around—the "tendencies" among the lines are not the result, but the cause of the "tendencies" among the magnets. The first such action of the lines is, Faraday says, to "coalesce"; the second, "to pass through the best conductors"; and the third, "to contract in length." The first is perhaps the key to Faraday's point of view. "Coalescence" denotes the joining together of two bodies into one combined system, the specific image being the conflux of two streams to form a single river. This is not equivalent to summation into a mathematical resultant. Faraday says specifically that coalescence does not mean simply "addition" (*XR* iii/431). It is only lines "favorably disposed" which coalesce, so the term evidently implies flow in a common direction. The "tendency to coalesce" then refers to the first mechanical motion observed when the system is assembled, namely that in which the magnets, turning, "set to each other." Two dipoles swing around so that a north pole is adjacent to a south pole, whereupon the lines of force are able to flow through first one and then the other. When two north poles are juxtaposed, on the other hand, and the magnets are restrained from "setting to," the lines of force conflict, and they do not coalesce—hence we see the magnets exhibiting a tendency to rectify this situation.

The distinction of this process from the formation of a single mathematical resultant is crucial for Faraday's view of physics. The sphondyloids, even when coalesced, belong to their proper magnets, and their union is temporary and reversible. The sphondyloids are conserved, not simply in total quantity, but as individual entities, and they merely coexist in the coalesced system—they are not added up out of existence. "Coalescence" is thus a particularly interesting example of Faraday's terminology. It reflects his deep-seated unwillingness to exchange an entity for an undifferentiated quantity.

Once "coalesced," the lines tend to flow through the best conductors. This returns us, of course, to the principle of differential action: diamagnetic materials will move out of the field, and paramagnetic materials into it, to permit maximum flow through the easiest channels. Now, however, this is phrased as a tendency, not of the bodies, but of the lines: the lines of force find their ways, and the magnets are carried with them. Finally, when all magnets have been oriented and shifted, there remains a stress in the line of force joining apparent north and south poles, a stress tending to draw the system together if constraints permit. In this aspect, the lines must be thought of as under tension; whenever in a pattern the lines extend from one magnet to another, they are sure signs of a tendency of the two bodies to draw together. Again, this is by no means magnetic "attraction"; it is an inner tendency to contraction, on the part of the system of lines of force.

A fourth principle, it seems to me, must be added to these, though it does not appear in Faraday's summary above. Faraday speaks elsewhere of "the tendency to lateral separation of the magnetic lines" (*XR* iii/436; cf. 419). This must be taken as applying to the coalesced pattern, as well as to the interaction of lines regarded as belonging to two independent sphondyloids. If one looks at the single pattern which results when two bar magnets are placed parallel with like poles together, the crowded set of parallel lines between them clearly exhibits this "tendency to lateral separation," which drives the two magnets apart.

Prepared with this set of principle, it is indeed possible to "read" in dynamic terms a set of iron-filing configurations, such as those collected by Faraday in the plate accompanying Series XXIX. The mathematical theories of Ampère and Poisson, concerned with quantities and phrased in terms of poles and fluids which Faraday finds nonexistent, are now replaced for Faraday by this vision of the distribution of a power of nature, arranged in distorted or coalesced sphondyloids. To one who has seen as Faraday has, the calculations are no longer necessary: the pattern speaks for itself of tendencies of all the bodies to "point" and to move.

On the whole, the above account has been in terms of systems of hard, permanent magnets and soft paramagnetic and diamagnetic materials. I have given relatively little attention to the systems of lines belonging to currents or to the interactions of currents, because Faraday himself tends to slight them. The circular field about a straight wire is referred to, as we have seen, and the equivalence of a helix to a bar magnet, but in general Faraday does not speak often of the fields of currents in other configurations. He has no art to match Ampère's for the prediction of field patterns by the summation of the contributions of individual current-elements. It appears to me that Faraday, recognizing at once the general approval of Ampère's exhaustive theory of current-current interactions, and his own inability to understand it, tended to leave this topic aside as an area of research he could not profitably pursue. We do not find him doing experiments with interacting linear conductors (except in his own specialty, the production of continuous rotations), or with Ampère's famous frames. We have seen that he once expressed a wish that the "lateral action" of currents might prove to occur by action-at-a-distance (page 20 above)—that is, by a "higher" process which he might leave to others to understand. When he became convinced that it, like electric induction, occurred through the action of a medium, his own task of explanation was much enlarged, and Ampère's phenomena became part of the general theory of magnetic lines of force. This he found it easier to pursue using magnets, Ampère himself having established the equivalence of the effects of magnets and currents. Faraday no doubt felt that he might thus arrive at a theory which would obviate Ampère's elaborate calculations. It is my impression that the lack of quantitative rigor of the rules Faraday gives for the interaction of lines of force does not strike him as a significant drawback, so long as none of the observed motions goes unexplained.

There remains the problem of achieving unity of the accounts of electricity and magnetism, phenomena which Faraday at one point envisioned as "direct" and "lateral" modes of induction, as we have seen. Ampère's hypothesis, that permanent magnetism can be explained as the effect of circulating molecular currents whose axes are parallel, Faraday finds very inviting as a step toward such a unified account: "there is no idea put forth ... which at all approaches in probability and beauty to that of Ampère" (*XR* iii/423). Again perhaps wishing that it might be true, he nonetheless finds himself forced to reject the hypothesis of molecular currents, because of a difficulty which he finds insuperable. He compares a helix carrying current, and a bar magnet, as agents by which a new magnet can be formed, and finds this irreconcilable difference: while the helix can produce a magnet whose strength is very much greater than its own, the bar magnet, supposed by Ampère's hypothesis to be equivalent, can produce a new magnet of strength no greater than its own. Faraday did not see the fallacy which lies in the terms of comparison: the coil, when approached by a piece of soft iron, will induce the strong field he observed *only* if the cell supplying it is able to sustain a constant current against an induced counter-emf, which means that power must be supplied. The bar magnet is equipped with no such external source of power. Hence, Faraday's rejection of Ampère's hypothesis is a misfortune consequent upon his limited understanding of Lenz's principle and its relation to the conservation of energy; in view of Faraday's limited knowledge of formal mechanical principles, one would hardly expect him to see this point.

A similar misfortune shadowed Faraday's judgement of the extension of Ampère's hypothesis by Weber, to include the case of diamagnetism. Here Faraday is the victim of a persistent fallacy. Weber, adopting a proposal which Faraday himself first made (*XR* iii/73), correctly theorizes that diamagnetism is the consequence of molecular currents *induced* by the approach of the diamagnetic body to the pole of the magnet; the induced currents must be, by Lenz's principle, of a sense to thrust the diamagnetic body out of the field. Faraday from the first thought of this hypothesis believes that such an induced current would be in a direction which is the reverse of normal induction. He thus insists that if bismuth is to have such currents induced, it must give results the reverse of those obtained with copper when each is formed into a helix and a bar magnet inserted:

> ... this view would be equivalent to the supposition, that as currents are induced in iron and magnetics parallel to those existing in the inducing magnet or battery wire; so in bismuth, heavy glass and diamagnetic bodies, the currents induced are in a contrary direction. (*XR* iii/73)

He experimented carefully, and found of course that induction is the same for all conductors. Nine years later, he is still convinced of the same difficulty:

> [T]hose who rest their ideas on *magnetic fluids* must assume that in all diamagnetic cases, and in them only, the fundamental idea of their mutual action must not only be set aside but inverted, so that the hypothesis would be at war with itself; and those who assume that

> *electric currents* are the cause of magnetic effects, would have to give up the law of their inducing action (as far as we know it) in all cases of diamagnetism. (*XR* iii/541)

Without trying to trace the source of the difficulty, which would seem to lie in his understanding of the direction of induction in ordinary cases, we may simply note that though he was corrected on the point by Weber, the fallacy is unshakable, and Faraday thereby missed the possibility of grasping a brilliant unification of electric and magnetic phenomena which was already clear, both theoretically and empirically, to Weber. Faraday was not always a good listener.

Ampère's hypothesis has the effect of giving emphasis to currents, and hence of making electricity the fundamental science, as indeed Faraday must have supposed it was when he titled his *Experimental Researches*, and when he wrote the introductory paragraphs to his Series XI. By contrast, perhaps his failure to see the possibilities of Ampère's and Weber's hypotheses helped to keep Faraday's thoughts in the later researches directed to magnetism, and to the magnetic lines of force as the fundamental principle of explanation. Certainly electric induction, to which Faraday had devoted so much attention in Series XI–XIV, is relatively seldom discussed in Volume Three, though it is clear that Faraday holds to the concepts of Series XI, and he finds them supported by disturbing phenomena observed in long cables, which he discusses in two later papers. At the close of the *Researches*, then, Faraday holds a view which embraces two systems of physical lines of force, magnetic and electric, with perhaps priority given to the former. They are related, speculatively, through the concept of the electrotonic state, of which he was never able to give a satisfactory interpretation.

We have seen the extent to which this qualitative physics of lines of force embraces, in its diagrammatic manner, a wide range of phenomena of electromagnetism. It is important to keep in mind at the same time that the basis of the representation of both electric and magnetic lines of force is a quantitative procedure, the one based on the moving wire, and the other on the proof plane and electroscope. Although Faraday uses his patterns only qualitatively, the possibility remains that these geometrical images contain in them a richness of mathematical principles and relations which Faraday himself was unable to realize. It is this scientific richness of Faraday's imagery which Maxwell discovered, and translated into appropriate mathematical forms in the *Treatise*.

Chapter 2.
The Translation of Faraday's Ideas Before Maxwell

THE "TRANSLATION" OF FARADAY

Maxwell's *Treatise* is perhaps best understood as a project of translation—the translation of Faraday's ideas. In Part iv of the *Treatise*, embedded in a discussion of Faraday's methods in electromagnetism, is this remark:

> It is mainly with the hope of *making these ideas the basis of a mathematical method* that I have undertaken this treatise. (*Tr* ii/176; emphasis added)

An account of Maxwell's own experience as a student of both Faraday's *Researches* and certain theoretical treatises on electricity is given in the Preface to the *Treatise*; it concludes:

> When I had *translated what I considered to be Faraday's ideas into a mathematical form,* I found that in general the results of the two methods coincided, so that the same phenomena were accounted for, and the same laws of action deduced by both methods. (*Tr* i/viii–ix; emphasis added)

It is clear, I think, from these passages and a number of others like them that Maxwell's *Treatise* has its origin in Faraday's *Researches*, and that its main purpose is to give these ideas a form which is at once methodical and mathematical; Maxwell himself regards this fundamental reformulation of Faraday's thought as a work of "translation."

Ultimately, this characterization as a translation may seem unfair to Maxwell's work, for the *Treatise* certainly goes far beyond Faraday's attainments, and it is difficult to be sure that even in essentials the *Treatise* has not grown away from Faraday's thought in certain fundamental respects. But its origin, and its overarching purpose, are by Maxwell's own testimony the interpretation of Faraday's thought, and I propose in the present study to explore this view of the *Treatise*, keeping in mind that the translation of a body of ideas such as Maxwell proposes is inevitably a creative work, and that Faraday's thought is, as we have already seen, both perplexing and rich in suggestion. Maxwell does not necessarily imply, in reformulating Faraday's ideas, that the original statement is defective; it may be the world that is at fault. We are being shown in the *Treatise*, in effect, how to read the *Experimental Researches*, and the end of the whole project may be to return us to Faraday, having shown us how to realize the wealth which Maxwell has found latent there:

> If by anything I have here written I may assist any student in *understanding Faraday's modes of thought and expression*, I shall regard it as the accomplishment of one of my principle aims—to communicate to others the same delight which I have found myself in reading Faraday's *Researches*. (*Tr* x/x–xi; emphasis added)

It is not only in the *Treatise* that Maxwell undertakes this translation of Faraday; the earlier electrical papers over a period of some fifteen years may be regarded as a series of essays in the same direction:

> The ideas which I have attempted to follow out are those of action through a medium from one portion to the contiguous portion. Those ideas were much employed by Faraday, and the development of them in a mathematical form, and the comparison of the results with known facts, have been my aim in several published papers. (*Tr* ii/158)

The *Treatise*, however, is the work that is to bring this project to completion, and the present study will be confined to the culmination of this effort in Part iv of the *Treatise*. The project would continue with the *Elementary Treatise on Electricity* and the second edition of the *Treatise*, both of which, though left incomplete at Maxwell's death, showed signs of a motion of thought still further in the direction of a commitment to Faraday's point of view:

> In the larger treatise I sometimes made use of methods which I do not think the best in themselves, but without which the student cannot follow the investigations of the founders of the Mathematical Theory of Electricity. I have since become more convinced of the superiority of methods akin to those of Faraday, and have therefore adopted them from the first.[1]

Maxwell from time to time shares with the reader his reflections on this project, and indeed it becomes clear that the very nature of the process of "translation" of Faraday's ideas is one of Maxwell's major concerns. It will be helpful to quote further from the Preface to the *Treatise*, in which he discusses Faraday's "way of conceiving phenomena":

> The general complexion of the treatise differs considerably from that of several excellent electrical works, published, most of them, in Germany. ... One reason of this is that before I began the study of electricity I resolved to read no mathematics on the subject till I had first read through Faraday's *Experimental Researches on* [sic!] *Electricity*. I was aware that there was supposed to be a difference between Faraday's way of conceiving phenomena and that of the mathematicians, so that neither he nor they were satisfied with each other's language. I had also the conviction that this discrepancy did not arise from either party being wrong. (*Tr* i/viii)

1. From the "Fragment of Author's Preface": James Clerk Maxwell, *Elementary Treatise on Electricity and Magnetism*, ed. William Garnett (Oxford: The Clarendon Press, 1881), p. viii.

Maxwell's reading of Faraday, which was at the same time the beginning of his study of electricity and magnetism, thus immediately led him to the problem of the "languages" in which two different points of view, each suspected of being true, were formulated. To test his conviction of the equal validity of both methods, he proceeded, as we have seen in an earlier quotation, to "translate" Faraday's ideas into the mathematical language, for comparison. In this he was following the lead of his slightly older friend and mentor, William Thomson.[2] He found, as Thomson and others were already finding, that Faraday's notions could be reconciled with mathematical physics in the standard form represented by the tradition of Coulomb, Poisson, and Ampère, in the sense that "the same phenomena were accounted for" by both methods.

This, of course, was a great step in the interpretation of Faraday, but it was not Maxwell's distinctive contribution. What came to interest Maxwell was not so much this equivalence, as the fundamental difference in point of view which accompanies the difference in the two "languages" involved—the language of the *Experimental Researches* and that of the mathematicians. Essentially, this is the difference between the concept of action through a medium, the point of view of what has since become "field physics," and that of action at a distance, which had been the foundation of most mathematical physics since Newton's *Principia*. Maxwell makes clear in the preface his interest in the distinction between those views despite his recognition that *both* have succeeded "in explaining the principal electromagnetic phenomena"; he contrasts Faraday's method with that of the mathematical theorists:

> The great success which these eminent men have attained in the application of mathematics to electrical phenomena, gives, as is natural, additional weight to their theoretical speculations, so that those who, as students of electricity, turn to them as the greatest authorities in mathematical electricity, would probably imbibe, along with their mathematical methods, their physical hypotheses [specifically, the hypothesis of action-at-a-distance].
>
> These physical hypotheses, however, are entirely alien from the way of looking at things which I adopt, and one object which I have in view is that some of those who wish to study electricity may, by reading this treatise, come to see that there is another way of treating the subject. ... In a philosophical point of view, moreover, it is exceedingly important that the two methods should be compared, both of which have succeeded in explaining the principal electromagnetic phenomena. ...
>
> I have therefore taken the part of an advocate rather than that of a judge, and have rather exemplified one method than attempted to give an impartial description of both. (*Tr* i/x–xi)

2. William Thomson became Sir William Thomson in 1866 and Lord Kelvin in 1892. Since this study will most often concern his work during the first period, it will perhaps suffice to allude to him as "Thomson."

In the terms of the present study, we recognize here Maxwell's commitment to a *rhetorical* enterprise, in which it is not the agreement with the phenomena that is at stake, but a "way of looking at things," reflected in a particular choice of language.

The task which Maxwell has set himself in the *Treatise*, then, is not simply to find a quantitative equivalent for Faraday's expressions, but to "translate" in a more exacting sense: to carry Faraday's "mode of thought and expression" over into a "mathematical form" that will be faithful to the method and point of view which were original with Faraday. Maxwell wishes to contrast this, as simply as possible, with the method associated with conventional mathematical forms. Maxwell's demand on himself is precisely the traditional requirement of the translator: he must find in one language the *appropriate* expressions for the thought originally formulated in another. This is an old problem, but in a new mode, since one of the "languages" involved is now mathematics itself. A special difficulty attaches to the project as well, because, once Faraday has been interpreted mathematically, the easy transformations to which mathematical expressions are subject can readily erase just the distinction Maxwell is determined to preserve. If one judges by criteria of measure and quantity alone, as he says, "the results of the two methods coincide." Maxwell, as a translator of Faraday's ideas, will be interested throughout the *Treatise* in an aspect of mathematical forms that is not measured by quantitative criteria alone.

Maxwell was not among the first to embark on the translation of Faraday; as mentioned above, he was preceded in particular by William Thomson, to whom he acknowledges a broad indebtedness in the Preface (*Tr* i/viii). But Thomson had on the whole taken the narrower view of the problem. The beginning of Thomson's writing on this subject is a paper of 1845, which was prepared at the request of Liouville for the latter's *Journal de mathématiques*, in order to resolve the question whether in fact Faraday's discoveries in electrostatic induction had cast legitimate doubt on the theories of Coulomb and Poisson.[3] Here the question was the more limited one, whether there was a *quantitative* disagreement between Faraday's views and the mathematical theory. Thomson's findings were that, properly understood, Faraday's methods and those of the mathematical physicists yielded the same predictions. He showed that this could be accomplished elegantly by a simple extension of the work of Poisson, applying to Faraday's new "dielectric" the theory of polarization which Poisson had developed for magnetism. By contrast with the view Maxwell was later to take, Thomson seems to seek one mathematical theory which would be indifferent to the distinction of physical hypothesis dividing Faraday and the French theorists:

3. Silvanus P. Thompson, *The Life of William Thomson* (London: Macmillan and Co., 1910), *1*, p. 128; (*EM*: 15n).

> [The] difference of his [Faraday's] ideas from those of Coulomb must arise solely from a different method of stating, and interpreting physically, the same laws: and farther, it may, I think, be shown that either method ... may be made the foundation of a mathematical theory which would lead to the elementary principles of the other as consequences. This theory would accordingly be the expression of the ultimate law of the phenomena, independently of any physical hypothesis. (*EM*: 26)

Following a tradition which stems from Newton's *Principia* and is reasserted, for example, in Ampère's *Théorie Mathématique*[4]—a tradition, indeed, which might seem the very essence of the discipline of mathematical physics—Thomson tends here to turn his back on any distinction of hypotheses which is not based strictly on the phenomenon, whereas Maxwell as we have already seen is fascinated by an aspect of Faraday's thought which does not affect the known phenomena at all. In discussing magnetic and electric "polarity" in a later paper, Thomson says:

> However different are the physical circumstances of magnetic and electric polarity, it appears that *the positive laws of the phenomena are the same, and therefore the mathematical theories are identical.* Either subject might be taken as an example of a very important branch of physical mathematics, which might be called "A Mathematical Theory of Polar Forces." (*EM*: 351; emphasis added)

Here Thomson is clearing the ground for the progress of *positive* exact science by identifying "physical mathematics" as directed strictly to the phenomena, and indifferent to hypothesis. If the "positive laws of the phenomena" are the same, there should exist a kind of rock-bottom mathematical theory strictly adapted to them, while any differences in the formulation of this theory would be uninteresting apart from considerations of convenience. Thomson, in this early period of fascination with mathematical physics as he has discovered it in Fourier and the French mathematicians, sees his own work as the recognition of mathematical equivalences and the reduction of merely apparent or hypothetical differences, while Maxwell from the beginning of his work on electromagnetism is interested in just the *difference* which characterizes Faraday's special point of view. In a manuscript "On Faraday's Lines of Force," evidently an early version of the printed paper (*SP* i/155 ff.), Maxwell discusses this distinction, as if he were replying to the passage we have quoted from Thomson:

> The phenomena of statical electricity and magnetism as studied by Coulomb, Poisson, Green &c. and those of dielectrics, paramagnetics

4. Quoi qu'il en soit de ces hypothèses et des autres suppositions qu'on peut faire pour expliquer ces phénomènes, ils seront toujours représentés par la formule que j'ai déduite des résultats de l'expérience, interpretés par le calcul." Ampère, *Théorie Mathématique*, p. 99; cf. p. 4.

> and diamagnetics and crystalline magnetic polarity as investigated by Faraday, Plücker and others, have evidently much similarity as to their mathematical expression. If we could abandon physical theory and use only mathematical formulae the methods would be identical. But though we might gain in generality of expression by this method, we should lose those distinct conceptions which a physical theory presents to the mind and the general laws of the science would be put out of the reach of any but professed mathematicians.[5]

Maxwell continues to recommend his own method of "physical analogy" as in effect preserving the "distinct conceptions" of physical hypotheses as well as the objectivity and precision which Thomson demands.

If Maxwell, as he says, writes as an "advocate," Thomson translates Faraday rather as a "judge," reducing the merely conceptual differences which have no foundation in the positive evidence of the phenomena. In the "advocate" and the "judge," Maxwell and Thomson, we see two quite distinct concepts of the translation of Faraday's ideas, and we may perhaps recognize the special mission which Maxwell undertakes when he enters the very active arena of electromagnetic theory in the middle of the nineteenth century.

The notion of "translation" demands further examination in this curious application to the problem of interpretation of Faraday's thought. Maxwell speaks of translation, but is this really a case of translation at all? It can be so only if the "thought" expressed, first in the *Experimental Researches*, and then again in partial differential equations, is truly common to both. In true translation, the thought must be the same, while only the languages in which it is expressed are diverse. If Faraday can be "translated" into mathematical guise, then he is in his own way doing mathematics in disguise. It is striking that both Thomson and Maxwell make this claim for Faraday: both assert that Faraday is genuinely a mathematician. Maxwell made it in effect his thesis in his first electromagnetic paper, "On Faraday's Lines of Force," to which we have referred above:

> The methods are generally those suggested by the processes of reasoning which are found in the researches of Faraday … and which, though they have been interpreted mathematically by Prof. Thomson and others, are very generally supposed to be of an indefinite and unmathematical character, when compared with those employed by the *professed mathematicians*. … [T]he limit of my design is to show how, by a strict application of the ideas and methods of Faraday, the connexion of the very different orders of phenomena which he has

5. From the collection of Maxwell manuscripts in the Cambridge University Library, MS.Add.7655. Items in the collection will be identified by reference to P. M. Harman, ed., *The Scientific Letters and Papers of James Clerk Maxwell* (3 vols.; Cambridge: Cambridge University Press 1990, 1995, 2002). For the present item, see H.84 and H.87.

> discovered may be clearly placed before the mathematical mind. I shall therefore avoid the introduction of anything which does not serve as a direct illustration of Faraday's methods, or of the mathematical deductions which may be made from them. In treating the simpler parts of the subject, *I shall use Faraday's mathematical methods* as well as his ideas. When the complexity of the subject requires it, I shall use analytical notation, still confining myself to the development of ideas originated by the same philosopher. (*SP* i/158; emphasis added)

This was Maxwell's manifesto to the "professed" mathematicians, whose rejection of Faraday's thought (as opposed to his experimental results) we shall consider later in the present chapter. Maxwell recalls this early insight—that Faraday, despite all appearances, really employs mathematical conceptions—when he writes the Preface to the *Treatise*:

> As I proceeded with the study of Faraday, I perceived that his method of conceiving the phenomena was also a mathematical one, though not exhibited in the conventional form of mathematical symbols. I also found that these methods were capable of being expressed in the ordinary mathematical forms, and thus compared with those of the professed mathematicians. (*Tr* i/ix)

The effect of this discovery is to detach the notion of mathematics from that of its "conventional form," so that the nature of mathematics itself is called into question. Maxwell is in fact deeply interested in this aspect of the inquiry. We shall turn to this question in Part II of the present study, in which we deal briefly with Maxwell's metaphysical convictions; suffice it to say for the moment that he assumes a concept of mathematics which is rooted in a philosophical tradition, but would not find widespread endorsement in the twentieth century. Mathematics, for Maxwell, is grounded in a thorough consciousness "of the fundamental forms of space, time, and force." This is the token by which he identifies Faraday as a mathematician:

> It was perhaps for the advantage of science that Faraday, though thoroughly conscious of the fundamental forms of space, time, and force, was not a professed mathematician. He was not tempted to enter into the many interesting researches in pure mathematics which his discoveries would have suggested if they had been exhibited in a mathematical form, and he did not feel called upon either to force his results into a shape acceptable to the mathematical taste of the time, or to express them in a form which mathematicians might attack. He was thus left at leisure to do his proper work, to coordinate his ideas with his facts, and to express them in natural, untechnical language. (*Tr* ii/176)

Two things immediately become clear in this passage: "mathematics" must be distinguished from its symbolic expression, and mathematics has its history, subject to fashions and the "taste of the time." Even a "translation" of Faraday's mathematics is not necessarily a transcription into conventional symbols or analytic form; the method of "physical analogy" which Maxwell introduces in "Faraday's Lines" is a third possibility, in which the mathematical conception of a perfect fluid replaces equations throughout most of the first part of that paper. When Maxwell does introduce analytic expressions in "Faraday's Lines," he does so with an apology virtually addressed to Faraday himself, as has been suggested.[6] The viewpoint of "Faraday's Lines" is really that Faraday has led the way into a new realm of mathematics, which has not been recognized by the mathematicians themselves. Maxwell's role is to restore the dialectic.

On the other hand, in Maxwell's view the isolation of Faraday's thought, and its almost total disconnection from an established body of doctrine and habit, may have been an important blessing: it may have been precisely the price of intellectual liberty. If to be a mathematician (as contrasted with a "professed" mathematician, of the sort who rejected Faraday's work) is to be "thoroughly conscious" of certain "fundamental forms," this may have nothing essentially to do with a technical mastery, which can be a tyranny. Maxwell sees Faraday as having been liberated from the tyranny of "mathematical taste." As we shall see, largely to make this point Maxwell devotes a whole chapter of the *Treatise* (Part iv, Chapter II) to a display of this traditional "taste," in giving an account of Ampère's theory of electrodynamics in a severely analytic form. Maxwell sees Faraday's freedom from technique and its demands as the *leisure* of the philosophical mind to do its own "proper work."[7] In turn, he appreciates the generation of unconventional mathematical forms from a more direct contact with nature—from the "coordination" of thought with fact—which Faraday was permitted to achieve through his innocence of formal mathematics. Out of the *Experimental Researches*, to which fact and nature lead, has emerged a new, more natural language, the language of new mathematics, to which Maxwell is to serve as our guide.

If Maxwell followed Thomson in the reading of Faraday, it may be that Thomson later followed Maxwell's lead in the recognition of Faraday's role as mathematician. Thomson reflects on this at the time of the collection of his own early papers into the *Reprint* of 1872:

> The singular combination of mathematical acuteness, with experimental research and profound physical speculation, which Faraday, though not a "mathematician," presented, is remarkably illustrated by his use of the expression, conducting power of a magnetic medium for lines of force. (*EM*: 489)

6. Joan Bromberg, *Maxwell's Concept of Displacement Current* (Ph.D. Thesis, University of Wisconsin, 1967: University Microfilms No. 67–475), p. 149.

7. Tyndall had made much the same point. John Tyndall, *Faraday as a Discoverer* (New York: 1872), p. 144.

> Faraday, without mathematics, divined the result of the mathematical investigation … and, what has proved of infinite value to the mathematicians themselves, he has given them an articulate language in which to express their results. Indeed, the whole language of the magnetic field and "lines of force" is Faraday's. It must be said for the mathematicians that they greedily accepted it, and have ever since been most zealous in using it to the best advantage. (*EM*: 581)

At the time Maxwell read the first part of "Faraday's Lines" to the Cambridge Philosophical Society in 1855, the mathematicians had, with the exception of Thomson himself, almost unanimously ignored or rejected Faraday's language; by the time Thomson writes the above, the "mathematicians" who have adopted Faraday's terms might principally include Maxwell, Thomson himself, and their mutual friend Peter Guthrie Tait.

If we were to accept the principle that Faraday is already doing mathematics, and at the same time follow Maxwell in denying that a translation must be into symbolic, analytic form—how would we then understand the "translation" process? Translation from *what* to *what*? The strategy of "Faraday's Lines," to which we have already alluded, I think suggests one answer, in terms of Maxwell's method of "physical analogy":[8]

> By a physical analogy I mean that partial similarity between the laws of one science and those of another which makes each of them illustrate the other. (*SP* i/156)

> [A]n impression still prevails, that there is something vague and unmathematical about the idea of lines of force, and that they will inevitably lead astray any student who is not guided by that familiar acquaintance with the laws of nature which distinguishes the great electrical philosopher [Faraday]. To illustrate the mathematical character and scientific usefulness of the method of lines of force I have developed certain of their properties in a series of propositions which I have proved in full without the introduction of technically mathematical expressions. (Manuscript version of "Faraday's Lines")[9]

> It is by the use of analogies of this kind that I have attempted to bring before the mind, in a convenient and manageable form, those mathematical ideas which are necessary to the study of the phenomena of electricity. (*SP* i/157)

In terms of one of the most significant words in Maxwell's methodological vocabulary—again, one which we will have occasion to discuss further in Part II of this study—his device is to *illustrate* the mathematical character of the lines of force, to "bring them before the mind" as mathematical ideas, by drawing an analogy between Faraday's account of electricity and

8. On the method of "physical analogy." see Joseph Turner, "Maxwell on the Method of Physical Analogy," *British Journal for the Philosophy of Science, 6* (1955–56) pp. 226 ff.

9. "On Faraday's Lines of Force" (H.491).

magnetism, and the admittedly mathematical concept of a perfect, incompressible fluid. By "illustration," Maxwell means "illumination," and the light is the intellectual light, "the first day's work" of which Bacon spoke, which is for the sake of the eye of the human understanding.[10] The geometrical fluid, incompressible and without inertia, can be grasped by the mind intuitively, but its motion can also be formulated analytically. The physical analogy, which reveals to the mathematician the inherent "generality and precision" of Faraday's method, serves as a kind of mathematical Rosetta stone opening Faraday's text to a mathematical reading.

Chapter 1 of this study has explored the sense in which the *Experimental Researches* is anything but a work of mathematical physics; Faraday's urge is to reveal in nature a plenum of physical powers, and in his work there is no real place for either quantitative relation or mathematical argument. But he sees his powers as "distributed" and moving; and in conceiving these shapes and their motions Faraday has proved himself "so thoroughly conscious of the fundamental forms of space, time, and force," Maxwell has discovered, that he inevitably presents them in relations that are consistently geometrical. It is in this way that the *Researches* become mathematically so ambivalent: in one sense the antithesis of mathematics, closer rather to alchemy; in another, field-theory almost fully formed.

Once "the mathematical character and scientific usefulness" of Faraday's method has been unveiled, the interesting possibility arises that the "translation" might run the other way: professed mathematics can be translated *into* the language of Faraday, and thereby gain the power and vividness of this new representation:

> In this outline of Faraday's electrical theories, as they appear from a mathematical point of view, I can do no more than simply state the mathematical methods by which I believe that electrical phenomena can be best comprehended and reduced to calculation, and my aim has been *to present the mathematical ideas to the mind in an embodied form*, as systems of lines or surfaces, and not as mere symbols, which neither convey the same ideas, nor readily adapt themselves to the phenomena to be explained. (*SP* i/187; emphasis added)

In a letter to Faraday, Maxwell described his own introduction to electrical studies; the passage corresponds to one quoted earlier from the preface to the *Treatise* (page 39, above), but with an interesting difference. In the Preface, he speaks of translating "what I considered to be Faraday's ideas into a mathematical form." In writing to Faraday, he had described the reverse of this process:

10. Francis Bacon, Preface to the *Great Instauration*. James Spedding and Robert Ellis (eds.), *The Works of Francis Bacon* (14 vols; London: 1858–74) *4*, p. 17.

> When I began to study electricity mathematically I avoided all the old traditions about forces acting at a distance, and after reading your papers as a first step to right thinking, *I read the others, interpreting as I went on, but never allowing myself to explain anything by these forces*. It is because I put off reading about electricity till I could do it without prejudice that I think I have been able to get hold of some of your ideas. (*L*: 245; emphasis added)

By this account, Maxwell began the study of electricity through the language of Faraday, as in effect his mother-tongue in the new science, and then read the "mathematical" texts, interpreting them *out of* the language of equations of action-at-a-distance in which they were written, *into* the new language of Faraday, the terminology of "contiguous particles" and lines of force in a medium. Faraday himself, as we shall see (page 62, below), in his own way did somewhat the same thing.

In a letter to Maxwell, written somewhat earlier than Maxwell's letter quoted above, Faraday concluded with this plea:

> There is one thing I would be glad to ask you. When a mathematician engaged in investigating physical actions and results has arrived at his conclusions, may they not be expressed in common language as fully, clearly, and definitely as in mathematical formulae? If so, would it not be a great boon to such as I to express them so?—translating them out of their hieroglyphics ... I think it must be so, because I have always found that you could convey to me a perfectly clear idea of your conclusions. (*L*: 206)

This is the use of the Rosetta stone to move out of the "hieroglyphics" of analytic mathematics, into a mode which Faraday can grasp; Faraday's own devices, of which Maxwell has endeavored to make himself master, are perhaps the best language into which the analytic results could be transformed.

I have proposed that the *Treatise* be understood as the culmination of a project of translation, and I have explored briefly the sense in which the method of "physical analogy" serves Maxwell as a first step in this process. The *Treatise*, on the other hand, is not, on the whole, written in the mode of physical analogy. The first phase has been passed through, the mathematical validity of Faraday's method has been established, and the objective of the *Treatise* has become somewhat different—not so much to illustrate Faraday's method, as to cast it, whole, into the form of a systematic treatise. The techniques of analytic representation of Faraday's ideas have been practiced and are already at Maxwell's command; the new problem is to exhibit these ideas for a student in a way that will first make them clear, and then show their coherence in a deductive account drawn from first principles. From this point of view, the "translation" to be accomplished in the *Treatise* is a methodological one, from an account written in the order of inquiry to one presented in the order of demonstration—from a series of research reports (the *Experimental Researches*) to a textbook (the *Treatise*). Here Whately's discussion is relevant: Faraday has written in the mode of the inquirer, which is the mode of philosophy, or logic; Maxwell is to write in the mode of teaching, which is that of rhetoric, in Whately's broadened sense of that term.[11] This is not to suggest that the *Treatise*, as a textbook, is in any sense light work; the "student" to whom Maxwell so often addresses the text, is everyman, or Maxwell himself. The task is to assure that Faraday's viewpoint indeed admits a cogent, coherent statement—and indeed there has been no dearth of subsequent critics to point out that the *Treatise* by no means fully accomplishes its object.

The new formulation in systematic order would not necessarily be better than Faraday's original, even if the new order were perfect; the *Researches* and the *Treatise* represent two complementary motions of mind. Maxwell was deeply impressed by the vividness and clarity which belong to "the nascent state" (*Tr* i/xi). The final result is to be, through the exercise of translation, an avenue into the insights and the language of the original text.

THE INTERPRETATION OF "LINES OF FORCE"

If we are to read the *Treatise* as a "translation" of Faraday, it will be important first to give some consideration to the efforts which had been made, to a large extent even before Maxwell began his study of electricity, to incorporate Faraday's *Experimental Researches* into the scientific dialogue of his time. We have already seen that Faraday's methods were of a sort which the best theorists would be likely to find alien in spirit, and very difficult to use in practice. Nonetheless his steady flow of discoveries, interwoven in the *Researches* with an insistent element of prediction and interpretation, could not be ignored. The result was a vast range of responses to Faraday throughout European scientific literature over a period of many years. There has probably never been a case quite like this, in which the normal course of intelligent scientific thought was so fundamentally disrupted by a voice speaking a different language, but saying what appeared to be very interesting things.

Some theorists chose to use his data and ignore the rest. This I think would be true of Franz Ernst Neumann, the Königsberg physicist (1798–1895), for example—and if so, it does not necessarily represent either a fault of judgement on Neumann's part or a misfortune for science as a whole. Maxwell is genuinely fascinated by the extent to which the same data can quite properly be treated in very different styles, and while these may indeed await some ultimate reconciliation, it is, according to him, to

11. See Author's Preface, page xx above.

the advantage of scientific thought if the alternatives are worked out separately and distinctly—by "advocates" rather than "judges." And it must be kept in mind that if read critically, Faraday's *Researches* appear as a chain of outrageous fallacies: a mathematically fastidious reader must somehow penetrate this barrier, to discover how fruitful such a fallacy as that of the "curved lines" can be in the hands of a brilliantly naive thinker. Neumann made no use of the "lines of force," but elegantly employed potential methods to draw Faraday's results on electromagnetic induction into the framework of Ampère's theory, thereby developing an extended theory of electrodynamics based entirely on action-at-a-distance. Maxwell ultimately found Neumann's concept of the mutual potential of two circuits to be of great value, even in the interpretation of Faraday.[12] But Neumann paid virtually no attention to the notions with which Faraday presented and explained his own discoveries.

On the other hand, Faraday did find warm friends, even among theorists, and one encounters in the literature of his time a few very perceptive appreciations of his work, as well as occasional experimental and theoretical investigations which carry it forward in very nearly its own terms.

For us, at present, this study of the scientific world's reaction to Faraday can be only an incidental investigation; it would be impossible here to attempt a thorough survey, or to draw the generalizations which could only emerge from a detailed study. No biographer of Faraday, or historian of science, has to my knowledge as yet really undertaken this most revealing inquiry. It is not likely, indeed, that the fruits of the study would take the form of generalizations; in each case, it would be the response of one individual scientific personality to a disturbing and challenging stimulus. It will be helpful here nonetheless to examine a select few of these cases, which will serve as touchstone to reveal the difficulties in the translation of Faraday's ideas. We should then understand better the unique contribution which Maxwell makes in accomplishing a systematic reformulation of the *Experimental Researches*.

Since one of the most striking phrases in Faraday's vocabulary is the famous "lines of force" which we have discussed at some length in Chapter 1, let us begin by looking briefly at the place this concept found in the literature of Faraday's time. We will meet right away a certain anomaly—but then, anomalies must abound when a mind as original as Faraday's plunges into a conversation for which he is in any formal sense quite unprepared. The anomaly in this case is that a good many people knew, in certain respects, more about the lines of force than Faraday ever came to know.

First of all, we must remember that the patterns traced by iron filings had had a long history: "I have even seen Samothracian iron dance, and

12. (*Tr* ii/190). Compare page 137, below.

iron filings go mad in a bronze bowl," Lucretius says,[13] and Niccolò Cabeo discussed the *lineae virtutis* in his *Philosophica Magnetica* (1629).[14] Earlier, Gilbert had observed the "verticity" with a detecting compass needle, thereby tracing the lines on the surface of a magnetic sphere.[15] Such phenomena had altogether different meanings for different philosophers, but the topic has a certain continuity, and enters the modern literature explicitly in many eighteenth-century texts. For example, Euler, in a text Faraday cites:

> Every magnet exhibits phenomena altogether similar. You have only to place one on a table covered with filings of steel, and you will see the filings arrange themselves round the loadstone AB, nearly as represented in fig. 6 ... in which every particle of the filings may be considered as a small magnetic needle, indicating, at every point round the loadstone, the magnetic direction.
>
> The arrangement assumed by the steel filings leaves no room to doubt that it is a subtile and invisible matter which runs through the particles of the steel, and disposes them in the direction which we here observe. It is equally clear that this subtile matter pervades the loadstone itself, entering at one of the poles, and going out at the other: so as to form, by its continual motion round the loadstone, a vortex.[16]

Not only did Euler help to draw attention to the patterns formed by the filings, but we see from such passages as this that he, like Descartes himself, regards them as delineating a physical process in the space around the magnet, a process which is in turn the principle of all magnetic phenomena. At the end of the long discussion of the magnet in the *Principles of Philosophy*, Descartes calls special attention to the patterns formed by filings, which he says confirm his theory of magnetic vortices, and he discusses certain patterns in detail.[17] We have seen (page 27) that Faraday considers Euler the source of his own "conduction" hypothesis, even though he does not necessarily agree literally with Euler concerning the actual flow of a fluid. The iron filing patterns have the same degree of physical significance for

13. Lucretius, *De Rerum Natura*, trans. W. H. D. Rouse (Cambridge, Mass.: Harvard University Press, 1947), vi, ll.1044–45; p. 519.

14. Edmund T. Whittaker, *A History of the Theories of Aether and Electricity: Volume I, The Classical Theories* (2nd ed., London: Thomas Nelson & Sons, 1951), p. 171.

15. William Gilbert, *On the Loadstone and Magnetic Bodies* (1600), trans. P. F. Motteley (1892) (Ann Arbor: Edwards Bros., 1938), p. 118; Book III, especially Chapter IX.

16. Leonhard Euler, *Letters of Euler on Different Subjects in Physics and Philosophy Addressed to a German Princess*, trans. Henry Hunter (2 vols.; London: 1802), *2*, p. 240. Cited by Faraday, (*XR* iii/528).

17. René Descartes, *Principes de la philosophie*, Book iv, §179, *Œuvres de Descartes*, ed. Charles Adam and Paul Tannery (Paris: Leopold Cerf, 1904) *9*, pp. 302–303.

Descartes, Euler, and Faraday, and to all three they suggest the idea of conduction. Not many of Faraday's readers would have shared such an opinion, but Euler's hypothesis was well known, and the patterns of the filings or compass directions appear in many treatises.[18]

What is perhaps more significant is the fact that the "magnetic curves" had become the object of analytic study, and were well understood as the trace of the vector resultant of forces acting from magnetic poles. This is not indeed a late development; concurrently with Euler's own interest, the equation of the curves due to a pair of poles was determined by the mathematician Johann Lambert at Berlin.[19] Dr. Peter Roget, though a physician by training, wrote extensively on the mathematics of the magnetic curves in his treatise on magnetism, published in 1832 by the Society for the Diffusion of Useful Knowledge, and this was included in turn in Sir David Brewster's article "Magnetism" in the seventh edition of the *Encyclopædia Britannica*.[20] Roget in fact lectured at the Royal Institution in 1831 on the subject of a set of rules he had constructed for tracing the forms of the magnetic curves mechanically, and he at the same time demonstrated a remarkable locus property of the curves, on which his tracing machine was based.[21] Although Roget later became Fullerian professor of physiology at the Royal Institution, and hence at least officially Faraday's colleague there, it seems clear from all we have said that Faraday did not share this fairly widespread familiarity with the magnetic curves as mathematical entities, precisely defined and understood. He does once, late in the *Experimental Researches*, compare the iron filing patterns with Roget's mathematical constructions; but he never reveals any recognition of the principle of vector composition on which the construction of the curves is based, even though this must have been entirely familiar to his scientific contemporaries, and is fully diagrammed in many works, including Roget's.

We see that the magnetic curves as representations of the forces of magnets were very well known and understood in Faraday's time—and yet we were not wrong in identifying the "lines of force" as one of Faraday's most distinctive and private terms. As *curves*, they were well known, and not

18. For example: George Adams, *An Essay on Electricity* (2nd ed.; London: 1785), which includes excellent diagrams of the patterns; and John Robinson, *A System of Mechanical Philosophy* (4 vols.; Edinburgh: 1822).

19. Auguste de la Rive, *A Treatise on Electricity*, trans. Charles V. Walker (3 vols.; London: 1853–58), *1*, p. 185. De la Rive reports Lambert's determination of the equation from the force law, and also the converse, the determination of the force law from the form of the curves. The latter, he says, was done by Charles Cellerier.

20. Peter Roget, *Treatise on Electricity, Galvanism, Magnetism, and Electro-magnetism* (London: Society for the Diffusion of Useful Knowledge, 1832), pp. 17 ff.

21. *Ibid.*, p. 20.

thought of particularly in terms of Faraday. Auguste de la Rive, though a friend of Faraday's and writing at a time when the latter's work was reaching fruition, was able to devote a long and thorough section of his *Treatise* to the magnetic curves without so much as mentioning Faraday—it is de Haldat, for example, who is credited with developing a technique for preserving the filing patterns, and with naming them (*phantômes magnétiques*).[22] What is unique is Faraday's extension of the role of the lines, until as we have seen they become crucial to his own understanding of electromagnetic induction, and emerge finally as physically operative agents in nature.

With the exception of Thomson, whose work is so fundamental to this study that it will be discussed separately below, I have found only one writer who formulates Faraday's theory of electromagnetic induction in Faraday's own terms, using the "lines of force." This was Richard van Rees (1797–1875), professor at Utrecht. In Faraday's Series XXVIII ("On Lines of Magnetic Force," 1852), which he had seen in German translation in the *Annalen der Physik*, and in Faraday's paper "On the Physical Character of the Lines of Magnetic Force" (1852), van Rees had discovered the role which Faraday was giving to the lines of force, and he evidently sensed the possible importance of this "new viewpoint." In a paper "On Faraday's Theory of the Magnetic Lines of Force" the following year, he examined with great care the mathematical possibilities and implications of Faraday's principles.[23] He has in view particularly sections 3094–3115 of Series XXVIII, in which Faraday concludes first that no current is induced in a loop which lies outside a cylindrical magnet rotating on its own axis (*XR* iii/338), and finally draws the general law of induction in terms of the lines of force, in a passage we have already quoted (page 13):

> They also prove, generally, that the quantity of electricity thrown into a current is directly as the amount of curves intersected. (*XR* iii/346)

Van Rees gives analytic form to this last statement, drawing for the purpose on the action-at-a-distance theory of the Continental tradition. He thinks of the situation of Figure 1a, in which a magnetic pole of strength μ gives rise to a magnetic intensity **H** at a distant point, through which an element **ds** of a circuit passes.[24] If the circuit is moved, so that the element

22. A. de la Rive, *op. cit.*, *1*, p. 184.

23. Richard van Rees, "Ueber die Faradaische Theorie der Magnetischen Kraftlinein," *Annalen der Physik*, *90* (1853), p. 415.

24. [H ↔ R]. The term "magnetic intensity" was used by Siméon Denis Poisson (1781–1840) to denote (in effect) the force per unit pole strength on a test pole at a point. Poisson, "Mémoire sur la theorie du magnetisme," *Mémoires de l'academie royale des sciences de l'institut de France*, *5* (1826, for the years 1821–22) p. 265. The theory of magnetic quantities and measurements was systematized by Karl Friedrich Gauss (1777–1855), "Intensitas Vis Magneticae Terrestris ad Mensuram Absolutam Revocata" (1832), *Werke* (12 vols.; Göttingen: 1863–1933), 5, pp. 81 ff.

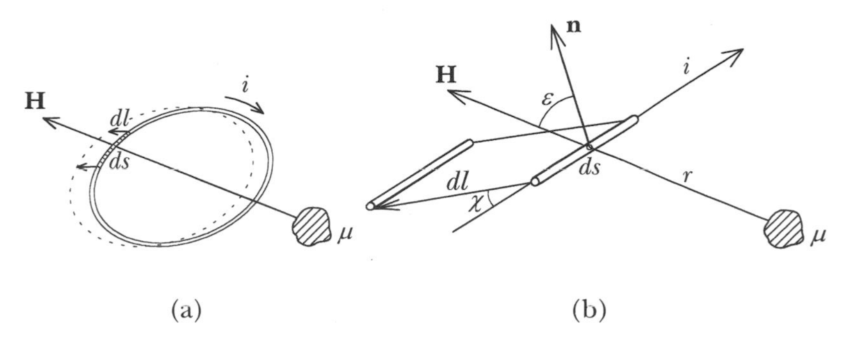

Figure 1. Van Rees' interpretation of Faraday

ds travels distance **dl** at the angle χ to its own length, as in Figure 1b, the area swept out is

$$ds\, dl \sin\chi\,. \qquad [ds \leftrightarrow \delta l]$$

He interprets Faraday's phrase "the amount of curves intersected" as the quantity:

$$(\mathrm{H} \cos\varepsilon)\,(dl\; ds \sin\chi)$$

in which ε is the angle between **H** and the normal to the plane (**ds, dl**). That is, he understands the "amount of curves" as the product of the area and the normal component of the intensity, or the quantity now termed the "flux." Hence van Rees writes for the entire circuit:

$$\iint \mathbf{H} \cos\varepsilon\, \sin\chi\; dl\, ds.$$

The "quantity of electricity thrown into a current" he translates as the integral of the electromotive force with respect to the time.[25] Hence Faraday's law quoted above is translated as:

$$\int E\, dt = \iint H \cos\varepsilon\, \sin\chi\; dl\, ds.$$

He points out that if one thinks of intersections at various angles with some given bundle of Faraday lines, this quantity will remain constant, and hence "obliquity of intersection makes no difference," as Faraday asserted (*XR* iii/345).

In 1846, Wilhelm Weber of Göttingen (1804–1890), originally a junior co-worker with Gauss, had asserted a sweeping new action-at-a-distance equation which included Coulomb's law, Ampère's phenomena of the attraction of currents, and Faraday's electromagnetic induction; this single

25. The term "electromotive force" to denote work per unit charge between points in an electric circuit derived from the earlier term "electroscopic force" used by Georg Simon Ohm (1787–1854) (Whittaker, *op. cit.*, p. 91). F. E. Neumann used the term "elektromotische Kraft" in his 1845 memoir, "Die mathematische Gesetze der inducierten elektrischen Ströme," *Abhandlungen der Berliner Akademie*, (1845), pp. 1 ff.; ed. Carl Neumann, *Ostwalds Klassiker der exakten Wissenschaft Nr. 10.*

fundamental equation (*Grundgesetz*) was subsequently more clearly reformulated.[26] In general, Weber's equation is this:[27]

$$f = \frac{q_1 q_2}{r^2}\left[1 - \left(\frac{dr}{dt}\right)^2 + 2r\left(\frac{d^2 r}{dt^2}\right)\right] \qquad \begin{matrix}[q_1 \leftrightarrow E] \\ [q_2 \leftrightarrow E'] \\ [r \leftrightarrow R]\end{matrix}$$

Here q_1 and q_2 are two charges, which are, in general, moving with respect to one another with the relative velocity dr/dt; to resolve static and current electricity into a single equation, Weber employs the hypothesis of Gustav Fechner (1801–1887) that a current consists of equal and opposite flows of positive and negative charges. Whether Weber's equation would indeed account for electromagnetic phenomena without paradox was the subject of a very long and interesting dialogue; it is discussed in the last chapter of the *Treatise* (*Tr* ii/486 ff.).

Very much as Liouville sought to establish that Faraday's phenomena of induction accorded with Poisson's theory (page 42 above), van Rees is anxious to show that Weber's law will account for Faraday's observation that there is no induction in a loop placed in the vicinity of a revolving cylindrical magnet. He applies Weber's law to this case, and by showing that the induced electromotive force is a complete differential, concludes that there will be no net induction. Faraday's facts, he concludes, are in accord with known laws, and thus in themselves do not demand a change of theory.[28]

Van Rees has at this point gone a long way toward grasping analytically the physics which Faraday articulated in such a different way. Is it possible, he wonders, that here is a new fundamental law, an alternative to Weber's? Admittedly, the precision van Rees demands has to be furnished by support from concepts out of another theory:

> [E]s kann schwerlich geläugnet werden, dass Faraday sich hier nicht mit der Klarheit und Genauigkeit ausgedrückt hat, welche erfondert wird, um aus den von ihm gegebenen Sätzen eine mathematische Entwicklung der Induktionserscheinungen abzuleiten, wenn sie nicht anders woher beleuchtet werden.[29]

26. Wilhelm Weber, *Elektrodynamische Maasbestimmungen über ein allgemeines Grundgesetz der elektrischen Wirkung* (1846–48) *Wilhelm Weber's Werke* (6 vols.; Berlin: 1892–94), *3*, pp. 25 ff., and *Elektrodynamische Maasbestimmungen: insbesondere Widerstandsmessungen* (1850), *Werke*, *3*, pp. 301 ff.

27. Weber, "On the Measurement of Electro-dynamic Forces" (1848), a partial translation from the *Elektrodynamische Maasbestimmungen*, in Richard Taylor, ed., *Scientific Memoirs* (5 vols.; London: 1837–52; reprinted New York: Johnson Reprint Corp., 1966), *5*, p. 519.

28. Van Rees, *op. cit.*, p. 427.

29. *Loc. cit.*

But van Rees has himself, as we have seen, supplied this additional light. He has made it possible to go further, and ask whether this new theory might not supplant the others:

> Es ... kann indessen gefragt werden, ob die Faradaischen Theorie die früheren Hypothesen nicht überflüssig mache.[30]

Van Rees immediately identifies a difficulty: Faraday has shown how to determine the induced current when a circuit cuts the lines of force, but has not shown at all how the lines arise, or what precise shape they will have. The theories of Ampère or Coulomb, by contrast, predict the form of the magnetic lines in good agreement with observation. Faraday's method is, then, incomplete; we have to have recourse to a higher principle, based on the inverse-square law. Faraday does not provide us with the inverse-square law, or any equivalent to it; van Rees detects that although Faraday does not deny this law, "allein es ist seiner Theorie fremd,"[31] Faraday's principle for diamagnetism, that a diamagnetic material moves from points of strong to points of weak magnetic force, has, van Rees finds, no connection with the rest of his theory—while in the Continental theory, it can be derived. He concludes that Faraday's theory is dependent on the higher principles embodied in other theories such as Weber's, and cannot replace them.

Given this negative conclusion, which indeed might seem hard to dispute on the evidence van Rees has carefully assembled, he then grants Faraday all the credit that can be summoned: Faraday has shown more fully than anyone before the properties of the lines of force; he has obtained laws for them, which can indeed be derived from other, higher theories, but which until Faraday's assertion of them were not known; and van Rees appreciates the fact that Faraday has put the analytic law in an "intuitive form" which simplifies its application—he has opened a new point of view, and shown the way to new discoveries.[32]

Van Rees, standing on mathematical *terra firma,* has leaned as far toward Faraday's strange new world as he sees possible; he retreats from it with what seems genuine regret, and with a sense of the possibilities that lie in the direction Faraday has indicated. Where is the gap that separates van Rees from Faraday? What prevents this effort at translation from yielding a coherent new field theory? We see that van Rees insists that a "higher" theory, action-at-a-distance, is needed to account for the origin of the lines of force; has he, then, missed an aspect of Faraday's thought? The answer is, I think, that he has not seized upon Faraday's concept of the lines as themselves physical—that is, as themselves the operative entities in

30. *Ibid.*, p. 432.

31. *Ibid.*, p. 433.

32. *Ibid.*, p. 436.

electromagnetism. Van Rees, in accord with tradition, sees them as consequences of entities elsewhere, poles and currents, which as sources produce the lines according to a law of action-at-a-distance. Faraday's view has dispelled such sources, the erstwhile "poles" becoming merely regions of concentration of the lines of force. But though Faraday has thrown out certain hints, he has not shown how to give precision to the configurations of lines of force. It is left to Maxwell to show, as van Rees could not, how the lines can give strict mathematical formulation without recourse to any theory as "higher" than Faraday's.

Van Rees had shown that precision could be supplied to certain of Faraday's phrases; others were not able to see this much. Leopoldo Nobili (1764–1835), who with Vincenzio Antinori (1792–1865) in Florence was one of the first experimenters to seize on early reports of Faraday's discovery of electromagnetic induction, and to pursue the inquiry empirically, speaks later of his own initial employment of Faraday's concept of "lines of force," with a note of apology:

> I too used the curves as an aid until 1824, in explaining the electrodynamic phenomena. ... This viewpoint is so to speak more concrete, but less philosophical than that of M. Ampère.[33]

This tendency, even among Faraday's admirers and friends, to apologize for the imprecision of his thought, is manifested particularly in the statements of John Tyndall. Often he expresses confidence in the profundity of Faraday's understanding but always with a qualification, as in this reflection in 1870:

> Faraday ... had his mind fixed upon his lines of magnetic force. To this conception, however, though it formed the guiding light of his researches, he never gave a mechanical form. Hence arose his difficulty in dealing with the phenomena exhibited by crystals in the magnetic field. His thoughts doubtless dwelt in the profoundest depths of the subject.[34]

At times, and in private, Tyndall's impatience with Faraday's "mistiness" and his dissatisfaction with the "line of force" as a scientific concept, became more outspoken, as in this 1855 letter to T. A. Hirst, published by Williams:

> The fact is that an experiment made with reference to the lines of force is virtually made with reference to the principle on which the lines of force themselves depend, and the deeper scientific mind

33. Leopoldo Nobili (1832), trans. as "Physikalische Theorie der elektro-dynamischen Vertheilung," *Annalen der Physik 27* (1833), p. 434. I have translated from the German version.

34. John Tyndall, *Researches on Diamagnetism and Magnecrystallic Action* (London: 1870), p. 183.

> looks through the lines of force after this principle. Faraday's achievements are due to his immense earnestness and great love for his subject and this very mistiness which serves to obscure the verity of matters may have its compensations by rendering the subject attractive. ... I heard Biot once say that he could not understand Faraday, and if you look for exact knowledge in his theories you will be disappointed—flashes of wonderful insight you meet here and there, but he has no exact knowledge himself, and in conversation with him he readily confesses this.[35]

This latter quotation undoubtedly represents Tyndall's true judgement of Faraday's methods and his "lines of force," at least at the time when Tyndall's own dialogue with Faraday on the subject of diamagnetism had reached its climax. It must also reflect the general opinion of Faraday among scientists of his time, and the hopelessness they felt in giving strict, scientific interpretation to such concepts as that of the "lines of force." The reaction of George Biddell Airy (1801–1891), Astronomer Royal and an accomplished mathematical physicist, reveals the degree of near exasperation which Faraday could elicit in a clear thinker:

> I declare that I can hardly imagine anyone who practically and numerically knows this agreement [between Ampère's theory and Faraday's in predicting electromagnetic phenomena], to hesitate an instant in the choice between this simple and precise action, on the one hand, and anything so vague and varying as lines of force, on the other.[36]

It is such reflections as these, which would certainly have reached Maxwell in Cambridge, to which he refers when he says of Faraday's researches, at the outset of "Faraday's Lines":

> [T]hough they have been interpreted mathematically by Prof. Thomson and others, [they] are very generally supposed to be of an indefinite and unmathematical character, when compared with those employed by the professed mathematicians. (*SP* i/157)

Faraday was, of course, aware of many of the reactions to his papers, and was very much concerned by them, coming as they often did from a mathematical realm he could not himself understand, but was obliged to respect. The Reverend John Barlow, to whom the Airy comment above was addressed, did Faraday the favor of sending it on to him, and Bence Jones, Faraday's biographer, prints his response:

> I return you Airy's second note. I think he must be involved in some mystery [!] about my views and papers; at all events, his notes mystify me. ... What [the law of the inverse square] has to do with my

35. L. P. Williams, *op. cit.*, p. 509.

36. Bence Jones, *Life and Letters of Faraday*, 2, p. 353.

> consideration, I cannot make out. I do not deny the law of action referred to in all like cases; nor is there any difference as to the mathematical results (at least, if I understand Thomson and van Rees) whether he takes the results according to my view or that of the French mathematicians. Why, then, talk about the inverse square law? I had to warn my audience against the sound of this law and its supposed opposition on my Friday evening, and Airy's note shows that the warning was needful.[37]

The difficulty of communication between Faraday and anyone who knows mathematical physics is evident, as we have seen in the first chapter. In a sense, Faraday is unable to help with the conversation from his side—the work of establishing any connection between the scientific tradition and the *Experimental Researches* had all to be done by others, and there were few who would sustain the effort to communicate with this innocent. Airy was shown the above reply—and returned it to Barlow "without comment."[38]

Yet as the allusion to Thomson and van Rees in the above reveals, he found any successful mathematical interpretation of his views encouraging, even though he could not understand it. He wrote to Tyndall, concerning van Rees's paper which we have discussed:

> Reading Matteucci carefully, and also an abstracted translation of van Rees's paper, is my weighty work. ... I think they encourage me to write another paper on lines of force, polarity, &c., for I was hardly prepared to find such strong support in the papers of van Rees and Thomson for the lines as correct representants of the power and its direction; and many old arguments are renewed in my mind by these papers.[39]

The work of Matteucci referred to here, as a further source of encouragement, is no doubt the *Cours spécial sur l'induction* of Carlo Matteucci of Pisa (1811–68), which Faraday sent on to Tyndall.[40] Matteucci was for a long time an ardent admirer of Faraday, who had corresponded with him over the years, and had made what amounted to a pilgrimage to England to meet him.[41] Matteucci had become a central figure in research in Italy, much of his work relating to Faraday's discoveries in electricity and magnetism. Riccardo Felici (1819–1902), his student, who produced perhaps the most definitive study of both the theoretical and experimental aspects of electromagnetism, identified this *Cours spécial* as the particular source of

37. *Ibid.*, p. 355.

38. *Ibid.*, p. 356.

39. *Ibid.*, p. 347.

40. Carlo Matteucci, *Cours spécial sur l'induction* (Paris: 1854); cf. Tyndall, *op. cit.*, p. 72.

41. N. Bianchi, *Carlo Matteucci e l'Italia del suo tempo* (Rome: 1874), pp. 29, 63–65, 89–94.

inspiration for his work.[42] Interestingly, Lewis Campbell, Maxwell's friend and biographer, remarks (without explanation) that Maxwell, in his travel to Italy in 1867, claimed that he had learned Italian especially for the sake of conversing with Matteucci (*L*: 238). Matteucci's reaction to Faraday's *Researches*, in his *Cours spécial*, is then of considerable interest for the present study, as an indication of what a sympathetic and influential physicist was able to make of Faraday's work at the distance of Pisa.

What was it in the *Cours spécial* which Faraday found such a source of encouragement? For one thing, I suspect, simply the fact that Faraday's view of the role of the lines of force in induction is fully reported—not, indeed, as a concept which Matteucci has himself adopted, but as a postscript to a long account of experimental results in induction:

> Je n'achéverai pas cette leçon sans dire quelques mots sur les *lignes* ou *courves de force magnétique*, et sur le rôle que Faraday attribue à ces lignes dans les phénomènes de l'induction. ... [Faraday] dans ses Mémoires se sert à chaque instant du mot *line of force*.[43]

Matteucci reports that for Faraday induction occurs only when a conductor moves across the lines; that a given section of the lines represents a constant sum of inductive force; and that a current is induced in a circuit when one side cuts more lines than the other. In connection with the discussion of diamagnetism, later in the book, he describes the methods of plotting the curves, and it is interesting to see that Matteucci himself uses the term *champ magnétique* freely, and takes the lines as delineating this field. He discusses Faraday's principle that diamagnets move toward areas in which the force is weaker, and in this connection points out that the shape of pole pieces of the magnet, by changing the distribution of the field, can make all the difference in the observed motions. He recognizes that Thomson has obtained the same results analytically (in an 1847 paper, to be discussed below). He joins Faraday in his criticism of Weber's hypothesis for diamagnetism.

Probably, however, the passage in the *Cours* which most moved Faraday "to write another paper on lines of force" was the conviction underlying Matteucci's closing remarks:

> Poursuivre par de nouvelles expériences et par les études théoriques la découverte de la relation qui doit exister entre ces deux effets [diamagnetism, and the rotatory power of magnetism on polarized light] et l'induction magnétique, expliquer cette rélation et rattacher

42. Because of its particular interest for Maxwell, Felici's work will be discussed in connection with the examination of Part iv, Chapter III of the *Treatise*, in Chapter 3 of the present study.

43. Matteucci, *op. cit.*, p. 84–85.

> par consequent la force inconnue de l'électricité et du magnétisme à celle de l'éther distribué dans les corps, tel est le travail ... qui donnera une grande impulsion à toutes les parties de la philosophie naturelle.[44]

Faraday must have recognized here a fellow spirit, one who was willing to try out his methods, who was an almost equally ardent experimenter, and who above all was ready to pursue a relation "qui doit exister" among the forces Faraday was exploring. The correspondence between Faraday and Matteucci reveals this feeling of rapport.

Matteucci asked Faraday, in a letter subsequent to the publication of the *Cours*, to explain the term "lines of force" for him. Faraday replied with care, in one of his most interesting efforts to make clear the meaning which the term had for him. He distinguishes, as usual, "physical lines" from the lines as "representations," and of the latter he says:

> The use of lines of magnetic force (without the physical) as true representations of nature, is to me delightful, and as yet never failing; and so long as I can read your facts, and those of Tyndall, Weber, and others by them, and find they all come into one harmonious whole, without any contradiction, I am content to let the erroneous expressions, by which they seem to differ, pass unnoticed.[45]

As representations, the lines appear to Faraday as a language in which facts can be "read," and into which other, conflicting statements can be translated and thereby reconciled. This letter was written in the midst of a tense dispute between Matteucci and Tyndall, which had been thrust into Faraday's lap. Faraday is in effect offering his volatile Italian correspondent the lines of force as peacemakers, a universal language in which nature can be represented without dispute. Faraday testifies that he finds harmony in the lines really in two senses, first as bringing unity to our view of the many forces of nature, and hence as a means to the unity Matteucci is seeking in the quotation from the close of the *Cours*, above; and second, as a neutral representation of facts, a language through which peace can be brought to the community of philosophers. To judge from the literature of the discussion of the "lines of force," however, their effect may often, unfortunately, have been quite the opposite.

INTERPRETATIONS OF FARADAY'S ELECTROSTATICS

The present study, directed as it is toward a reading of Part iv of Maxwell's *Treatise* in which a theory of electromagnetic phenomena is developed, does not emphasize electrostatics. Thus in exploring certain of

44. *Ibid.*, p. 274.

45. H. Bence Jones, *op. cit.*, p. 367.

Faraday's views, we have concentrated on those of Volume Three of the *Experimental Researches*, in which magnetism holds the center of the stage, and we did not enter upon those earlier series in which Faraday's whole preoccupation is with electrostatics. Nonetheless, there are two instances of the "translation" of Faraday's ideas prior to Maxwell's work, which lie in the realm of electrostatics, but are so revealing of both the possibilities and the difficulties inherent in the encounter of mathematical minds with Faraday's thought, that they demand our consideration. A brief digression into electrostatics will be rewarded by a fuller understanding of the magnitude of Maxwell's task.

Let us turn briefly then to the reception of Faraday's ideas on electrostatics, where the problems are somewhat different from those we have thus far encountered. Here Faraday is not bursting upon the world with new discoveries introducing a totally new science, as he did in the first *Experimental Researches* on electromagnetic induction, but is rather proposing a reformation of an established science. Electrostatics had already been carried to the highest levels of mathematical analysis, and physical theories of many sorts had already been advanced to account, quantitatively as well as qualitatively, for the phenomena. Certainly Faraday's discovery of specific inductive capacity, announced in Series XI (1837), was a new and crucial development, but it has the effect of extending and modifying an old science, rather than introducing a new one. As has been mentioned, it was quickly seen (by others) that the new development fitted with no difficulty at all into an already existing and quite complete mathematical theory. I mentioned earlier that Thomson reconciled Faraday's findings concerning specific inductive capacity with Poisson's theory of polarization, developed for a magnetic medium but equally applicable, as Thomson showed, to electrostatics (page 42, above). Ottaviano Mossotti of Turin (1791–1863) did essentially the same thing; he had only to translate magnetism into electricity, as Maxwell suggests (*Tr* i/70).[46] Poisson's theory, however, though it saves Faraday's observations, does not altogether represent his ideas.

Faraday, beginning with Series XI, undertook to interpret the phenomena of electrostatics as effects of a polarized medium, for which the "dielectric" materials he studied in this series became the paradigm. A body is not electrically "charged" with "fluid" on its surface, as action-at-a-distance theories had in one way or another supposed, but becomes rather the terminus of a state of strain in the surrounding medium. This state of the medium he calls that of "induction," the strain being conveyed through a sequence of "contiguous" particles. Perhaps we can best suggest Faraday's thought on

46. The apparent problem presented by the new results in electrostatics, especially those of Snow Harris, is described in Antoine Becquerel, *Traité de l'électricité* (7 vols.; Paris: 1834–40), 5, Part 2 (1840), pp. 63 ff. Cf. pages 92 ff., below.

electrostatics with some quotations from what is in effect an ode to induction, with which he concludes Series XI:

> Thus induction appears to be essentially an action of contiguous particles, through the intermediation of which the electric force, originating or appearing at a certain place, is propagated to or sustained at a distance, appearing there as a force of the same kind exactly equal in amount, but opposite in its direction and tendencies. ...
>
> Induction appears to consist in a certain polarized state of the particles, into which they are thrown by the electrified body sustaining the action...
>
> The principle of induction is of the utmost generality in electric action. It constitutes charge in every ordinary case, and probably in every case; it appears to be the cause of all excitement, and to precede every current. The degree to which the particles are affected in this their forced state, before discharge of one kind or another supervenes, appears to constitute what we call *intensity*.
>
> When a Leyden jar is *charged*, the particles of the glass are forced into this polarized and constrained condition... *Discharge* is the return of the particles to their natural state...
>
> All charge of conductors is on their surface, because being essentially inductive, it is there only that the medium capable of sustaining the necessary inductive state begins. (*XR* i/409–10)

Even in these few paragraphs, the full effect of which can only be felt by reading the original to the end without pause or omission, Faraday's idea gathers height and power and comes crashing in upon him like a wave upon the shore: what he finds himself saying, explicitly even if not quite intentionally, is that charge—i.e., "electricity"—is *nothing but* a strained state of a medium. "Charge" appears to be on the surface of "conductors" because it is "essentially" (i.e. in its essence, really nothing but) inductive, and the surface of the conductor is the place at which charge, understood in this way as a *state* of the surrounding medium, "*begins*." Again the "lines of force" are useful—borrowed here from magnetism, but later, in the researches of Volume Three, feeding back into it (*XR* i/411).

Faraday is here in strong dissent from the orthodox or, as he often says, "high" mathematical theory of action-at-a-distance. Yet his electrostatic views do bear a certain relationship to others which had already been formulated mathematically, in terms of atmospheres of "ether" or "caloric" envisioned about centers of force; in this sense, one might expect greater success for Faraday's dialogue with his contemporaries about electrostatics, than for that about magnetism and electromagnetism, in which his "lines of force" play a more dominant role.

Faraday did seize upon an earlier view of electrostatics as sympathetic to his own. This was a restatement by Mossotti, in terms of an electric

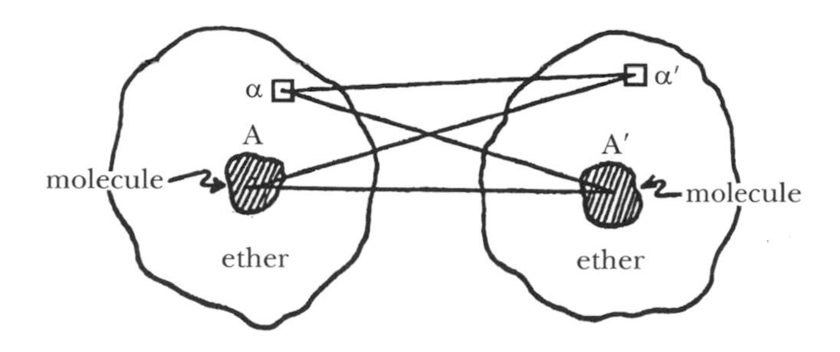

Figure 2. The forces assumed by Mossotti (1836)

atmosphere of the action-at-a-distance theory of Franz Aepinus (1724–1802).[47] Faraday greeted Mossotti's paper, "On the Forces Which Regulate the Internal Constitution of Bodies," with great excitement when it came to him in 1836, and he communicated it promptly for translation and publication in the first volume of Taylor's *Scientific Memoirs*.[48] Mossotti, reciprocally, found support for his theory in Faraday's Series XI, and developed Faraday's ideas analytically in 1846, as in the earlier paper he had given analytic form to Aepinus' theory. Despite the promise of this interesting exchange, it is hindered by the fact that Mossotti works chiefly in analytic mathematics, with the result that Faraday can read only his introductions and conclusions.

Let us review briefly the contents of the paper which Faraday received with such delight in 1836. Fortunately for Faraday, a long discursive section precedes the mathematical analysis; in it Mossotti reviews the project of Aepinus and others of accounting for the observed properties of bodies in terms of static equilibria of forces between molecules. Static electricity, he says, is a special aid in this effort; it seems that nature has given us the phenomena of electrostatics, in order "by separating the forces which she employs, to present herself in all her simplicity."[49] He sees a theory of static electricity as a clue to a molecular theory, which will ultimately include all the phenomena which bodies exhibit, including universal gravitation. He turns directly to the hypothesis of Franklin as formulated mathematically by Aepinus, but now, unlike Aepinus, he takes the "single fluid" of Franklin as a compressible ether surrounding the electrified body, and assumes a set of forces among two bodies and their surrounding ethers as indicated in Figure 2.

47. Whittaker, *op. cit.*, pp. 51 ff.

48. Ottaviano Mossotti, "On the Forces Which Regulate the Internal Constitution of Bodies," R. Taylor, ed., *Scientific Memoirs, 1* (1837), pp. 448 ff., L. P. Williams, *op. cit.*, pp. 294–96.

49. Mossotti, *op. cit.*, p. 448.

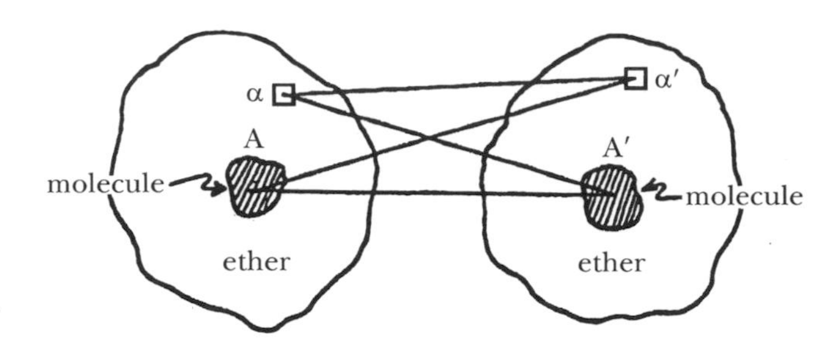

Figure 2 (repeated)

The symbols in the figure have the following signification:

A, A' : elements of the molecules
α, α' : elements of the corresponding ethers.

It was Aepinus' suggestion that not only would the particles of the ether repel one another, but that those of two molecules would repel likewise, despite Newton's demonstration of a universal force of attraction among masses. On the other hand, a force of attraction would bind molecules of matter to particles of ether, and if this ether-molecule attraction were to exceed the force of repulsion at large distances, universal gravitation would result—no longer as a fundamental law, but as a synthesis out of the same forces which caused electric and other phenomena of bodies. The forces assumed by Mossotti may be summarized as follows, again in relation to Figure 2:

$\alpha\alpha'$: ether/ether repulsion
$\alpha A'$: ether/molecular attraction
$\alpha' A$: ether/molecular attraction
AA' : molecular/molecular repulsion.

At very short distances, repulsion must dominate in order to account for the observed resistance of bodies to compression.

Mossotti undertakes to carry out this suggested hypothesis of Aepinus', making it the basis of a universal, mathematical theory of the phenomena of bodies:

> I have supposed that a number of material molecules are plunged into a boundless aether, and that the atoms of the aether are subject to the actions of the forces required by the theory of Aepinus, and then endeavoured to ascertain the conditions of equilibrium of the aether and the molecules. Considering the aether as a continuous mass, and the molecules as isolated bodies, I found that, if the latter be spherical, they are surrounded by an atmosphere the density of which decreases according to a function of the distance which contains an exponential factor. The differential equation which determines the density being linear, is satisfied by any sum of these functions answering to any

> number of molecules. Whence it follows that their atmospheres may overlay or penetrate each other without disturbing the equilibrium of the aether. ... [T]he reciprocal action of two molecules and of their surrounding atmospheres is independent of the presence of the others, and possesses all the characteristics of molecular action. At first it is repulsive, and contains an exponential factor which is capable of making it decrease very rapidly: it vanishes soon after ... so that they would remain in a state of steady equilibrium. At a greater distance the molecules would attract each other, and their attraction would increase with their distance up to a certain point, at which it would attain a maximum: beyond this point it would diminish, and at a sensible distance would decrease directly as the product of their mass, and inversely as the square of their distance.[50]

Mossotti's paper is in a certain sense an essay in field theory, and it will be worth our while to examine with some care the principal equations Mossotti develops. They will serve as a paradigm of the *genre* of "caloric" theories, as one kind of approach to a theory of the electrostatic field; here the term "caloric" is used in the sense given it by Pierre Simon Laplace (1749–1827). For Laplace it is a space-filling elastic fluid obedient to a strict mathematical law of force, and capable of accounting for the observed properties of bodies under compression.[51] Because the caloric theory of heat became untenable with the recognition of the interchange of heat and mechanical work, it is an easy error for the student of the history of science to think of "caloric" theory as if it were of an intrinsically imperfect or archaic type. It is important to recognize, therefore, that in the early years of the nineteenth century caloric theory belonged to the most modern and promising mode of physical theory; it was particularly adapted to the mathematical physic of force laws and potentials which had been brilliantly developed by Laplace, initially for the purposes of celestial mechanics.[52]

As Figure 2 shows, every region in the ether is subject to two types of force, the ether-ether repulsion and the ether-molecular attraction, while every element of a molecule experiences both an ether-molecular

50. *Ibid.*, pp. 450–51.

51. Pierre Simon Laplace, *Traité de mécanique céleste*, Book XII (1823); *Œuvres de Laplace* (14 vols.; Paris: Gauthiers-Villars, 1843–1912), *5*, pp. 104 ff. For an account of Laplace's caloric theory, see Stephen G. Brush, *Kinetic Theory*: *Volume 1, the Nature of Gases and of Heat* (Oxford: Pergamon Press, 1965), pp. 11 ff.

52. On caloric theory in general, see E. Mendoza's introduction to his edition of Sadi Carnot's *Reflections on the Motive Power of Heat* (New York: Dover Publications, 1960), pp. xv ff., and John T. Merz, *A History of European Thought in the Nineteenth Century* (4 vols.; New York: Dover Publications, 1965), *1*, p. 342n. Note for example John Dalton's use of caloric in *A New System of Chemical Philosophy* (1808) (New York: The Citadel Press, 1964), pp. 114, 131, based directly on Proposition 23, Book II, of Newton's *Principia*. James R. Partington, *A History of Chemistry* (4 vols.; London: Macmillan, 1961–1972), *3*, p. 767.

attraction and a molecular-molecular repulsion. Each of these forces is assumed to *act at a distance*, and to vary inversely with the square of the distance; the force at any point on an element of the ether, or on an element of a molecule, is the vector sum of all the forces on it, due to all the sources in the universe. Mossotti expresses these forces as potentials; with some revision of symbols, we may write these four potential equations to correspond to the four forces listed above:

POTENTIALS AT A POINT P IN THE ETHER:

Potential of ether/ether repulsion:

$$F = f\int_{\infty} \frac{q' d\tau'}{r}$$

Potential of ether/molecular attraction:

$$G_i = g\int_{\infty} \frac{\omega_i' d\tau'}{r}$$

POTENTIALS AT A POINT Q IN A MOLECULE:

Potential of molecular/molecular repulsion:

$$H_i = \gamma\int_{\infty} \frac{\omega_i' d\tau'}{r}$$

$$[H_i \leftrightarrow \Gamma_i]$$

Potential of ether/molecular attraction:

$$\Phi = g\int_{\infty} \frac{q' d\tau'}{r}$$

In each integral, $d\tau'$ denotes a volume element at the source-point, r denotes the distance from source point to the point at which the potential is measured, and the integral is a volume integral over all the source points. The remaining symbols have the following meanings:

f : coefficient of the ether/ether repulsion (i.e., force per unit mass at unit distance).

g : coefficient of the ether-molecular attraction.

γ : coefficient of the molecular-molecular repulsion.

q : ether-density at point P in the ether, at which the potential of the ether is measured.

q' : ether-density at point P' in the ether, a source-point contributing to the potential at P or Q.

ω : molecular density at point Q in a molecule, at which the molecular potential is measured.

ω' : molecular density at point Q_i' in the ith molecule, a source-point contributing to the potential at P or Q.

p : the fluid pressure at any point P in the ether.

$$[p \leftrightarrow \varepsilon] \quad \left[r \leftrightarrow \sqrt{(x'-\xi)^2 + (y'-\eta)^2 + (z'-\zeta)^2}\right]$$

$$[d\tau' \leftrightarrow dx', dy', dz', \text{etc.}]$$

Mossotti's question now is, given these equations of potential, what can we say about the equilibrium density configuration of the ether? He first proves the general result that the equilibrium density of the ether is, apart from a constant, directly proportional to a potential expression $\Sigma G_i - F$ —that is, that the *the ether is in effect a reified potential.* The steps in the reasoning to this conclusion are these:

(1) For mechanical equilibrium in the ether, the pressure-gradient must just balance distance forces, which are given by the negative gradients of potentials F and G; bearing in mind that that F is repulsive and G attractive, this may be written:

$$\nabla p = q\left(-\nabla F + \nabla \sum_{i=1}^{n} G_i\right).$$

(2) Mossotti now borrows from the mathematical theory of caloric in Laplace's *Méchanique céleste* the following result for the relation between pressure and density for the ethereal fluid:

$$p = \frac{1}{2} K q^2$$

in which K is a coefficient characteristic of the fluid.[53]

(3) Combining steps (1) and (2), we get:

$$q = \frac{1}{K}\left(-F + \sum_{i=1}^{n} G_i\right).$$

and we see that the ether-density is in effect itself a potential. This is what Newton had recognized about the elastic ether in his own speculations; though of course he did not speak of a "potential," he saw that the gradient of the ether-density would be a force by which one might attempt to account for the courses of the planets.[54] This helps us to understand the *idea* of caloric: as an elastic atmosphere in the sense of Newton, Dalton, Laplace, and Mossotti, caloric (whether by this or another name) is a "fluid" (i.e., a gas, not a liquid) the gradient of whose density is proportional to a force. *It is a scalar potential, taken as a physical substance.*

Differentiating the potential expression above, Mossotti gets Poisson's equation; it turns out that the density is at once potential and source:

$$\nabla^2 q = \frac{4\pi}{K} f q \text{ .}^{55}$$

This is the partial differential equation of caloric. Mossotti first obtains a

53. Laplace, *op. cit. 5*, p. 129; Brush, *op. cit.*, p. 12.

54. Isaac Newton, *Opticks* (4th ed., 1730) (New York: Dover Publications, 1952), Query 21, pp. 350–52.

55. Mossotti, *op. cit.*, equation VI, p. 455.

general solution by throwing the equation into spherical coordinates, and then (following Poisson) expressing the solution as a sum of orthogonal functions in a way quite familiar to students of potential theory, or, for that matter, to students of quantum mechanics in the Schrödinger form—the problem and the methods of solution are essentially the same as those of caloric theory. Mossotti turns directly from the general solution to the special case in which the molecules are spherical, and he obtains an equation for the ether-density about a molecule, on the assumption that the molecules are far apart in relation to their radius δ:

$$q = q_0 + \frac{4\pi}{3K}\sum_{i=1}^{n}\frac{g\omega_i' + fq_i}{r_i}\delta_i^3 e^{-\alpha r_i}.$$

in which q_i now denotes the density of the ether at the surface of the ith molecule, and α is defined as follows:

$$\alpha = \sqrt{\frac{4\pi}{K}}.$$ [56]

This crucial quantity, in effect relating the self-interaction of the ether to its compressibility, must be very large, Mossotti says, in which case the density given above will fall off very rapidly about any given molecule.

Mossotti concludes by deriving the force upon a molecule from this expression for the ether-density; here the mechanical equilibrium is based on the principle that the ether-pressure on the surface of a molecule must balance the distance-forces upon it:

$$\int_S p\,dS = \omega\int\left(\nabla\Phi - \sum_{i=1}^{n}\nabla H_i\right)d\tau,$$

in which the left-hand integral is over the surface of the molecule, p being the pressure upon it everywhere, and the right hand integral, expressing the body-forces on it, is taken over its volume.[57] Combining this principle with the ether-density already calculated about any given molecule, and assuming (with Aepinus) that $f = g$ and that γ is slightly less than g, Mossotti obtains this expression for the net equilibrium force upon a molecule as a function of the intermolecular spacing r_{12}:

$$f = gv_1(\omega_1 + q_1)\cdot v_2(\omega_2 + q_2)\frac{1+\alpha r_{12}}{r_{12}^2}e^{-\alpha r_{12}} - (g-\gamma)v_1\omega_1\cdot v_2\omega_2\frac{1}{r_{12}^2},$$ [58]

in which a positive sign denotes repulsion. This can be written more schematically:

56. *Ibid.*, p. 463.

57. *Ibid.*, p. 453.

58. *Ibid.*, p. 466.

$$f = g M_1 M_2 \left(\frac{1 + \alpha r_{12}}{r_{12}^{\;2}} \right) \cdot e^{-\alpha r_{12}} - (g - \gamma) \frac{m_1 m_2}{r_{12}^{\;2}} \; .$$

Here the second term represents the gravitational attraction; the first a short-range repulsion, the effective masses M_1 and M_2 including a component for the ether-densities q_1 and q_2 at the surfaces of the molecules. From this force expression, Mossotti obtains the equilibrium intermolecular spacing (at which point the first and second terms are equal, and the force zero), and shows that it will be a stable equilibrium. He also calculates the point of maximum attraction (cohesion), between the neutral point and the asymptotic gravitational decrease of force with distance.

Finally, Mossotti carries out the solution for intermolecular spacing in the case of a regular tetrahedral group of molecules, and shows in this case the increase of spacing with an increment in ether-density. He draws from this result the conclusion that the ether in question is indeed "caloric":

> For what else, in fact, is an increase or diminution of temperature in respect to a body, than a new state in which its molecules ... form, in consequence of their being more or less widely separated, a greater or less volume.[59]

It has been proposed that Faraday was deeply attracted to the natural philosophy of Roger Joseph Boscovich (1711–1787). The claim is not so much that Faraday had ever read Boscovich's *Theory of Natural Philosophy* as that there was something about the popular account of Boscovich's theory which appealed to him.[60] Boscovich's basic concept is, indeed, almost as remote from Faraday's own viewpoint as one could get, inasmuch as Boscovich takes as fundamental the idea of a *mathematical function*, a law of force of a more complex form than those monotonic functions of distance with which Newton deals in the *Principia*.[61] The suggestion is nonetheless Newton's, originating in certain "Queries" appended to the *Opticks* in which Newton is sketching a theory of diffraction in terms of a wave-like

59. *Ibid.*, p. 469.

60. Roger Joseph Boscovich, *A Theory of Natural Philosophy* (1763), (Cambridge, Massachusetts: M.I.T. Press, 1966). For the claim that Faraday is deeply imbued with Boscovich's thought, see L. P. Williams, *op. cit.*, pp. 78, 130, *et. al.* A corrective to this view is asserted by J. Brookes Spencer, "Boscovich's Theory and its Relation to Faraday's Researches: an Analytic Approach," *Archive for History of Exact Sciences*, *4* (1967), pp. 184 ff.

61. Spencer, *op. cit.*, pp. 201–202: "...the fundamental basis for the exclusion of gravitation from Faraday's physical thought [that it appears to be action-at-a-distance with no physical mechanism] is also basic for the fantastic comprehensiveness of Boscovich's theory. In other words that aspect of Faraday's speculative papers, gravitation, which is related in its fundamentals to the Boscovich theory is precisely the one aspect which is a conundrum for Faraday."

photon trajectory.[62] The required function oscillates with distance, and it struck Faraday and other readers as well that the function assumed by Boscovich might be just that synthesized by Mossotti out of the combined action of the four forces he assumes. Faraday says in a discussion of Boscovich's theory:

> This, at first sight, seems to fall in very harmoniously with Mossotti's mathematical investigations and reference of the phenomena of electricity, cohesion, gravitation, &c. to one force in matter; and also again with the old adage, "matter cannot act where it is not." (*XR* ii/293)

The editors of the *Philosophical Magazine* similarly, in printing an account of Mossotti's paper, inserted a note calling the interested reader's attention to Boscovich as related reading.[63]

Faraday's interest in Boscovich's theory may serve as a clue to the significance of Mossotti's paper for him. It is not, certainly, as a mathematical function that Boscovich's hypothesis appeals to him; but the following passage from a famous "speculation" of Faraday's makes clear the way in which Faraday adapts Boscovich to his own mode of thought:

> [The atoms of Boscovich,] if I understand aright, are mere centers of forces or powers, not particles of matter, in which the powers themselves reside. If, in the ordinary view of atoms, we call the particle of matter away from the powers *a*, and the system of powers or forces in and around it *m*, then in Boscovich's theory *a* disappears, or is a mere mathematical point, whilst in the usual notion it is a little unchangeable, impenetrable piece of matter, and *m* is an apostrophe of force grouped around it. ...
>
> To my mind ... the *a* or nucleus vanishes, and the substance consists of the power or *m*; and indeed what notion can we form of the nucleus independent of its powers? All our perception and knowledge of the atom, and even our fancy, is limited to ideas of its powers: what thought remains on which to hang the imagination of an *a* independent of the acknowledged forces?[64]

We see that Faraday simply substitutes for the mathematical function assumed by Boscovich's *Theory* his own concept of a "system of powers or forces," and he makes explicit his view that this is not a question of a mere

62. Newton, *op. cit.*, Queries 1–4, p. 339; Query 30, p. 395. Boscovich himself cites the latter reference (Boscovich, *op. cit.*, p. 19), but feels that his theory goes far beyond Newton's suggestion.

63. *London, Dublin, and Edinburgh Philosophical Magazine, 10* (1837), p. 357. (This publication will hereafter be referred to as the *Philosophical Magazine*.)

64. "A Speculation Touching Electric Conduction and the Nature of Matter" (1844) (*XR* ii/290–91).

mathematical quantity, but of a substance: in fact *the* substance of which the bodies we encounter in nature are composed.

Boscovich assumes just one mathematical function; Faraday in his own interpretation of Boscovich has converted this to one "power," and he has linked it in turn to Mossotti, who he sees as having reduced various phenomena, including electricity, "to one force in matter." Mossotti, too, sees his own work as leading to a unified account:

> It is a subject which appears to me entitled to the greatest attention, because the discovery of the laws of molecular action must lead mathematicians to establish *molecular mechanism* on a single principle, just as the discovery of the law of universal attraction led them to erect on a single basis the most splendid monument of human intellect, the *mechanism of the heavens*.[65]

But Mossotti here celebrates Newton's achievement in the *Principia*, and the unity he seeks is, like that of Newton and Boscovich, exhibited in a single mathematical force law. (In Mossotti's case, it would seem rather to be three force laws of one single type, all three of the distance forces which he assumes obeying the inverse-square law.)

We see, then, that Faraday and Mossotti are not working with the same concepts; and yet they do share certain basic common ground. Mossotti, with Faraday, is convinced of this; when he later extended his theory to include Faraday's dielectric, he wrote:

> [L]es vues de l'illustre physicien s'accordent avec celles que nous avons adoptées dans nôtre mémoires sur la constitution des corps.[66]

They are in accord in giving an essential role to a space-filling substance; for Mossotti, this is the "ether," which in turn proves to be "caloric"; for Faraday, it is the "power" as a system in space. They both seek a unifying principle for the forces of nature. But for Mossotti, the medium has only a limited role; true, it acts partly by contiguity through the exertion of pressure, but even its agency is chiefly through action-at-a-distance, both on remote parts of its own substance and on the molecules themselves. Furthermore, the molecules continue to act on each other at a distance, just as they did in Newton's *Principia* itself. Mossotti helps very little, then, in bringing the operations of nature into accord with Faraday's "old adage."

Nonetheless, as mentioned earlier, Mossotti was convinced that his theory was well adapted to give analytic form to the ideas of Faraday's Series XI. In fact, he sees the interpretation of Faraday, as so many did, as a problem of *translation*; the first task is just this:

65. Mossotti, *op. cit.*, pp. 450–51.

66. Mossotti, "Recherches théoriques sur l'induction électrostatique, envisagée d'après les idées de Faraday," *Bibliothèque universelle archives des sciences physiques*, *6* (1847), p. 197.

> 1° Trouver le moyen de traduire, par une expression analytique, l'action du corps diélectrique pour l'introduire dans l'équation différentielle de l'équilibre.[67]

He makes very clear the importance he attaches to Faraday's *Researches*, and his conviction that he is in fact interpreting Faraday's ideas:

> Parmi les problèmes généraux que présente la théorie mathématique de l'électricité statique, il n'en est peut-être pas de plus interessant que celui qui a donné naissance à la proposition suivante, suggerée à la Société italienne des sciences, comme sujet d'un prix à conférer: "En prenant pour point de départ les idées de Faraday sur l'induction électrostatique, donner une théorie physico-mathématique de la distribution de l'électricité sur les corps de forme diverse."[68]

Mossotti is frank to explain that he is using the analytic theory of Poisson, just as, indeed, Thomson was doing at approximately the same time. Mossotti, however, adapts Poisson to his own purposes. In place of the polarization which Poisson envisioned as a shift of a magnetic fluid within a molecule to which it is bound, Mossotti substitutes a counterpart in terms of the ether theory of his own 1836 paper. He modifies his theory by extension, without in fact altering any of its fundamental assumptions. The dielectric is now envisioned as a molecular structure in which the atmospheres of the molecules are distorted—rarefied on one side, and compressed on the other. The atmosphere can in fact be torn from the molecule; in this, he points out, lies the difference between magnetism and electricity from a mathematical point of view, there being no currents of magnetism to correspond to the electric current. At the same time, he makes more explicit his view that the electric current is indeed the motion of the caloric atmosphere. But this one difference in no way affects the application of Poisson's theory of magnetism to the present case, and Mossotti arrives without difficulty at a mathematical theory of the dielectric phenomena Faraday has observed, understanding "polarization" as a distortion of the caloric atmosphere about each molecule of the dielectric.

Again, however, Mossotti's interpretation is really far from the views toward which Faraday's Series XI is tending: Mossotti sees the charge on a conductor as a layer of fluid, a tightly-bound, dense layer of the caloric atmosphere. It acts *at a distance* on an opposite layer; the action is merely modified by the net displacement of the bound ether of the dielectric. This is the view now generally known as the Faraday-Mossotti hypothesis.[69] Distance-forces retain their fundamental role, and charge remains a

67. Mossotti, *op. cit.*, p. 194.

68. *Ibid.*, p. 193.

69. Alfred O'Rahilly, *Electromagnetics* (1938): republished as *Electromagnetic Theory* (2 vols.; New York: Dover Publications, 1965), pp. 77 ff.

concentration of a fluid, even though this belongs to the space rather than to the conductor itself. It by no means agrees with Faraday, who has concluded, as we have seen, that "induction ... *constitutes charge*." In Faraday's view, charge has become the state of the medium, and there is no action-at-a-distance. it is interesting that Mossotti not only does not catch this in his translation, but is not aware of the fact that he has fallen so far short. Yet his rendition, I think, represents an effort of one of the most sympathetic, and most competent, interpreters of Faraday.

Maxwell, as an interested student of those earlier efforts at the translation of Faraday, points out the extent to which Mossotti's attempt falls short. He says of it in the *Treatise*:

> Since, as we have seen, the theory of direct action at a distance is mathematically identical with that of action by means of a medium, the actual phenomena may be explained by the one theory as well as by the other, provided suitable hypotheses be introduced when any difficulty occurs. Thus, Mossotti has deduced the mathematical theory of dielectrics from the ordinary theory of attraction merely by giving an electric instead of a magnetic interpretation to the symbols. (*Tr* i/70)

Mossotti's remains essentially "the ordinary theory of attraction," merely applied to Faraday's dielectric phenomena. It has not captured the revelation of thought to which this led Faraday, in which a power present in the medium replaces distant actions between centers of force.

One other interpreter of Faraday's electrostatics demands our attention. Peter Riess (1805–1883), a theorist and experimenter in Berlin who produced one of the "several excellent electrical works" to which Maxwell alludes in the Preface to the *Treatise*,[70] made an especially valiant and revealing effort to bring Faraday to grips with the concepts of electrostatic induction as anyone reasonably familiar with the mathematical tradition would understand them. In 1842, he had written an article arguing that Faraday's reasoning in Series XI was fallacious, and that his experiments in "specific inductive capacity" in no way affected the status of the standard theory.[71] In this, Riess is of course correct, as both Kelvin and Mossotti were later to show in detail. Apparently having thus satisfied himself that Faraday had introduced nothing essentially new, Riess omitted discussion of Faraday's views in his 1853 treatise, *Reibungselektrizität*. Shortly after its publication, however, he was moved to repeat his argument against Faraday, evidently more than a little exasperated by the necessity of reiterating what should have been apparent to all. This second article against Faraday appeared in the *Annalen der Physik* in 1854, and then in translation in the

70. Peter Riess, *Reibungselektrizität* (2 vols., Berlin, 1853); (*Tr* i/viii, x).

71. Riess, "Die lehre von der Elektrizität," *Repertorium der Physik*, *6* (1842), p. 219.

Philosophical Magazine in 1855, where Faraday read it.[72] Faraday promptly repeated an experimental result which had been brought into question, and then wrote a reply to Riess, who in turn responded to this letter. The exchange (which took place in English) was printed as well in the *Philosophical Magazine*, so that we have, in all, a most interesting attempt to converse earnestly with Faraday.[73] The result was, on the whole, a disaster.

To abbreviate this account of the exchange, I shall epitomize their positions on one or two of the issues raised, without attempting to recount the sequence of their remarks. Essentially, Riess is interested in one central question: is the basic principle on which traditional action-at-a-distance theory is based jeopardized by Faraday's experimental results in electrostatic induction? Riess identifies this principle as follows: the electrostatic action of an electrified body *E* on another, *A*, is not affected by the presence of any third body *B*. *E* may act on *B*, and *B* may in turn act on *A*, but the original action of *E* on *A* is not thereby "hindered or weakened." Otherwise put, the issue is this: can the resultant effect be compounded as a linear combination of a set of individual effects, each of which can be calculated separately according to the theory of Coulomb and Poisson?[74]

In the exchange with Riess, Faraday never really meets this issue. One might conclude that he simply does not grasp the question, but in fact he had expressed an opinion earlier, in the *Experimental Researches*:

> A striking character of the electric power is that it is limited and exclusive, and that the two forces being always present are exactly equal in amount. (*XR* i/535)

Riess cites this passage, and insists that Faraday has at least led others (de la Rive and Melloni are cited as instances) to believe that the effect of *E* on *A* is diminished by the presence of *B*. But Faraday is looking at the whole effect, and he is concerned with what he calls "power." We have already seen that he does not necessarily mean by this term either "energy" or "force" (page 24, above). What is "limited and exclusive" in the present situation is, I believe, the "power" as he measures it with a proof plane, or as he envisions it in terms of the lines of force (*XR* i/386). A typical paragraph suggesting this interpretation is this, likewise from Series XI:

> As another proof that the whole of these actions were inductive I may state a result which was exactly what might be expected, namely, that if uninsulated conducting matter was brought round and near the excited [charged] shell-lac stem, then the inductive force was directed towards it, and could not be found [i.e., with a proof plane]

72. Riess, "On the Action of Non-conducting Bodies in Electrical Conduction" (1854), *Philosophical Magazine, 9* (1855), p. 401.

73. *Philosophical Magazine, 11* (1856), pp. 1 ff., 10 ff.

74. *Philosophical Magazine, 11* (1856), p. 10.

> on the top of the hemisphere. Removing this matter the lines of force resumed their former direction. (*XR* i/383–84)

What Faraday is considering is the total number of lines of force terminating on a surface, a quantity which is, in a given medium, directly proportional to the charge on it; if the charge is fixed (as it is, in this example, upon the shellac stem), then indeed the total number of lines is conserved. If some are drawn off, then the number remaining to terminate on the body under induction (the hemisphere in this case) is in fact reduced.

Faraday, then, is thinking of his lines, and in effect is calling their number, which is estimated by measurements with the proof plane, the "power." Riess is demanding an answer concerning force, only wishing Faraday to admit that the force of one charged body on another is unaffected by the presence of a third. Faraday, who has denied the validity of action-at-a-distance theories to his electrostatics of "curved lines" (*XR* i/386), will not and perhaps cannot yield the simple "yes" which would permit the dialogue to progress fruitfully.

Riess tries desperately to get to the bottom of Faraday's notion of "curved lines," which figure so largely in Series XI. They are, as a great many passages in the *Experimental Researches* confirm, a decisive indication for Faraday of the operation of a medium:

> Amongst those results deduced from the molecular view of induction ... the expected action in curved lines is, I think, the most important at present; for, if shown to take place in an unexceptionable manner, I do not see how the old theory of action at a distance and in straight lines can stand, or how the conclusion that ordinary induction is an action of contiguous particles can be resisted. (*XR* i/380)

Riess, to force the issue, carries out a procedure analogous to the "curved line" procedures of Series XI (*XR* i/381 ff.). He produces "curved lines" in a simple way, by placing a grounded metal sheet between a charged body and a proof plane; the fact that the proof plane is unresponsive immediately in the region shadowed by the metal sheet, while one farther from the source and the screen is more affected, is evidence of the type Faraday has cited for his "curved lines," and yet Riess believes no one could argue that they are significant of the action of a medium in a case so easily explained by conventional theory. Presumably Faraday should grant that curved lines are not necessarily evidence of the participation of a medium. Faraday however does not yield the point; it is difficult not to describe his response as a quibble about the arrangement of the intervening conductor.

Riess, still eager to find the distinctive position which Faraday is asserting, says that Faraday denies action-at-a-distance. (The quotation above might be regarded as sufficient grounds for this conclusion.) But Faraday expresses himself as offended at having been thus misunderstood, and points to passages in the *Researches* in which he has, it is quite true, reserved judgement about the *vacuum.* This reservation, which would seem quite irrelevant, so softens the ground of the argument that Riess finds it hard to sustain the discussion. Faraday could have helped by keeping his attention on the point in question, in which the vacuum is not at issue. Does he not deny action-at-a-distance in all cases in which a dielectric, even air, is present?

Riess, in summarizing his own quite valid conclusion from the discussion, no doubt speaks for virtually all competent scientists in 1856. Nothing Faraday has said seems to him have touched the principle which Riess initially asserted; therefore the old theory stands. Old theories in any science, he says, should not be abandoned without reason. Sound evidence demanded the abandonment of the emission theory of light, but:

> I see not the like in the old theory of electricity. It assumes, indeed, the action at a distance, and I agree entirely with you that such a action is extremely difficult to conceive; but admit we not the like in the great theory of gravitation?[75]

For Faraday, the "great theory of gravitation" is, as Maxwell said of him, "a sacred mystery" which leaves him totally unsatisfied.[76] Riess is not persuading Faraday, but he is making good sense to the mathematical mind:

> These premises granted, the [action-at-a-distance] theory accounts for the phenomena of static electricity in the simplest manner. All these phenomena [those of Series XI] are instances of the arrangement of electricity upon the surface of bodies, and their arrangement is made dependent upon the equilibrium of a number of forces which the electric particles exert mutually on each other. Thus the electrostatic problems are changed into problems of pure mechanics. ... The advantage of this method is very great; it gives the result of each experiment as the sum of single actions, which the mind conceives without difficulty, and leaves to mathematics the pains to sum up the single effects and to give the amount of the sum. ... Therefore I have long ago defended this theory against its—indeed not very dangerous—antagonists, and I could not abstain from continuing the defence, when an adversary arose in the man whom I venerate as the greatest natural philosopher of the age.[77]

I believe the last statement is quite sincere, and reflects accurately the despair which Faraday's ideas caused for those who recognized in him a brilliantly intelligent experimenter. Riess closes the conversation rather beautifully:

75. *Loc. cit.*

76. Maxwell thus characterized Faraday's view (*SP* ii/318).

77. *Philosophical Magazine, 11* (1856), pp. 10–11.

> I have little hope to persuade you, my dear Sir, to modify your views ... and, I confess, if I could I would scarcely wish it. The great philosopher works best with his own tools, whose imperfections he avoids by dexterous application. But these tools, so efficacious in his hand, are not only useless but very dangerous in the hands of others.[78]

Riess is quite correct, and his position, which was almost universally shared, is entirely understandable. Yet Riess's name has virtually disappeared from our records—mention is hardly to be found of him in our encyclopedias or our histories. The *Reibungselektrizität* is one of the "excellent electrical works, published, most of them, in Germany" on which Maxwell has turned his back at the outset of the *Treatise*, in the passage quoted earlier from his preface. It is easy to refute Faraday, and extremely difficult to understand him. Yet Maxwell enters the discussion of electrical theory precisely in order to show what unexpected possibilities lie in the direction Faraday is indicating.

MAGNETISM AND DIAMAGNETISM

In examining these efforts to interpret, first Faraday's ideas of "lines of force" with their role in electromagnetic induction, and then his concept of the action of the dielectric medium in electrostatic induction, we have taken a fair sample of the frustrations which surrounded Faraday's encounters with other scientific minds. But there was a third major problem which, as Williams's biography of Faraday has emphasized, brought the "field" concept to a new level of development in his own mind, and at the same time plunged him into even more intense dialectic with other theorists and experimenters. This was the question of diamagnetism and magnecrystallic action.

The history of the discoveries of Faraday and others in this area has been recounted by Williams in such detail that it seems unnecessary to review it here; Faraday's own principal ideas have been discussed in Chapter 1, above. Let us turn, therefore, directly to the issues as they shaped up in the discussion. Faraday, as we have mentioned (page 12, above), himself introduced the hypothesis of induced "polarity" as an explanation of diamagnetic repulsion. But Weber had no sooner adopted Faraday's point of view and begun to reveal its fruitfulness through supporting experiments and the construction of a coherent theory, than Faraday recognized in it the face of his old *bête noire,* action-at-a-distance, and disowned both poles and polarity. Through Tyndall's notion of "elective polarity," and Weber's hypothesis of induced molecular polarity, Faraday was brought face-to-face with his own term "polarity"—and while the ensuing discussion was hardly

78. *Ibid.*, p. 17.

more constructive than the others we have examined, it at least offered an extended opportunity for Faraday to articulate for himself and the world the alternative to "polarity." This alternative involved some of the augmented work for the "lines of force" which we have already seen in the review of Faraday's methods in Chapter 1. Many scientists contributed to the complex discussion—Edmond Becquerel, Auguste de la Rive, Julius Plücker, Mateucci, Thomson, Weber, and Tyndall—but Tyndall formulated the case for "polarity" with particular care in a long series of six monographs, supporting his case all the while with new experiments, and directing the whole effort, it sometimes seems, toward the persuasion of Faraday. It will suffice, therefore, if we trace here the exchange of thought between Faraday and Tyndall.

Tyndall was personally drawing closer to Faraday during the course of the debate, becoming, in the midst of it, Faraday's colleague at the Royal Institution; years later, he was to become Faraday's successor. Furthermore, Tyndall spoke the language of experiment, and thus had the best chance to find common ground with Faraday; Tyndall's papers contain little mathematics. He certainly held Faraday's respect—though as we have seen, it is clear on the other side that Tyndall was confident that he saw far deeper than Faraday, whom he regards with what Williams has called "paternal tolerance."[79] (Tyndall was twenty-nine years Faraday's junior; it was, he said, his "wish and aspiration to play the part of Schiller to this Goethe"—yet he indeed thought of the aging Faraday as "boy-like."[80]) No one who had studied mathematical physics—not even Maxwell—could indeed be altogether innocent of this sense of knowing much which lay beyond Faraday's ken. But Tyndall makes a great effort to understand Faraday, to show him by experiment and to win him around by argument—and finally, I think, in the Sixth Discourse, simply to *teach* him. The power of Faraday's grip on the scientific minds of his time is shown, not so much by others' ability to interpret and use Faraday's ideas, as by their patient effort to persuade him away from them. Perhaps there were others who, like Riess, would not have wished to do that; but even these sought with earnest effort to convince him that there existed a tenable alternative to his view.

Faraday is sensitive to the patience his obduracy demands of others, but he is never persuaded that there in fact exists any alternative to his views; he writes to Tyndall in 1855, late in the discussion:

> Some persons may feel surprised that I dwell upon points which are perfectly and mathematically explained by the hypothesis of two magnetic fluids, i.e., by action-at-a-distance theory. ... My reason is, that being satisfied by the phenomena of diamagnetism, &c. that the

79. L. P. Williams, *op. cit.*, p. 509.

80. John Tyndall, *Faraday as a Discoverer* (New York: 1873), pp. 169, 163.

> hypothesis *cannot be true*, all these and such like phenomena acquire a new character and a high importance which they had not before.[81]

In truth, what is "perfectly and mathematically explained" for others is not explained *at all* for Faraday, and so the effort to reason with him is essentially endless. It is, nonetheless, not idle. It represents the struggle of the human mind, on the one hand to articulate an almost totally new point of view, and on the other, to comprehend it.

This particular discussion between Faraday and Tyndall is of special interest to the present study because it may have incubated Maxwell's own impulse to come forward as Faraday's spokesman. Maxwell read the first part of his "Faraday's Lines" to the Cambridge Philosophical Society in December, 1855, at a time when the interchange between the two men had reached its height; he sent copies of it to Tyndall, Thomson, and Faraday as the principal participants. If the *Treatise* is indeed the culmination of Maxwell's effort to achieve a consistent translation of Faraday's thought, then it was perhaps born of this debate on the theory of diamagnetism.

By the time Tyndall wrote his first paper on diamagnetism and the properties of crystals under the action of magnets in 1850 (jointly authored with Karl Hermann Knoblauch of Marburg, where Tyndall had studied),[82] the discussion of diamagnetic polarity was already underway. Faraday had proposed the hypothesis in his second series dealing directly with diamagnetism (Series XXI, 1846); Weber had seized upon it in a paper of 1848, which Faraday read in translation in 1849.[83] Faraday's alternative account, according to which a diamagnetic body tends to move from strong to weak regions of magnetic force (page 28 above), was suggested in the same series as the principle of induced polarity. This second principle was of immediate interest to Thomson, who soon afterwards, early in 1847, produced a mathematical translation of it, once again employing Poisson's 1824 memoir on magnetism as the source of the necessary formal theory.[84] We shall return to a discussion of Thomson's interpretation of this principle in the next section of the present chapter (page 99 below); for the moment, it will suffice to note that Faraday's Series XXI gave rise to two quick responses on

81. John Tyndall, *Researches on Diamagnetism and Magnecrystallic Action* (London: 1870), p. 220 (emphasis added).

82. John Tyndall and K. H. Knoblauch, "The Magneto-optic Properties of Crystals and the Relation of Magnetism and Diamagnetism to Molecular Arrangement" (1850), in John Tyndall, *op. cit.*, pp. 1 ff.

83. Wilhelm Weber, "On the Excitation and Action of Diamagnetism According to the Laws of Induced Currents" (1848), translated in R. Taylor, ed., *Scientific Memoirs*, *5* (1852), pp. 477 ff. Cf. Faraday, *Diary*, *5*, p. 153.

84. William Thomson, "On the Forces Experienced by Small Spheres under Magnetic Influence..." (1847) (*EM*: 499 ff.; note especially p. 504).

the part of mathematical physicists, Weber's in terms of "polarity," and Thomson's in terms of the tendency of the body to move in the magnetic field.

Weber's 1848 account included experimental results which he felt strongly supported the hypothesis of diamagnetic polarity due to induced currents in the molecules, and he credited Faraday fully with this brilliant extension of Ampère's electrodynamics. By the time this appeared in English translation (1849), Faraday was occupied with the new principles of magnecrystallic action in terms of "conduction" of the lines developed in Series XXII, read as the Bakerian Lecture at the Royal Society in December, 1848 (see pages 26 ff., above). Although he does not there disown his former hypothesis of induced polarity, it is clear that the new possibilities for the magnetic lines of force are much closer to his heart. When Faraday, having read Weber's paper, then repeated the experiments with which Weber thought to have confirmed the hypothesis, he exposed an apparent fallacy: instead of the supposed molecular currents, gross circulating currents were causing the forces Weber had observed and taken for a diamagnetic effect.[85] This becomes the occasion for Faraday's Series XXIII of 1850, questioning the "supposed magnetic polarity." But though the failure of Weber's experimental demonstration is indeed the immediate occasion for Series XXIII, one feels that opportunity, or nature, has played into Faraday's hand: he would by now much prefer to find an explanation of another sort.

As I have mentioned earlier, it should be made clear that whatever difficulties Faraday himself came to feel concerning it, his disowned hypothesis proved splendidly successful in the hands of Weber, Tyndall, and others (pages 36–37, above).[86] The experimental problems in demonstrating diamagnetic polarity were in fact severe, but in 1852 Weber published a new set of investigations which, at least in Tyndall's final judgement in 1870, met all the objections which had been raised especially by Matteucci.[87] Later, Tyndall had Weber design for him (and Weber's own instrument maker construct) an apparatus in which the original problem of spurious circulating currents could not arise, Faraday's original heavy glass itself being substituted for an electrically conducting diamagnetic object. Tyndall's

85. (*XR* iii/144, 157). L. P. Williams, *op. cit.*, pp. 431–33.

86. Williams unfortunately appears to leave the impression that Faraday's judgement of Weber's experiment was decisive (*loc. cit.*).

87. W. Weber, *Elektrodynamische Maasbestimmungen: insbesondere über Diamagnetismus* (Leipzig: 1852); trans. by Tyndall with the title "On the Connexion of Diamagnetism with Magnetism and Electricity," in J. Tyndall and W. Francis, eds., *Scientific Memoirs: Natural Philosophy* (1853) (New York: Johnson Reprint Corp., 1966), pp. 163 ff. For Tyndall's judgement, see Tyndall, *Researches on Diamagnetism*, p. 157.

successful result with this apparatus was shown by him to Faraday personally at the Royal Institution, and was published in 1855.[88] With that, any reasonable doubt concerning the validity of the experimental result must have ended, so that both theoretically and experimentally, the hypothesis of induced diamagnetic polarity had stood all tests. It is not clear, nevertheless, that Faraday was ever convinced by the evidence before his eyes;[89] his justly famed sensitivity to the voice of nature occasionally took the form of a fine sense of what that voice *ought* to say!

Even with the evidence offered by Tyndall and Weber admitted as decisive, however, it would only show the admissibility of "diamagnetic polarity" as a possible explanation of the diamagnetic force; it would not in itself decide against Faraday's principle as an alternative. The discussion therefore was carried on as well on higher ground, in which the issue became the tenability of Faraday's mode of viewing diamagnetic and magnecrystallic actions, and, in a still broader sense, the form which physical theory should take.

The first paper of Tyndall's series, that of 1850 cited earlier, objects to Faraday's rule of magnecrystallic force. Tyndall writes:

> Let us look a little further into the nature of this magne-crystallic force, which, as is stated, is neither attraction nor repulsion, but gives position only. The magne-crystallic axis, says Mr. Faraday, *tends* to place itself parallel to the magnetic resultant passing through the crystal; and in the case of a bismuth plate, the recession from the pole and the taking up of the equatorial position is not due to repulsion, but to the *endeavour* the bismuth makes to establish the parallelism before-mentioned. Leaving attraction and repulsion out of the picture, we find it extremely difficult to affix a definite meaning to the words 'tends' and 'endeavour.' 'The force is due,' says Mr. Faraday, 'to that power of the particles which makes them cohere in regular order, and gives the mass its aggregation, which we call at times the attraction of aggregation, and so often speak of as acting at insensible distances.' We are not sure we fully grasp the meaning of the philosopher in the present instance; for the difficulty of supposing that what is here called the attraction of aggregation, considered apart from magnetic attraction or repulsion, can possibly cause the rotation of the *entire mass* round an axis, and the taking up of a fixed position by the mass, with regard to surrounding objects, appears to us insurmountable.[90]

What is at issue in the last sentence is not really the particular case of the orientation of a crystal in relation to a distant magnet, so much as the

88. Tyndall, "On the Nature of the Force by which Bodies are Repelled from the Poles of a Magnet" (1855), in Tyndall, *Researches on Diamagnetism*, p. 157.

89. Tyndall, *op. cit.*, p. 75.

90. *Ibid.*, pp. 21–22.

more general question of a possible physics. *Could* local actions ("at insensible distances") account for a bodily rotation ("the taking of a fixed position") in relation to other bodies which are placed at a distance ("with regard to surrounding objects")? When Tyndall cannot grasp Faraday's *meaning* here, it is because he *cannot see the possibility of a field physics.* As we have noted, Thomson had already given mathematical form to those "tendencies" and "endeavours," before this impasse of Tyndall's was reached.

Tyndall's third memoir (1851) continues the discussion.[91] Here we see Tyndall learning to use Faraday's methods, and yet questioning their significance. He has learned to speak of the "magnetic field"; Faraday's term appears in the paper's title with Tyndall's own term, "polarity" ("On the Polarity of Bismuth, Including an Examination of the Magnetic Field"), and indeed the entire study is based on the configuration of the magnetic field between pole pieces, the field being modified by the introduction of a helical coil or a piece of iron. Tyndall carefully investigates Faraday's principle of motion within that field. He shows that a bismuth exploring sphere does indeed move to positions of lower field strength, as Faraday's law required.[92]

To pursue the significance of this method of Faraday's, Tyndall envisions a dialogue with a hypothetical advocate of induced polarity, Tyndall himself undertaking to speak for Faraday. Having been shown the satisfactory operation of Faraday's rule in a simple case, the hypothetical friend objects to Tyndall-Faraday:

> 'I grant you,' he may urge, 'that in a simple magnetic field, consisting of the space before and around a single pole [an arrangement Tyndall has described for shaping the field], what you assume is correct, that a magnetic sphere will pass from weaker to stronger places of action; but for a field into which several distinct poles throw their forces, the law by no means sufficiently expresses the state of things. If we place together two poles of equal strengths but of opposite qualities close to a mass of iron, it is an experimental fact that there is almost no attraction; and if they operate upon a mass of bismuth, there is no repulsion. Why? Do the magnetic rays, to express the thing popularly, annul each other by a species of interference *before they reach the body*; or does one pole induce in a body the certain condition upon which the second pole acts in a sense contrary to the former, both poles thus exactly destroying each other? If the former, then I grant you that the magnetic field is rendered weaker ... but if the latter, then we must regard the field as possessing two systems of forces; and it is to the peculiar inductive property of the body, in virtue of which one system neutralizes the other, that we must

91. Tyndall, "On the Polarity of Bismuth, Including an Examination of the Magnetic Field" (1851), in Tyndall, *Researches on Diamagnetism*, pp. 76 ff.

92. *Ibid.*, p. 83.

> attribute the absence of attraction or repulsion. Once grant this, however, and the question of diamagnetic polarity, insofar as you are concerned, is settled in the affirmative.'
>
> 'Which will you choose? ... [E]ither you must refer the weakening ... to magnetic interference, or you must conclude, that the induced state ... which causes the bismuth to be repelled by the magnet ... causes it to be *attracted* by the coil [the discussion has shifted to a slightly different case], the resultant being the difference of both forces.'[93]

Tyndall's "friend," then, demands an account of the combination or interaction of two fields originating from separate magnets—or else, a retreat on the part of their advocate to a position in which the significant processes occur, as the "friend" would have them, as a composition of polar forces induced in the diamagnetic material (here, bismuth) and not as a process of any sort among the fields themselves. If this admission is made, presumably the account in terms of fields loses its interest, and the hypothesis of "diamagnetic polarity" carries the burden of explanation. Faraday, of course, would seek some way to comprehend the process in terms of the two systems of power; it is perhaps significant that his thoughts on the "coalescence" of systems of lines, which we discussed in Chapter 1 (pages 33–34, above), were formulated in 1852, shortly after this challenge.

In Tyndall's paper, the question is not left here; the author soon drops his mask to remind the reader that decisive evidence exists, showing that induction does indeed occur in diamagnetic as well as in paramagnetic bodies. The empirical proof is that the body is affected by an electromagnet, not in proportion to the current in the latter, but to the *square* of the current, as an *induction* process would require. This supports Tyndall's view—the "friend's" view, in the dialogue—that the crux of diamagnetic phenomena is the induction of poles in the diamagnetic body, and hence, essentially an action-at-a-distance process of the traditional type.

The conclusion for Tyndall is this: Faraday's method of accounting for forces by field configurations is a valid working rule in simple cases, but it raises fundamental questions in more complex instances. As a philosophical principle, it can be dismissed as a result of the square-law evidence that diamagnetism is an induction process. The status of Faraday's methods is readily assigned:

> In order to present magnetic phenomena intelligibly to the mind, a material imagery has been resorted to by philosophers. Thus we have the 'magnetic fluids' of Poisson and the 'lines of force' of Mr. Faraday.[94]

93. *Ibid.*, pp. 85–86.

94. *Ibid.*, p. 89. *Cf.* de la Rive, *Treatise, 2*, p. 75.

Even though the lines of force are no more than a "mental imagery," the fact that Tyndall has begun to learn to work with them is an important step in the direction of field theory, and the interpretation of Faraday.

Perhaps the most significant advance in the discussion came, however, with Faraday's essay "On Some Points of Magnetic Philosophy" (1855) (*XR* iii/528 ff.). He deals in one section of the paper with the concept of "magnetic polarity," which had become the sticking-point of the discussion:

> The meaning of this phrase is rapidly becoming more and more uncertain. In the ordinary view, polarity does not necessarily touch much upon the idea of lines of physical force [!]; yet in the one natural truth it must either be essential to, and identified with it, or else absolutely incompatible with, and opposed to it. Coulomb's view makes polarity to depend upon the resultant in direction of the action of two separated and distant portions of two magnetic fluids upon other like separated portions, which are either originally separate, as in a magnet, or are induced to separate, as in soft iron. ... My view of polarity is founded upon the character in direction of the force itself, whatever the cause ... and asserts that when an electro-conducting body ... moving in a constant direction near or between bodies acting magnetically on themselves or each other, has a current in a constant direction produced in it, the magnetic polarity is the same; if the motion or the current be reversed, the contrary polarity is indicated. The indication is true either for the exterior or the interior of magnetic bodies whenever the electric current is produced, and depends upon the unknown but essential dual or antithetical nature of the force which we call magnetism. (*XR* iii/533)

If the prodding of Tyndall and others has led Faraday to this statement, it has indeed advanced the discussion. For Faraday, "polarity" has here emerged as a local characteristic of the field; it is now explicitly "polarity" *without poles*, a question of the sense of the lines, as determined by the direction of current flow in the moving wire. We might compare with this a statement of 1838, in which "polarity" was defined in immediate relation to the "contiguous particles" of electric induction. There:

> With respect to the term *polarity* also, I mean at present only a disposition of force by which the same molecule acquires opposite powers on different parts. (*XR* ii/411)

However Faraday may have intended that earlier statement in his own mind, its tendency had been to suggest a *particulate* polarity, however finely divided, on the model of Poisson's 1824 hypothesis of magnetic polarity. It left an opening for the molecular theory of Mossotti with a major role for action at a distance, and it plunged Faraday into an elementary conundrum when he was challenged on his notion of "contiguity" by Robert Hare of Philadelphia.[95]

95. An exchange of letters between Robert Hare and Faraday on this subject is printed in the *Experimental Researches* (*XR* ii/251 ff.). Williams, *op. cit.*, pp. 306 ff.

By contrast, Faraday's new statement of 1855 appears to suggest something very different: a directional *continuum*. If this is indeed what Faraday has come to mean by the term "polarity," it is no wonder that he writes to Matteucci, concerning his dialogue with Tyndall:

> I cannot help thinking that there are many apparent differences amongst us, which are not differences in reality. I differ from Tyndall a good deal in phrases, but when I talk to him I do not find that we differ in facts. That phrase polarity in its present undefined state is a great mystifier.[96]

Faraday reflects, in the paper "On Some Points of Magnetic Philosophy," upon the effect of this discussion of "polarity" in which he has found himself enmeshed:

> The views of polar action and of magnetism itself, as formerly entertained, have been powerfully agitated by the discovery of diamagnetism. I was soon driven from my first supposition, that the N pole of a magnet induced like or N polarity in the near part of a piece of bismuth or phosphorus; but as that view has been sustained by very eminent men, who tie up with it the existence of magnetic fields or closed electric circuits as the source of magnetic power, it claims continued examination, for it will most likely be a touchstone and developer of real scientific truth, whichever way the arguments may prevail. (*XR* iii/535)

In the same article, he tries out still another position as an alternative escape route from "polarity"—this is the suggestion that diamagnetism and paramagnetism are merely differential actions with regard to a universal magnetic medium (pages 29–32, above). The result was a new line of discussion among Thomson, Tyndall, Faraday and others concerning the possible existence of such a magnetic medium, in the course of which Tyndall proves at least to weigh the possibility with care. He offers what he conceives to be a decisive objection, however, while Thomson responds with strong support for the possibility, in principle, of such a magnetic medium.

The result of the entire discussion for Tyndall himself is indicated by the character of his sixth memoir (1856).[97] Faraday's speculations may be stimulating, but the argument in this final memoir is to establish the scientific value of the traditional polar theory. Step by step, using hypothetical examples, he works through an exposition of the theory in terms of induced poles. He describes his purpose in this way:

> The importance of the principle [of diamagnetic polarity] is demonstrated by the fruitfulness of its consequences; for by it we obtain a

96. Tyndall, *op. cit.*, p. 181.

97. Tyndall, "On the Relation of Diamagnetic Polarity to Magnecrystallic Action" (1856), in Tyndall, *Researches on Diamagnetism*, pp. 184 ff.

> clear insight of effects which, without it, would remain standing enigmas in science, being connected by no known tie with the ordinary laws of mechanics. Many of the phenomena of magnecrystallic action are of this paradoxical character. For the sake of those who see no clear connection between these and the other effects of magnetism, as well as for the sake of completeness, I will here endeavour to indicate in a simple manner … the bearing of the question of polarity upon that of magne-crystallic action.[98]

Among the persons who "see no connection," surely Faraday is chief. This becomes evident as Tyndall formulates the concept of polarity: he is speaking, through the endlessly hospitable pages of the *Philosophical Magazine*, directly to Faraday. He begins with a quotation from Series XXI:

> 'Hence a proof that neither attraction nor repulsion governs the set. … This force, then, is distinct in its character and effects from the magnetic and diamagnetic forms of force.' [*XR* iii/113]
>
> These experiments present grave mechanical difficulties, and are quite sufficient to justify the conclusion drawn from them, namely, that the force which produces them is *neither* attractive *nor* repulsive. We will now endeavour to apply the idea of a force which is *both* attractive *and* repulsive, or in other words of a *polar* force, to the solution of the difficulty.[99]

Working through the cases carefully, using geometrical diagrams—in one case even applying to Faraday that enormously needed medicine, a properly constructed vector diagram of a line of force—Tyndall shows the power of the polar theory to account for all the types of phenomena which magnecrystallic experiments had made familiar. All can be accomplished by a theory of pairs of opposite forces acting at a distance to produce couples, and hence the observed turning effects. All "enigmas" are systematically removed, and the ties "with the ordinary laws of mechanics" are made clear. He concludes:

> The whole domain of magne-crystallic action is thus transferred from a region of mechanical enigmas to the one in which our knowledge is as clear and sure as it is regarding the most elementary phenomena of magnetic action.[100]

This is the offer made to Faraday: peace, and a safe haven.

Faraday has already given his response; he does not want such precise mathematical knowledge, but

> …the true but unknown natural magnetic action. Indeed, what we really want, is not a variety of different methods of representing the forces, but the one true physical signification of that which is

98. *Ibid.*, p. 184.

99. *Ibid.*, p. 188.

100. *Ibid.*, p. 198.

> rendered apparent to us by the phænomena, and the laws governing them. (*XR* iii/530; compare page 10, above.)

We have seen how far the interpreters of Faraday were able to go, and where they have stopped. Faraday is resolved to go beyond them, beyond the whole world, indeed. There are only two who would take the next step with him—not merely to adjust his ideas to the old, or stretch the old to include his—but to undertake a new physics, the physics of Faraday's imaginings. These two are Thomson and Maxwell. We turn first to Thomson, in the section which follows.

THOMSON'S ANALYTIC TRANSLATIONS OF FARADAY'S IDEAS

Having examined the efforts of a number of interpreters for whom Faraday's mode of thought was difficult and resistant to translation, we now turn to one, Thomson, who had at his command precisely the methods and the insight that were needed. If the translation of a difficult text in some way requires the opening of a sealed doorway between two realms of discourse, it was Thomson in this case who possessed the magic combination. "Combination" it indeed had to be, for Thomson's art in this brilliant, formative period of electromagnetic theory was the identification of relevant patterns in a variety of different sciences, and the clear perception of their common ground in a single, inclusive mathematical theory. This bold exploitation of analogies among the sciences was not Thomson's work alone; the same, almost breath-taking vision of the reach of formal analogy was given to a number of scientists in this same period, but it was Thomson who brought the new power to bear on Faraday's *Experimental Researches*, and, above all, it was Thomson who taught Maxwell to use it in his turn.

How it happened that this panorama of sciences broke into view in the second quarter to the nineteenth century is a question we cannot explore here. In general, what is in question primarily is the mathematical physics of continua, and the formulation of their properties through a variety of mathematical methods—among them, differential geometry and the solution of partial differential equations under boundary conditions. A great reservoir of mathematical methods for dealing with continua had been accumulating for a long while, and Thomson was among those who recognized its possibilities. The history of this development of the mathematical physics of continua would wind its way through many of the richest areas of European thought of the eighteenth and nineteenth centuries; Truesdell and Whittaker have dealt with it competently, though finally it may be a history which no one is able to write.[101] Ultimately, it is the legacy of Descartes

101. Clifford Truesdell, *The Kinematics of Vorticity* (Bloomington, Indiana: Indiana University Press, 1954), and, by the same author, "Rational Fluid Mechanics, 1687–1765," in *Leonhardi Euleri Opera Omnia*, Series 2, Vol. 12 (Lausanne: 1954), pp. ix ff.

and Leibniz—the effort to understand the world as a plenum. It passes through Newton, who wrote the second Book of the *Principia* to construct Descartes' world and thereby to show that it is not ours, and Euler, who brought the Cartesian vision under mathematical control.

The essential thing is the analytic mathematics of three dimensions. Even when this is developed specifically for the solution of problems of action-at-a-distance theory, a *mathematical continuum* is nonetheless taking shape, which can be brought to bear on a hypothetical physical continuum at will. We have seen how easily a mathematical field can be transformed into a physical field in the work of Mossotti, whose molecular ether turned out to be a *reification* of a scalar potential function (page 69, above). Maxwell speaks of this kind of shift:

> The most remarkable example of the use of a mathematical abstraction of this kind in mechanical science is the function used by Laplace in the theory of attractions and to which Greene [George Green] in his Essay on Electricity has given the name of the Potential Function. We have no reason to believe that anything answering to this function has a physical existence in the various parts of space, but it contributes not a little to the clearness of our conceptions to direct our attention to the potential function *as if it were a real property of the space in which it exists.*[102]

As we shall see, Thomson discovered a perfect mating of boundary problems in action-at-a-distance attraction theory, having nothing physically to do with a continuum, and Fourier's theory of heat flow, which is throughout a question of flow of a caloric fluid under thermal pressure in a resisting medium.

The *scientifically* crucial step is to learn to *think* about a continuum. This calls for the definition of functions in space, which satisfy the condition specified by a differential equation at every point, and have given values, or derivatives, over boundaries. Once the thought has been given strict formulation, perhaps as a purely geometrical or analytic problem, it can be brought to bear on fluids, solids, or simply potential fields with great versatility. It is not, indeed, that the idea develops first, and its application follows, chronologically; rather, we see the idea for the most part developing in application. But if Laplace develops it for an *attraction* theory of planetary motion, Thomson can nonetheless apply it to a magnetic *medium*. It is this Protean power of a universal analytic mathematics which Thomson so brilliantly directs.

How Thomson happens to be the one who sees the relevance of all this to Faraday, is again a question which cannot be opened here—only a new biographer of Thomson could undertake to deal with it. On the other

102. Maxwell, "Mathematical Theory of Polar Forces," (H.36: "On the Potential Function"); emphasis added.

hand, Thomson was born into the modern Continental mathematics, which had only recently begun to be taught and learned in England. Scotland had its own direct access to France, and Thomson was enamored of French analytic mathematics in his undergraduate years; his special attachment to Fourier is well known, and is perhaps the generator of much of the rest.[103] From the beginning of his scientific work, he certainly sensed his own possession of riches which his contemporaries had not appreciated. He wrote as no ordinary laborer in an academic vineyard, but (like Faraday in the world of "fact") as a discoverer in the world of theory. He showed at the outset a new way of using Fourier to grasp difficult problems in attraction theory; a little later, when he had discovered Green's *Essay*, he produced it for Liouville and the French mathematicians with equal zest.[104] He seems to have felt, in his early productive years, a special flair for discovering unexpected relations among theories; and having tasted the success of this, he was just the man to seize upon mathematical techniques which were already available in continuum mathematics and put them to use in the formal articulation of Faraday's ideas.

We are told that Thomson began with an antipathy to Faraday's mode of physics, of the sort we have seen in other students of the mathematical tradition, and that he was confronted with the possibilities of Faraday's methods by David Thomson, a cousin of Faraday's and a graduate of Glasgow and of Trinity College, Cambridge. David Thomson had returned to Glasgow to substitute for William Meikleham, professor of natural philosophy, during 1840–41. This was William Thomson's last year at Glasgow before his leaving to enter Cambridge.[105] William Thomson said, according to his biographer, that he was "inoculated with Faraday fire" by David Thomson.[106] The "fire" was probably not only a reaction to the *Researches* themselves, but an impulse to reveal their essential truth to an unbelieving world—not the truth of the "facts," which no one doubted, but the validity of Faraday's thoughts, which, as we have seen, even the best-willed mathematicians had difficulty justifying. There is often the sense of the heroic about Thomson's early papers: not the least, in his discovery of the underlying mathematical validity of Faraday's "lines of force," which the most eminent mathematicians of the day had declared "vague" and "unclear."

The first occasion for that discovery was provided almost immediately upon Thomson's graduation from Cambridge, in the form of an international appeal which no doubt roused the best in Thomson's spirit. Thomson described the situation in a letter to his father in 1845. Poisson, who had

103. S. P. Thomson, *op. cit.*, pp. 13 ff.

104. *Ibid.*, p. 113–119. Compare (*EM*: 17n–18n).

105. P. J. Anderson, "David Thomson." in the *Dictionary of National Biography*.

106. S. P. Thomson, *op. cit.*, p. 19.

died in 1840, had asked Joseph Liouville to "do what he could" for the mathematical theory of electricity which Poisson had developed in two memoirs of 1812 and 1824.[107] According to Thomson's letter, the mathematical theory of electricity would have been set as a prize problem by the Institut des Sciences in 1845, had it not been for doubts about its validity, which had been raised as a result of the researches of Faraday and William Snow Harris. Arago, in particular, was concerned about the "objections" of Faraday, presumably those of Series XI, in which indeed, as we have seen, Faraday doubts the ability of the mathematical theory to account for specific inductive capacity, and particularly the "curved lines" of force.[108] Thomson had, he says, already seen through those supposed objections, and hence was quite ready to comply when Liouville asked him to write a paper on the subject:

> He asked me to write a short paper for the Institute, explaining the phenomena of ordinary electricity observed by Faraday, and supposed to be objections fatal to the mathematical theory. I told Liouville what I had always thought on the subject of these objections (i.e., that they are simple verifications), and as he takes a great interest in the subject, he asked me to write a paper on it.[109]

The result was a paper which was in fact published in 1845 in Liouville's *Journal de mathématiques*, given in a modified form to the British Association meeting the same year, and finally published in the *Cambridge and Dublin Mathematical Journal*.[110] It must have been eminently satisfying to the French scientists who awaited it, as it not only showed no incompatibility of Faraday's results with the theory of Coulomb and Poisson, but elegantly proceeds to show, as we have noted, that a complete theory of specific inductive capacity, together with the mysterious curved lines, is already contained in Poisson's 1824 memoir on what would appear to be another subject, that of induced *magnetism*. Thomson, who has already sensed the power of analogies among the sciences in applying heat theory to electrostatics, concludes that the real topic opened by this memoir, which intends to be about one thing but turns out now to be about something else as well, is "in the most general manner the state of a body polarized by influence"

107. *Ibid.*, pp. 128–129.

108. *loc. cit.*; compare (*XR* i/386).

109. S. P. Thomson, *op. cit.*, pp. 128–129.

110. William Thomson, "On the Elementary Laws of Statical Electricity," *Cambridge and Dublin Mathematical Journal, 1* (1845), pp. 75 ff. This is a translation, with additions, from *Journal de mathématiques, 10* (1845), pp. 209 ff. Compare *British Association Reports* (1845), pp. 11–12. In the form of the first reference, it is reprinted at (*EM*: 15 ff). In general, for a convenient guide to the many transformations of Thomson's papers, see the bibliography appended to S. P. Thomson, *op. cit., 2,* pp. 1223 ff.

(*EM*: 32). Thomson moves directly to the abstract mathematical theory which encompasses both magnetic induction and electric induction, equally; these are the "positive laws of the phenomena" which we noted earlier, in distinguishing Thomson as "advocate" as opposed to Maxwell, the "judge" (page 44, above). Faraday's view has now become mathematically possible, because of its demonstrated formal equivalence to Poisson's, which is evidently consistent. But Thomson does not concern himself at this point with Faraday's ideas, or attempt to shape the mathematical theory to reflect Faraday's viewpoint rather than that of action-at-a-distance mathematics.

Let us look now at the theory—really, as he says, Poisson's—which Thomson offers as an expression of these "polar forces" which serve to account for the dielectric phenomena Faraday has observed. The mathematical theory is not set out in Thomson's 1845 paper—the reader is merely directed to Poisson—and although Thomson embarked on a series of papers which he later collected under the heading "On the Mathematical Theory of Electricity in Equilibrium," these never quite reached the subject of induction in dielectrics with which the project began, and hence "polar forces" are not discussed.[111]

The theory of polar forces does appear in its due place in a sequence of papers which began in 1849 as "A Mathematical Theory of Magnetism"; Poisson's theory of *induced* forces is reviewed in an 1851 contribution to this series (*EM*: 471 ff.) Equally relevant to the theory of "polar forces," and far more original, is an extensive discussion of polar *sources* continuously distributed in three dimensions, in a paper of 1850. (*EM*: 382 ff.).[112]

Finally, perhaps the most interesting aspect of Thomson's various contributions to "polar" theory appeared in 1847 in a paper entitled "On the Forces Experienced by Small Spheres under Magnetic Influence," in which the mechanical force on a magnetic or diamagnetic sphere placed in a magnetic field is expressed in terms of the configuration of the field; the result directly corresponds to Faraday's principle of magnetic and diamagnetic motion (*EM*: 499 ff.). The same topic is pursued in a further paper in 1850 (*EM*: 505 ff.).

111. Strictly, the "Mathematical Theory of Electricity in Equilibrium" in the *Reprint* includes three "diversions," the first two of which deal with basic principles, while the third is a "Geometrical Investigation ... of Electricity on Spherical Conductors," the papers thus collected spanning the period from 1845–50. The assignment of papers to the series would seem, however, a rather desperate attempt to impose a unity.

112. The "Mathematical Theory of Magnetism" as collected seems to have ten chapters, though it is hardly more than a formal exercise to analyze Thomson's intentions in this respect. The theory of polar sources is Chapter V, "On Solenoidal and Lamellar Distributions of Magnetism" (1850), (*EM*: 382 ff.).

Thomson's publications came in flashes and bursts, and although there was often the suggestion of a grand scheme, which in the case of electricity and magnetism he tried to recover in the 1872 *Reprint of Papers* to which we frequently refer, the result is nonetheless a profusion of promises and insights which are difficult to review systematically.

Let us begin with the theory of polar *sources* (*EM*: 382 ff.). We must note clearly at the outset that this is completely an action-at-a-distance theory, *not* a field theory; its whole purpose is to determine the forces upon distant magnetic poles due to any distribution of dipole sources. In this, it follows Poisson, and takes a view very different from Faraday's. On the other hand, the theory becomes of great interest to us because the source itself is regarded as continually distributed. Apropos of the polar source, Thomson makes a step which might seem small, but which in fact represents an important advance in thought beyond Poisson, from a fine-grained dipole distribution of two fluids, to a polar continuum. Thomson stresses the artificiality of the notion of magnetic fluids, and adds his voice to Faraday's in denying the existence of "poles" (*EM*: 350). This as we saw was very nearly the sticking-point in the elusive dialogue between Tyndall and Faraday, Tyndall insisting on polarity in the sense of a separation of seats of power, Faraday having progressed to "the character in direction of the force itself" (page 86, above).

By taking "polarity" to the mathematical limit, Thomson has apprehended one aspect of the development of thought through which Faraday has passed, beginning with a notion of polarity not essentially different from Poisson's, and ending with the concept of the continuous system of the lines of force. In this final form, "polarity" is no longer a question of a separation of sources or poles, even on a molecular level, but a *direction* through a continuum, best indicated by the moving wire. Thomson has caught in formal symbols Faraday's polarity without poles.[113] Once the theory is thus developed for polar *sources*, which as such are foreign to Faraday's thought, it will be ready for later application to a polar *medium*.

Thomson defines, as the measure of this continuous polarity, the "intensity of magnetization," which is in effect the dipole moment per unit volume of a continuous source, and which we may denote I (*EM*: 361). He calculates the magnetic potential V and the "resulting magnetic force" (our magnetic field intensity, $\mathbf{H}$,) at an observation point (ξ, η, ζ), due to a given distribution of continuous polar sources. His results are these (*EM*: 369–71):

113. Thomson apologized in 1849 for his continued use of this term: "The terms 'polarity' and 'poles' are still retained, but the use of them, which has very generally been made, is nearly as vague as the idea from which they had their origin. ... Notwithstanding this vagueness, however, the terms poles and polarity are extremely convenient." (*EM*: 350).

$$V = \int \mathbf{I}\cdot\nabla\left(\frac{1}{r}\right)d\tau = -\int \frac{\mathbf{I}\cdot\hat{\mathbf{e}}_r}{r^2}d\tau$$

$$H = -\nabla_\xi V = \int \mathbf{I}\nabla^2\left(\frac{1}{r}\right)d\tau.$$

The integrations are over the source volume, of which $d\tau$ denotes an element, while the gradient $\nabla_\xi V$ is taken at the observation point; $\hat{\mathbf{e}}_r$ is a unit vector in the direction of r. Note that all these analytic results are essentially Poisson's, and thus come strictly from an action-at-a-distance theory.[114]

Here Thomson is writing on magnetism, and he does not interrupt to point out the relation to Faraday's electrostatics; yet it is clear that the thought of the polar source-continuum is part of the abstract "theory of polar forces" to which he alluded earlier in the same paper, where he applies this abstract ("positive") theory as fully to electrostatics as to magnetism. At the simplest level, Thomson's 1845 discovery in the paper for Liouville was that Faraday's dielectric medium, while under induction, becomes a polar source, and that if you regard it as such a source, *acting at a distance* according to Poisson's theory, you obtain effects which are precisely those which Faraday has observed:

> As far as can be gathered from the experiments which have yet been made [primarily Faraday's], it seems probable that a dielectric, subjected to electrical influence, becomes excited in such a manner that every portion of it, however small, possesses *polarity* exactly analogous to the magnetic influence of a magnet. By means of a certain hypothesis regarding the nature of magnetic action, ... Poisson has investigated the mathematical laws of the distribution of magnetism, and of magnetic attractions and repulsions. These laws seem to represent in the most general manner the state of a body polarized by influence. (*EM*: 32)

These "magnetic attractions and repulsions" are very different from Faraday's idea in Series XI, in which the polarized medium *replaces* action-at-a-distance. Yet we begin to see the two ideas converging. Thomson hints at such a convergence, a polar theory in Faraday's terms, in concluding the 1845 paper:

> It is, no doubt, possible that such forces at a distance may be discovered to be produced entirely by the action of contiguous particles of some intervening medium. (*EM*: 37)

In developing the mathematics of polar sources which act directly at a distance, he is formulating ideas equally applicable to the description of a medium in a true field theory where there is no action-at-a-distance. He

114. Poisson's theory of magnetic induction is summarized by Whittaker, *op. cit.*, pp. 63 ff. Compare page 54n. above.

may not explicitly say this—he may not even see it—but the idea is certainly latent in the 1849–50 paper "On the Mathematical Theory of Magnetism."

There are two powerful theorems concerning distribution of polar sources, one expressed in Thomson's paper, the other hinted at. The first is again Poisson's, a famous equivalence theorem asserting that a continuous distribution of dipole sources is equivalent to a spatial distribution of monopole sources throughout the same volume, together with a surface distribution of monopole sources over the bounding surface. It is shown by transforming the equation of potential, through integration by parts, into the following form (*EM*: 368):

$$V = -\int_{\tau} \frac{\nabla \cdot \mathbf{I}}{r} d\tau + \int_{S} \frac{\mathbf{I} \cdot d\mathbf{S}}{r}.$$

If we make the following definitions, the significance of this form becomes apparent:

Let μ' denote a monopole density per unit volume, and μ'' denote a monopole density per unit area, over the volume and surface respectively of the continuous source distribution. The potential then becomes that of two distributions of what Thomson enjoyed calling an "imaginary magnetic matter" (*EM*: 375):

$$V = -\int_{\tau} \frac{\mu'}{r} d\tau + \int_{S} \frac{\mu''}{r} dS.$$

Applied to a dielectric, this immediately accounts for the apparent surface charge on the dielectric under induction. Most significantly for Thomson in his reading of Faraday, it shows that even when the "existence" of poles is demonstrated by experiment, they may as well be accounted for by the action of a vector continuum of sources, as by any real magnetic fluid located at the supposed pole positions. It undercuts the significance of experiments which sought to demonstrate empirically the existence of induced diamagnetic polarity. The term "imaginary" matter no doubt suggests for Thomson the questionability of the "fluids" of action-at-a-distance theory, and the essential idleness into which debates, such as that between Tyndall and Faraday over diamagnetic polarity, could slip.

The second great step in comprehending these vector distributions is the distinction between two basic configurations. As Thomson points out (*EM*: 388n), the analogy between polar distributions and fluid flow is so strong that it is natural to analyze polar sources in terms of types of fluid motion; certainly the ideas which Thomson uses here come to him out of a long tradition of fluid theory. He identifies and names the two forms of polar distribution as follows:

> A magnetic solenoid is an infinitely thin bar of any form, longitudinally magnetized with an intensity varying inversely as the area of the normal section in different parts. (*EM*: 383)

> A magnetic shell is an infinitely thin sheet of any form, normally magnetized with an intensity varying inversely as the thickness in different parts. (*EM*: 384)

> If a finite magnet of any form be capable of division into an infinite number of solenoids which are either closed or have their ends in the bounding surface, the distribution of magnetism is said to be solenoidal, and the substance is said to be solenoidally magnetized.
>
> If a finite magnet of any form be capable of division into an infinite number of magnetic shells which are either closed or have their edges in the bounding surface, this distribution of magnetism in it is said to be lamellar, ... and the substance is said to be lamellarly magnetized. (*EM*: 384)

Thomson points out that the term "solenoid" is Ampère's, who used it to refer to a limiting concept of current flow in a dense set of vanishingly small parallel circular currents, stacked in effect like pancakes on a common axis; the "canal-like" (σωληνοειδής) magnetism resulting from a collection of such solenoids would be of the sort Thomson describes.[115]

The theory of magnetic shells, for which Thomson no doubt had in mind the French term *lamelles*, had been developed by Ampère and Poisson; the potential of a magnetic shell had been expressed by Gauss very elegantly in terms of the solid angle subtended at the observation point by the shell, as Thomson points out here (*EM*: 386n).

We see that Thomson, in making this classification of polar distributions, is drawing upon the mathematical theory of magnetism and electrodynamics; but he is also working closely with George Gabriel Stokes (1819–1903) on the theory of fluid flow. Between 1847 and 1849 he and Stokes collaborated to produce a series of "Notes on Hydrodynamics," published in the *Cambridge and Dublin Mathematical Journal*, half of them by Thomson, and half by Stokes.[116] In 1847, Thomson wrote to his father about this collaboration:

> I have been getting out various interesting pieces of work, along with Stokes, connected with some problems in electricity, fluid motion, etc., that I have been thinking on for years, and I am now seeing my way better than I could ever have done by myself, or with any other person than Stokes.[117]

115. Ampère, *op. cit.*, p. 74. The term may echo the role of the *parties canalées* of the first element, in the Cartesian account of magnetism (Descartes, *op. cit.*, *9*, pp. 271 ff.), in which the screw shape reflects the polarity. Compare also Euler, who uses the term of the Cartesian "pores": "[T]hese pores ... constitute tubes or canals." Euler, *Letters to a German Princess*, *2*, p. 244.

116. William Thomson, *Mathematical and Physical Papers* (6 vols.; Cambridge: Cambridge University Press, 1882–1911), *1*, p. 107n.

117. S. P. Thomson, *op. cit.*, pp. 204–5.

The theorem toward which Thomson's classification of distributions is tending—which, indeed, it in effect presupposes—was formulated almost concurrently by Stokes. In an 1849 monograph "On the Dynamical Theory of Diffraction," Stokes had shown that any steady fluid motion having the necessary continuity can be expressed as the sum of a lamellar and a solenoidal motion, derived respectively from a scalar and a vector potential u and $\mathbf{p}$, as follows:[118]

$$\mathbf{v} = -\nabla u + (\nabla \times \mathbf{p}),$$

$\mathbf{v}$ denoting the velocity of flow. In application to the theory of magnetism, there must then exist some U and $\mathbf{P}$ such that:

$$\mathbf{I} = -\nabla U + (\nabla \times \mathbf{P}).$$

Although this theorem is not stated in Thomson's 1850 paper on magnetic sources, he clearly has it in mind. The intimate relation between Thomson's work on electricity and magnetism during this period, and both his own and Stokes' on fluids, suggests that the concept of flow or conduction, which as we have seen was such an attractive paradigm for Faraday, was also a guiding principle in Thomson's thought concerning magnetism. It might be added that the relation in turn to the problem of the luminiferous ether, the topic of Stokes' monograph from which the fluid flow theorem is borrowed, suggest that in Thomson's interpretations of Faraday, the question of the optical ether was probably never altogether out of mind.

The criteria for solenoidal and lamellar distributions of polar sources are stated by Thomson in expressions equivalent to these:

solenoidal: $\nabla \cdot \mathbf{I} = 0$,

lamellar: $\nabla \times \mathbf{I} = 0$.

He also identifies a "complex-lamellar distribution," in which the magnetism is resolved into shells of varying intensity, and shows that for this case:

complex-lamellar: $\mathbf{I} \cdot (\nabla \times \mathbf{I}) = 0$.

The solenoidal distribution corresponds, then, to flow without divergence—that is, to the flow of an incompressible fluid.[119] Lamellar flow is

118. George Gabriel Stokes, *Mathematical and Physical Papers* (5 vols.; Cambridge: 1880–85), *2*, pp. 256–57; Truesdell, *Vorticity*, p. 27.

119. In speaking of the operations of vector analysis, I shall use the terms *gradient*, *divergence*, and *curl*. In the *Treatise*, Maxwell uses other terms, respectively: *space-variation* (*Tr* i/16), *convergence* (for the negative of divergence) (*Tr* i/30), and *rotation* (*Tr* i/30). Maxwell's interest in these concepts and his rôle in the assignment of names to them, are well known; the best evidence is in the published correspondence with Tait, in C. G. Knott, *Life and Scientific Work of Peter Guthrie Tait* (Cambridge: Cambridge University Press, 1911), pp. 143 ff. Apart from their quaternion or vector representation, the mathematical operations were, as we see in Thomson's paper, borrowed from classical potential theory and fluid theory, where most of the ideas of electromagnetic theory were incubated.

that without curl, and hence derivable from a scalar potential. Complex-lamellar flow is flow perpendicular to its own curl. A fourth category, a form of incompressible flow, had been derived by Stokes in 1842; it is the so-called "Beltrami" flow, in which the velocity is collinear with its curl.[120]

What we see, then, is a rapidly developing theory of the kinematic geometry of a vector field, here specifically applied to the classification of continuous magnetic *sources*, but immediately applicable in other ways already suggesting themselves to Thomson—to the possible dielectric medium, or to the vector field represented by Faraday's lines of force. In a sense, Ampère's "solenoids" *are* Faraday's "lines of force."

Actually, the theorem in Thomson's "Mathematical Theory of Magnetism" which is most immediately relevant to Faraday's *Experimental Researches* is one which refers to the force *experienced by* a polar distribution *placed in* a magnetic field, referred to earlier (page 93 above; *EM*: 499 ff.). Thomson's analytic expression for this force proves to be a direct counterpart of Faraday's principle of diamagnetic motion, the principle that a diamagnetic body tends to move from points of stronger to those of weaker magnetic force (page 28 above). This, we saw, was the form of expression which Tyndall found particularly unpalatable (page 84 above). A valid objection to the principle as a part of the general view of magnetism was certainly that it did not appear to be related to any other principles. Thomson *shows this relation* by deriving Faraday's principle from the Poisson theory of polarity as he has reformulated it. The translation of Faraday's principle becomes this equation (*EM*: 378):

$$\mathbf{f} = \int_{\tau} (\nabla \cdot \mathbf{I}) \mathbf{H} \, d\tau \, .$$

Here **H** is the external magnetic field in which the body experiencing the force is placed, and the integration is over the volume of the body. As the expression is written above, **I** refers to the state of polarization of the body; there is no reference to induction in the equation, as the body is assumed rigidly polarized. But there is no difficulty in extending the theory to include induction. Poisson had already done this. Assuming a linear relation between the external field and the magnetization produced by induction, he wrote in effect $\mathbf{I} = i\mathbf{H}$, in which i is a coefficient that is unity for an insulated conductor in the electric case, and nearly so for soft iron in the magnetic case. For a diamagnetic material, i is negative, so that the induced polarity is reversed in sense with respect to the external field. In the integral, then, if the material is homogeneous we will have simply:

$$\mathbf{f} = i \int_{\tau} (\nabla \cdot \mathbf{H}) \mathbf{H} d\tau .$$

Very conveniently, Poisson had worked out in full the case of induction in a sphere, so that his theory is immediately applicable to the popular form

120. Truesdell, *op. cit.*, p. 77.

of apparatus for experimentation with magnetism and diamagnetism, in which the magnet of a sensitive Coulomb torsion balance is terminated in spherical pole-pieces. This is the case which Thomson discussed in 1847, pointing out that it applies also to the explanation of the attractions by amber in electrostatics (and hence, we might add, supplies the law of the root phenomenon in the science of electricity, the electrostatic attraction observed by the Greeks, from which all else has sprung!).

Thomson triumphs, on Faraday's behalf, in a footnote at the beginning of the 1847 article:

> This has not been made the subject of a special investigation by any writer, so far as I am aware, although the nature of the result, in the case of magnetism, appears to be *entirely understood by Mr. Faraday*. Thus, from §2418 of his *Experimental Researches* [containing that passage quoted on page 28, above] ... we might infer that a small sphere or cube of soft iron would in some cases be "urged along, and in others obliquely or directly cross the lines of magnetic force." (*EM*: 499; emphasis added)

It seems that the more Thomson thought about it, the more he was impressed by Faraday's insight in the case of this remarkable law; he told the Glasgow Philosophical Society in 1870:

> One of the most brilliant steps made in philosophical exposition of which any instance existed in the history of science, was that in which Faraday stated, in three or four words, intensely full of meaning, the law of the magnetic attraction or repulsion experienced by inductively magnetized bodies. (*EM*: 580)

Promptly after the 1850 paper from which we have quoted, Thomson wrote again on the subject of Faraday's law of the motion of magnets and diamagnets; this time, however, he assumes the mathematical theory which he has given in the earlier paper, and simply attempts to make the phenomenon and its meaning clear. This second paper, "Remarks on the Forces Experienced by Inductively Magnetized Ferromagnetic or Diamagnetic Non-Crystalline Substances," appeared in the *Philosophical Magazine* for October, 1850, and I think we may properly understand it as a reply to Tyndall's paper which had appeared in the same journal in July of that year, in which Tyndall had announced his inability to conceive Faraday's meaning in the statement of this law:

> Leaving attraction and repulsion out of the question, we find it extremely difficult to affix a definite meaning to the words 'tends' and 'endeavour'... (see page 83, above)

Thomson remarks that Faraday's principle has not "been sufficiently attended to by subsequent experimenters" (among them perhaps Tyndall), and gives the following account of the sense in which the law can be reconciled with the concept of "attraction"—in other words, Thomson addresses himself directly to the issue on which Tyndall has taken his stand:

> The preceding conclusions enable us to define clearly the sense in which the terms "attraction" and "repulsion" may be applied to the action exerted by a magnet on a ferromagnet. ... A small sphere of ferromagnetic substance, placed in the neighbourhood of a magnet, experiences in general, a force, but the term *attraction*, according to its derivation, means a *force towards*; and if we apply it in any case, we must be able to supply an object for the preposition. Now, in this case the force is towards places of stronger "magnetic force"; and hence the action ... may be called an *attraction* if we understand *towards places of stronger force*. Places of stronger force are generally nearer the magnet than places of weaker force, and hence small pieces of soft iron are generally urged, on the whole, towards a magnet (in consequence of which no doubt the term "attraction" came originally to be applied); but, as will be seen below, this is by no means universally the case. (*EM*: 508)

What we really see happening here, in this critical examination of the term "attraction," is not only the undoing of an ancient presupposition (that iron is "attracted" to a magnet), but the suggestion of a radically different view of physics—a physics, not in terms of a motion toward or from a center, but rather, of a local tendency, determined by the configuration of the field in the immediate vicinity of the body. This of course is not absolutely new—vortex physics of the seventeenth and eighteenth centuries was an earlier rebellion against the concept of "attraction," and the rôle of the caloric atmosphere in Mossotti's hypothesis is analogous, so far as it goes; as we have seen, Faraday feels a bond of sympathy with both of these traditions. But as a precise and complete theory of the motion of magnetic and diamagnetic materials, it is a major step toward a physics of the "field."

The concepts of fluid theory had served Thomson well, long before he drew together the "Mathematical Theory of Magnetism" which we have been discussing. Fourier's *Théorie analytique de la chaleur* is a text on fluid flow—the flow of that particularly interesting fluid, caloric, in a somewhat different guise from that in which it appeared in the theories of elastic atmospheres, such as that of Mossotti. Here, it is a question of "heat flow," in which caloric moves from regions of higher temperature to those of lower temperature as an *incompressible fluid without inertia.* Thomson was a careful and enthusiastic student of Fourier and, as we have pointed out, his first papers on electricity in 1842 and 1843 bring the theory of heat flow strikingly to bear on the solution of problems of the equilibrium of electrically charged bodies (*EM*: 1 ff., 126 ff.).

In relation to the interpretation of Faraday, which was indeed probably not yet in Thomson's mind in these first papers, the impressive point is that the configuration of caloric in its steady flow through a heat-conducting body *is* the very pattern of Faraday's lines of force—so that it would be hardly an exaggeration to claim that as Poisson wrote the theory of Faraday's dielectrics, and to a degree laid the foundation for his mysterious principle of magnetic and diamagnetic action, so Fourier wrote the theory of Faraday's lines of force.

We find Thomson first dealing directly with the lines of force as such in an 1843 paper "On the Equations of Motion of Heat Referred to Curvilinear Coordinates."[121] The plan of the paper is to extend Fourier's theory by introducing the curvilinear coordinates which Thomson has learned from the writings of Gabriel Lamé.[122] This makes it possible to determine the *isothermal surfaces* and their orthogonal trajectories, which are in fact the stream lines of caloric flow, for problems with symmetry other than rectangular. Again, it is fascinating to see how closely the ground is prepared for this by Stokes, who in an 1842 paper, "On the Steady Motion of Incompressible Fluids," established the context for Thomson's specific problem; caloric is one of the incompressible fluids of Stokes' paper.[123] Stokes is interested in establishing the *lines of motion* in the fluid for various possible cases of incompressible flow. Working in two dimensions for simplicity, the differential equation of the line of flow, or stream line, simply expresses the fact that it is everywhere tangential to the velocity vector **v**:

$$\frac{dy}{dx} = \frac{v_y}{v_x}.$$

This, of course, is at the same time the differential equation of Faraday's lines of force, though Stokes makes no allusion to any such thing in his paper. The particular case of "caloric" among fluids is distinguished by the following pair of criteria:

> It is incompressible. $\nabla \cdot \mathbf{v} = 0$,
>
> It moves only under an applied pressure (it has no inertia), and therefore the force on it, to which the velocity is assumed proportional, is derivable from a potential; hence the velocity has no curl. $\nabla \times \mathbf{v} = 0$.

Notice that Fourier's caloric obeys Aristotle's law of motion: it moves only insofar as it is pushed! Hence there exists a potential, U, of which the velocity of flow is the negative gradient, and this potential will satisfy Laplace's equation in regions where there are no sources or sinks of caloric:

$$\mathbf{v} = -\nabla U; \qquad \nabla^2 U = 0.$$

121. William Thomson, *Mathematical and Physical Papers, 1*, p. 25.

122. *Ibid.*, p. 28.

123. Stokes, *Mathematical and Physical Papers, 1*, pp. 1 ff.

Stokes had experimented in fact with stream lines in actual fluids, in his rooms at Cambridge, in what Thomson called the first university research laboratory (forgetting Newton's furnace!).[124]

Thomson succeeds in this 1843 paper in showing that the equation of flow of heat is always solvable in the case of cylindrical symmetry; he has a method of integrating the differential equation of motion to obtain the equations of the lines of force in that case. This proves to be an extremely valuable theorem in a number of interesting instances that have the required symmetry, and Thomson subsequently put it to good use as he became increasingly interested in the forms of the lines of flow. He had a beautiful set of diagrams constructed for the case of a vanishingly short dipole lying parallel to the lines of force of a uniform external field. He used these drawings first with his natural philosophy class in Glasgow, in about 1849, and then exhibited them to the British Association in 1852 (*EM*: 493n, 519–23). By that time, this diagram had taken on its full significance in relation to Faraday's concepts, as the representation of his understanding of magnetic and diamagnetic motion. In describing it to the Association, Thomson made the most of its versatility, applying it to each of the sciences to which it provides a visual key; the same diagram represented, in quick succession: a magnet placed parallel or antiparallel to a magnetic field; the electric lines of force about an insulated spherical conductor; the lines of motion of a liquid flowing around a fixed spherical solid; a ferromagnetic or diamagnetic substance under induction in a uniform magnetic field; and the lines of motion of heat about spherical cavities filled with substances of differing thermal conductivity.

This was, as it was intended to be, both a brilliant display and a pointed lesson. It exhibited to the skeptics the lines of force as mathematically precise figures; and it must have revealed to some the extent to which Faraday had hit upon a mathematical idea belonging, not only to the two sciences to which he himself applied the lines, but to a great range of mathematical physics. In short, Thomson exhibited the lines of force in 1852 as a fundamental concept in our understanding of nature in general. He made sure that the relevance of all this to Faraday was not missed; he pointed out that the same diagrams were to be seen in Faraday's lectures at the Royal Institution.

We have traced an interesting turn of thought. Attention was originally directed to the function which mathematicians had been learning to employ with great success, for many years: the potential function. The solution of thermal problems was sought in a partitioning of space by means of isothermal surfaces, delineating in this way the function which satisfied Laplace's equation and the given boundary conditions. This ability to work with the potential, which as a scalar is far easier to handle than the vector,

124. S. P. Thomson, *op. cit.*, p. 298.

force, had been an outcome of one of the great revolutions in mathematical physics. Faraday, on the other hand, thinks in terms of the lines of force which he encounters in his laboratory, with their directional character, and hence his imagery runs along the orthogonals to the mathematicians' isothermals. The difference may seem minor, when one considers the intimate relation of the two functions, the potential and its gradient. But there is this point, which Thomson himself suggests in one of the early papers: the flow lines run from source to sink, in a way which either represents an actual propagation, as in the case of heat and of liquid motion, or strongly *suggests* such a transmission:

> Now the laws of motion for heat which Fourier lays down in his *Théorie analytique de la chaleur*, are of that simple elementary kind which constitute a mathematical theory properly so called; and therefore, when we find corresponding laws to be true for the phenomena presented by electrified bodies, we may make them the foundation of the mathematical theory of electricity: and this may be done if we consider them merely as actual truths, without adopting any physical hypothesis, *although the idea they naturally suggest is that of the propagation of some effect by means of the mutual action of contiguous particles.* (*EM*: 29; emphasis added)

> It is, no doubt, possible that such forces at a distance may be discovered to be produced entirely by the action of contiguous particles of some intervening medium, *and we have an analogy for this in the case of heat, where certain effects which follow the same laws are undoubtedly propagated from particle to particle.* (*EM*: 37; emphasis added)

I think it must be granted that the heat flow analogy is not a neutral instrument, but carries with it the suggestive power of metaphor. The shift we see in Thomson from the original search for isothermals to the depiction of the field entirely in terms of lines of flow is significant of an increasing interest in a possible *propagation* of effects along those lines, through an intervening medium. Thomson's growing conviction of the value of the lines of force in explanation of the phenomena is attested by his use of them in the synthetic mode of geometrical argument he adopted for a set of "Elementary Demonstrations of Propositions in the Theory of Magnetic Force" (1855) (*EM*: 531 ff.). Much more could be said about the development of what Thomson later dubbed the "hydrokinetic analogy," especially in relation to the important concept of what Faraday called the "conducting power" (pages 26 ff., above), and which Thomson formulated in terms of "permeability" (*EM*: 586). But these developments came later, as did the enthusiasm for vortex motion, which became a highly important concept for Maxwell as well. I have tried here very generally to take the measure of Thomson's contribution to the interpretation of Faraday in the period before Maxwell himself began work on the problem. We begin to see, I believe, some of the major strides which Thomson had taken, and yet also the fragmentary character of his approach. There is however, a further aspect of his early work to which we must now turn.

Fluids served splendidly to give precise form to the lines of force, and no doubt served to suggest the possibility of propagation through a medium, but as a possible *means* of explaining such propagation, solids probably looked more promising. A simple solid is more complex, and hence can bear more functions, than a simple fluid. Thomson made one early essay at an elastic-solid analogy to electric and magnetic phenomena; again, although it was proposed merely as a "representation," it evidently tempted him in the direction of a physical hypothesis. He wrote about it to Faraday:

> I enclose the paper which I mentioned to you about giving an analogy for the electric and magnetic forces by means of the *strain* propagated through an elastic solid. What I have written is merely a sketch of the mathematical analogy. I did not venture even to hint at the possibility of making it the foundation of a physical theory of the propagation of electric and magnetic forces which, if established at all, would express as a necessary result the connection between electrical and magnetic forces, and would show how the purely *statical* phenomena of magnetism may originate either from electricity in motion, or from an inert mass such as a magnet. If such a theory could be discovered, it would also, when taken in connection with the undulatory theory of light, in all probability explain the effect of magnetism on polarized light.[125]

As the letter suggests, the venture into an elastic solid analogy was prompted by the exciting Series XIX discovery of the effect of magnetism on rotating the plane of polarization of light, though the elastic solid analogy was initially suggested by Series XI:

> Mr. Faraday, in the eleventh series of his *Experimental Researches in Electricity*, has set forth a theory of Electrostatical Induction, which suggests the idea that there may be a problem in the theory of elastic solids corresponding to every problem connected with the distribution of electricity of conductors, or with the forces of attraction and repulsion exercised by electrified bodies. The clue to a similar representation of magnetic and galvanic forces is afforded by Mr. Faraday's recent discovery of the Faraday effect.[126]

So begins the paper of 1847, "On a Mechanical Representation of Electric, Magnetic, and Galvanic Forces." This theory as well—it seems to go almost without saying—is grounded on a Stokes' slightly earlier paper, "On the Theories of the Internal Friction of Fluids in Motion, and of the

125. S. P. Thomson, *op. cit.*, p. 203.

126. William Thomson, *Mathematical and Physical Papers, 1*, p. 76.

Equilibrium and Motions of Elastic Solids" (1845, published in 1849).[127] What Thomson draws from Stokes is the mathematical characterization of the equilibrium of an elastic solid, for which Stokes' theory gives the following equation:

$$K\nabla(\nabla\cdot\mathbf{u})+\nabla^2\mathbf{u} = 0$$

in which the vector **u** represents a small displacement, and K is a constant characteristic of the solid. Except for a deliberately omitted term, this is the equation now generally known as Navier's.[128] Note that the first term may be given a certain intuitive significance if the equation is written as follows:

$$\nabla[-K(\nabla\cdot\mathbf{u})] = \nabla^2\mathbf{u}\ .$$

The negative of the divergence of the displacement vector (or, to use Maxwell's term, the "convergence" of **u**) represents a compression; if we were dealing with a fluid, this would simply give rise to a pressure, in which case the quantity $-K(\nabla\cdot\mathbf{u})$ would denote that pressure, and the left-hand side of the equation would represent the gradient of that pressure. This would then be a driving force, very much as in Mossotti's ether. We are, however, dealing with a solid, in which there is no unique pressure at a point, but rather a stress with components which must in general be represented by a tensor. It is nonetheless useful to keep in mind the comparison with fluid pressure. It turns out that the *sum* of the normal stresses upon planes set up perpendicular to the coordinate axes is proportional to the divergence of the displacement; one can thus define an *average* stress, the "mean normal stress," $\bar{S}$, and write:

$$\bar{S} = -K(\nabla\cdot\mathbf{u})$$

as if we were dealing with a hydrostatic pressure in a fluid. The coefficient K is termed the "bulk modulus of elasticity."[129] It is not quite the same as Stokes' coefficient, but evidently the quantity $-K(\nabla\cdot\mathbf{u})$ in Stokes' equation is proportional to the mean normal pressure, and the left-hand term is the gradient of this average pressure. The right-hand term, $\nabla^2\mathbf{u}$, measures the character of the displacement of the solid at any point; this displacement can be analyzed as an elementary rotation without distortion, together with a relative displacement of the points within an element of the medium. It might be added that Stokes' is a two-constant theory of solids, but the second constant would appear only if we had occasion to consider the density of the solid. Since in Thomson's application of the theory he is neither accelerating it nor submitting it to the action of gravity, only one constant

127. Stokes, *Mathematical and Physical Papers, 1*, pp. 75 ff.

128. Robert R. Long, *Mechanics of Solids and Fluids* (Englewood Cliffs, NJ: Prentice-Hall, 1961), p. 81; Truesdell, *op. cit.*, p. 41.

129. Long, *op. cit.*, p. 78.

arises. The behavior of the solid in compression and in shear are both involved in the one constant K.

Since the quantity $\nabla^2\mathbf{u}$ is shown by the equilibrium equation to be the gradient of a scalar, namely $-K(\nabla\cdot\mathbf{u})$, we know that the displacement in any particular case must be such that $\nabla^2\mathbf{u}\cdot\mathbf{ds}$ is a complete differential ($\mathbf{ds}$ denoting an element of a path of integration). We do *not* know, however, whether $\mathbf{u}$ itself derives from a potential: it may or may not. Thomson now considers three types of strain, which he identifies in turn with electricity, magnetism, and electromagnetism. The corresponding assumptions may best be tabulated; either:

(1) $\mathbf{u}\cdot\mathbf{ds}$	is a complete differential, or
(2) $(\nabla\times\mathbf{u})\cdot\mathbf{ds}$	is a complete differential, or
(3) neither	is a complete differential.

(1) The first case, in which the displacement $\mathbf{u}$ itself is the gradient of a potential, clearly permits us to identify $\mathbf{u}$ with the electric intensity, setting:

$$\mathscr{E} = \mathbf{u}, \quad \text{so that} \quad \mathscr{E} = -\nabla U.$$

(2) In the second case, we may identify, not $\mathbf{u}$, but curl $\mathbf{u}$ with the magnetic intensity:

$$\mathbf{H} = \nabla\times\mathbf{u}\,.$$

$\mathbf{H}$ will then be derivable from a potential, as indeed the magnetic field *is* so derivable, in a region free of sources; the potential is then that of a distribution of dipole sources, given on page 95, above:

$$V = \frac{\mathbf{p}\cdot\hat{\mathbf{e}}_r}{r^2}.$$

Proceeding in a way suggested by the mechanics of solids and fluids (pages 98 ff.), Thomson has in fact taken the very important step here of introducing the *vector potential* of the dipole distribution, though he does not so name it. In the guise of the displacement, he has introduced the quantity

$$u = \frac{\mathbf{p}\times\hat{\mathbf{e}}_r}{r},$$

a vector from which the force is derived, not by taking the gradient as in the case of a scalar potential, but by taking the curl:

$$\mathbf{H} = \nabla\times\mathbf{u}\,.$$

In the elastic-solid analogy, this is not a merely formal operation; the quantity $\nabla\times\mathbf{u}$ measures the angle of rigid rotation of an element of the solid.[130] Thomson has therefore taken the angle of rotation of an element of the solid at any point as the measure of the magnetic intensity there.

130. *Ibid.*, p. 52.

(3) Finally, in the third case in which neither of these conditions is met, there still remain displacements of the solid which satisfy the equilibrium equation—that is, which are possible equilibrium configurations of an elastic solid. It remains true for these that $\nabla^2\mathbf{u}$ itself must be the gradient of a scalar. Thomson uses this case to represent the magnetic intensity due to a current element $i\,\mathbf{ds}$ as given by Ampère's theory. He obtains this result by defining the potential of which $\nabla^2\mathbf{u}$ is to be the gradient, as follows:

$$\nabla^2\mathbf{u} = -\nabla\left(\frac{i\,\mathbf{ds}\cdot\hat{\mathbf{e}}_r}{r^2}\right).$$

This gives, for the displacement $\mathbf{u}$ itself:

$$\mathbf{u} = \frac{1}{2}\nabla\left(\frac{i\,\mathbf{ds}\cdot\hat{\mathbf{e}}_r}{r^2}\right) - \frac{i\,\mathbf{ds}}{r}.$$

The curl of this will be:

$$\mathbf{H} = \nabla\times\mathbf{u} = \frac{i\,\mathbf{ds}\times\hat{\mathbf{e}}_r}{r^2},$$

which is just the magnetic intensity of a current element. Since identically curl grad $\mathbf{v} = 0$ for any $\mathbf{v}$, the first term in the expression for $\mathbf{u}$ does not contribute to $\mathbf{H}$; and if we therefore ignore this first term, we see that Thomson has in effect introduced the vector potential of a current element in the form now familiar:

$$\mathbf{dA} = \frac{i\,\mathbf{ds}}{r}$$

(cf. *Tr* ii/257, here setting $\mu = 1$).

To summarize, then: Thomson has proposed an elastic solid continuum as an analogue to a field in which electric and magnetic intensities are present. The analogy does not go beyond a mathematical description of the equilibrium state; though it is *suggestive* in terms of stresses, it does not undertake to account for forces or motions, but merely to give a mathematical analogy to the configurations of the quantities **H** and $\mathscr{E}$ under equilibrium conditions. Thomson takes the displacement itself as representing the electric intensity, and the angular displacement everywhere as representing the magnetic intensity, whether due to dipole sources (case 2), or to current elements (case 3). Since the angular displacement is measured by the curl of the displacement, in the magnetic case **u** itself functions in effect as a vector potential of the magnetic intensity.

We have seen that both Thomson and Mossotti have introduced continua and their displacements into the discussion of the region about charged bodies; Mossotti worked with an elastic fluid, whose equilibrium state is characterized everywhere merely by a compression and a corresponding hydrostatic pressure, while Thomson's solid could sustain more complex stresses, corresponding to both distortions and rotations. Thomson's then is able to represent both the electric and the magnetic fields. Mossotti's on the other hand is not a representation, but a mechanical hypothesis; its role is to exert pressures on bodies which, together with other, quite independent forces, will account for the observed electrostatic effects, as well as gravity and molecular phenomena. The density configuration in Mossotti's ether is a potential, but it is not the electric potential, and it is not intended to be; his ether is not a mathematical model, but a hypothesized instrument of nature.

Thomson, by contrast, is not at this point making hypotheses, but is discovering analogies which will help to solve problems. In this sense, Mossotti is much closer to Faraday's own line of thought. Faraday, I have argued, is concerned with entities, not with analogies as such, or with the solution of problems.

Yet Thomson's analogy, though it is no more than a "sketch," as he tells Faraday in the letter quoted above, inevitably becomes a "hint" of a physical theory. It is a representation which points strongly toward a unified mechanical theory, and as such, it has achieved one enormous advance from the point of view of field theory. Here all the effects are produced without the invocation of action-at-a-distance. The "body force" term (ordinarily representing the direct action of gravity on the parts of the solid) was omitted from the equation of equilibrium (page 106), with the result that the equilibrium is entirely one of contact action. In establishing the equilibrium configuration of displacements, each part of the solid continuum acts only on the contiguous parts.

If this is no more than a hint, a hint may sometimes suffice. Maxwell turned to Thomson as his tutor in electricity at a time when he could find "no other man to apply to," and it is clear that this "allegorical representation," to which Thomson referred him, set his mind turning.[131] Maxwell had worked with elastic solids before he turned to electricity; his second published paper, five years before, had described the configurations of stresses in transparent solids as revealed by polarized light. These stresses which Maxwell observed are reported in his first diagrams of field patterns and lines of force *in solids* before he ever began to think about magnets (*SP* i/68). He was therefore prepared to take a cue from Thomson, and to bring the concept of material continua to bear upon Faraday's sphondyloids of magnetic power and systems of lines of force. Thomson's fluid representations became Maxwell's "physical analogy" in "Faraday's Lines of Force" (1855–56), and Thomson's elastic-solid analogy became the germ of Maxwell's "Physical Lines of Force" (1861–62).

131. Letters of Maxwell to Thomson, Nov. 13, 1854, and Sept. 13, 1855; in Joseph Larmor, *Origins of Clerk Maxwell's Electric Ideas* (Cambridge: Cambridge University Press, 1937).

In each case, Maxwell fundamentally converted the material Thomson supplied, and yet in the suggestion of ways in which Faraday might be understood—as reading lessons in the *Experimental Researches*—Thomson's interpretations of Faraday may mark the inception of Maxwell's enterprise.

Chapter 3. Maxwell's Initial Statement of Faraday's Electromagnetism

THE "DIRECT" AND "INDIRECT" METHODS

The first two chapters of this study have been concerned with preliminaries: first, with an examination of Faraday's concept of science, along with some of the ideas about electromagnetism that are woven into his *Experimental Researches*; and second, with a few of the efforts to understand and "translate" Faraday's ideas which had been made by others, before Maxwell. We turn now to the central concern of this study, Maxwell's *Treatise* itself, and to the translation of Faraday's electromagnetism which Maxwell achieved in Part iv of that work—the part he entitled simply, "Electromagnetism." To a large extent, the mode of the present study must now become that of a commentary, following Maxwell's argument in Chapters I–X of Part iv of the *Treatise*. To avoid excessive quotation, I shall make frequent references directly to the *Treatise*, trusting that the reader has a copy of the text close at hand.

It must be acknowledged again that the *Treatise* is not Maxwell's first endeavor to achieve an adequate translation of Faraday's concepts, but rather (if I am right) the culmination of a series of papers directed in various ways to that end. No attempt is being made here to tell the "history" of the translations of Faraday: the earlier examples have been examined for the sake of a better understanding of the problem which the interpretation of Faraday presents, and thereby to cast in a sharper light Maxwell's achievement in its final form in the *Treatise*. To compare the several facets of Maxwell's successive interpretations of Faraday over the decade and a half which preceded the writing of the *Treatise* would be an illuminating study, but a different one.[1]

A reader of Part iv of the *Treatise* is forcibly struck by a strange feature of Maxwell's plan: the development comes to an abrupt halt at the end of Chapter IV, as if a deep breath were being drawn, and then a new start is

1. Maxwell wrote four principal papers on electromagnetic theory prior to the *Treatise*. These were "On Faraday's Lines of Force" (1855–56) (*SP* i/155 ff.); "On Physical Lines of Force" (1861–62) (*SP* i/451 ff.); "A Dynamical Theory of the Electromagnetic Field" (1864–65) (*SP* i/526 ff.); and a "Note on the Electromagnetic Theory of Light," included in a paper of 1868 (*SP* ii/137 ff.). There is a considerable literature of commentary on Maxwell's electromagnetic writings; see for example Boltzmann, Hertz, Duhem, Whittaker, Bork, Bromberg, Scott, and Tricker in the Bibliography.

made with Chapter V. This second start marks the beginning of the exposition of Maxwell's "Dynamical Theory of Electromagnetism," to which Chapter V itself, "On the Equations of Motion of a Connected System," is a necessary preliminary. In commenting on Maxwell's text, this study will make a division at the same point. The present chapter will deal with Maxwell's Chapters I–IV, which sets forth an initial statement of Faraday's electromagnetism, then contrasts it with Ampère's electrodynamics, an examplar of action-at-a-distance theory. Part II of the present study, beginning with the next chapter, will follow the development of Maxwell's "dynamical theory" and will discuss it as a further stage in the effort to give systematic form to Faraday's insights concerning the electromagnetic field.

We turn now to the discussion of Chapter I of Part iv of the *Treatise*, the chapter entitled "Electromagnetic Force." Its topic is the mechanical force exerted by one circuit on another, a realm of phenomena for which Ampère had given a brilliant and exhaustive theory in action-at-a-distance form before Faraday even began his *Experimental Researches*; Maxwell reserves treatment of Faraday's original topic, electromagnetic induction, for Chapter III of Part iv. He thus begins his exposition of Faraday's method in electromagnetism on ground that is sacred to the mathematical theory, and the chapter is a deliberate recasting of theory into a new mold. Maxwell is certainly sensitive to the implicit confrontation of methods which is lurking as he develops his account of these electromagnetic phenomena from a point of view appropriate to Faraday's concepts, and at a certain point in the chapter he stops to make this confrontation explicit. He does so by distinguishing formally between what he calls the "direct" and the "indirect" methods in physics. The former is Ampère's, and is that of action-at-a-distance theory in general. The latter is Faraday's; and I believe Maxwell hopes to represent it in a way that captures the essence of Faraday's approach. If so, Maxwell's definition of the "indirect" method may be taken as a characterization of the method of "field" physics.

According to Maxwell, there are three stages in the method of Faraday, while there is only one in that of Ampère. For Ampère, whose theory Maxwell will discuss at length in Chapter II of Part iv of the *Treatise*, it is only necessary to calculate "the direct action of a portion of one circuit on a portion of another" (*Tr* ii/152). In the method of Faraday, the *field* intervenes. Maxwell identifies the stages of the *indirect method* in this way:

> [I] We shew, first, that a circuit produces the same effect on a magnet as a magnetic shell, or, in other words, we determine the nature of the magnetic field due to the circuit.
>
> [II] We shew, secondly, that a circuit when placed in any magnetic field experiences the same force as a magnetic shell. We thus determine the force acting on the circuit placed in any magnetic field. (*Tr* ii/152)

In steps I and II (the enumeration is mine), we treat two interacting circuits separately, considering the relationship of each to a *field*, the one producing it as source, the other acted upon by it. It is only in the third step that we bring the two circuits together, through the mediation of the field:

> [III] Lastly, by supposing the magnetic field to be due to a second electric circuit we determine the action of one [whole] circuit on the whole or any portion of the other. (*Tr* ii/152)

For Maxwell, these two methods represent not so much two ways of structuring a physical theory, as two views of the world: the "methods" are ultimately nature's, not the scientist's. Thus the comparison and contrast of these two ways, and ultimately the resolution of this methodological alternative, raises questions not merely of convenience but of truth; in Maxwell's view, it is a problem of central concern for natural philosophy. Maxwell's long concern with the alternative we meet here in the *Treatise* is suggested by his account, in an early letter to Thomson, of his own first steps in approaching the study of electricity:

> I tried to make out the theory of attractions of currents [Ampère's] but tho' I could see how the effects could be determined I was not satisfied with the *form of the theory which treats of elementary currents* & their reciprocal actions. ... I read Ampère's investigations this term and greatly admired them but thought there was a kind of ostensive demonstration about them wh: must have been got up, after Ampère had convinced himself, in order to suit his views of philosophical inquiry, and as an example of what it ought to be. ... Now I have heard you speak of "magnetic lines of force" & Faraday seems to make great use of them, but others seem to prefer the notion of attraction of elements of currents *directly*.[2]

The sentence which follows has in it the germ of the three parts of the "indirect method" as we have seen it described in the *Treatise*; we can number them according to the stages he distinguished there:

> Now I thought that [I] as every current generated magnetic lines & [II] was acted upon in a manner determined by the lines through wh: it passed that [III] something might be done by considering "magnetic polarization" as a property of the "magnetic field" or space and developing the geometrical ideas according to this view.[3]

Between that early proposal, which led directly to "Faraday's Lines of Force" in 1855, and the *Treatise*, many approaches to the problem have intervened, but the underlying effort to devise a theory in which distant elements do not act upon one another *directly* has been constant.

2. Joseph Larmor, *Origins*, p. 8.

3. *loc. cit.*

The indirect method is to be the key to the translation of Faraday. Here the crucial point is to be able to work mathematically with the *field*—that is, to characterize rigorously the state of the space or region in which the conductors are located, and to express equally precisely the relation of each conductor to the state or configuration of that region in which it is placed. We have seen the roles of the "lines of force" and the "sphondyloids of power" in Faraday's own accounts: for him, finally, they became the operative entities in nature. Maxwell's "indirect method" must supply a counterpart to the "lines" and the "sphondyloids."

In setting up a mathematical theory appropriate to Faraday's view, Maxwell does not make the *forces* in the field, represented by the lines as paths through a vector field, his central consideration. What Maxwell chooses as the formal device most truly representative of Faraday's concept is, strangely, a quantity drawn from an entirely separate context of thought. This is the *mutual potential* of two circuits, a quantity developed on the Continent by such men as Franz Ernst Neumann, and I believe almost completely unknown in this form to Faraday. Conversely, as we have mentioned, Neumann took no more from Faraday than his empirical result on electromagnetic induction (page 51, above).

We thus meet the apparent anomaly that in Chapter I of Part iv of the *Treatise* Maxwell presents Faraday's *indirect* method by means of a mathematical device incubated by theorists of the *direct* method. Maxwell does so, I think, because he feels that in the concept of the mutual potential he has identified what is really at the core of Faraday's insight; and he wishes, from the beginning of the account of electromagnetic theory in the *Treatise*, to develop in the reader's mind what will ultimately prove the most appropriate forms of thought. Only gradually will Maxwell unfold for the reader the strict relation between Faraday's own terms and images, and this new and abstract mathematical interpretation which Maxwell is proposing to bring to them.

THE THEORY OF MAGNETIC SHELLS

The first two chapters of Part iv of the *Treatise* are conceived by Maxwell as a pair, in order to contrast as sharply as possible the methods of Faraday and Ampère, and thereby to make more vivid the nature of Faraday's method itself. The first chapter is devoted to Faraday's view, and the second exclusively to Ampère's. At the beginning of the third chapter, Maxwell looks back to survey these two paradigms of antithetical modes in science, weighing in terms of the contrasting scientific styles of these two men, the significance of the transformation from action-at-a-distance theory to a physics of fields.

In Chapter I—the "Faraday" chapter of the pair—Maxwell uses a single overall strategy, based on the formal equivalence of electric circuits to magnetic shells. This move achieves a notable economy in the composition of the *Treatise*; because he has already developed shell theory fully in Part iii, virtually all the mathematics he will need in the discussion of electromagnetics has already been presented. Not only the mathematics but, more significantly, certain basic ideas are now already familiar; he need only point out their application in the new context.

Maxwell would not, however, have seized on this device for the sake of economy if the ideas developed in magnetic shell theory were not at the same time those which he felt were appropriate to the interpretation of Faraday. Again, however, we meet an anomaly: this equivalence on which the design of the *Treatise* rests, and which will be the avenue into the translation of Faraday, is borrowed virtually intact from Ampère's action-at-a-distance theory.

Magnetic shells are appropriate to Faraday's concepts primarily because the substitution of a shell for a circuit commits us from the outset to a view of the *circuit as a whole.* As we shall see in further detail later, the contrast of Faraday's method with Ampère's yields two distinctions: not only does Faraday attribute the active role in electromagnetism to an intervening field, but he sees this field as an *entity*, a "system" of power. As we have discussed in Chapter 1, the sphondyloid belonging to a magnet is for Faraday an individual substance—subject indeed to spatial distortions, but quantitatively conserved in its interactions with the sphondyloids of other magnets. To catch this spirit of Faraday's thought, a theory of electromagnetism must treat the field of a circuit as belonging to the *whole* circuit, thereby establishing a single measure of the "power" of the system in its integrity.

By contrast, Ampère works in the spirit of analysis that stemmed from Newton's *Principia*, and which by Ampère's time had achieved magnificent formulation in the works of the Continental mathematicians. He thus takes as fundamental a law of force between *current elements*, just as Newton had taken a law of force between elementary masses as fundamental in the theory of gravitation. In his second, "Ampère" chapter, Maxwell will develop an analytic argument which leads to such a force law, but we find nothing comparable in Chapter I. Currents are here taken *primarily* as whole geometrical circuits, equivalent to corresponding shells, and functioning as the seats of corresponding geometrical fields. It is always possible to work back from wholes to parts by differentiating, and Maxwell will make a limited use of that technique in this chapter. But the essential point is that in his statement of Faraday's view, Maxwell takes whole circuits as the "elements" of the theory.

In place, then, of the law of force between current elements (in the manner of Ampère), we will find an expression for the interactions of circuits, in the guise of shells. The single quantity which relates two such shells, from which both forces and torques on the shells can be derived, is their *mutual potential energy*, precisely the quantity to which the theory of shells led in the opening chapters of Part iii. Indeed, when Maxwell introduced this quantity at the end of Chapter III of Part iv, he hinted at its future importance "in the theory of electric currents" (*Tr* ii/46). Still further applications of the potential method will unfold later, in relation to the theory of induction.

We see, then, that Maxwell approaches a theory of electromagnetism patterned upon Faraday's ideas by employing a device which immediately treats circuits as wholes, and which assigns a single scalar quantity to their mutual energy. This mutual energy will translate Faraday's term "power," in the phrase "system of power." It is interesting that, in interpreting Faraday, Maxwell is led to give central place to the scalar *energy*, rather than the vector *force*—to the single quantity which refers to the field as a whole, rather than to the specific quantity which belongs to a point. This places emphasis on the "power" from which, in Faraday's terms, the individual "tendencies to move" can be inferred. It is, indeed, an aspect of a larger transformation of viewpoint within physics which is very much part of Maxwell's own thought, and which has to do, as we shall see in a later chapter of this study, with his overall conception of "dynamical theory." The essence of Faraday's view, as Maxwell interprets it, accords with the reformulation of physics in terms of energy, and the derivation of *forces* from more fundamental *potentials*.

Since magnetic shell theory is Maxwell's principal instrument in the mathematical translation of Faraday's electromagnetism, we must begin with a synopsis of this theory as it was developed in Chapter III of Part iii of the *Treatise*. This theory was not original with Maxwell, and indeed he gives as an overall chapter reference Thomson's article, which we have already noted, in which the theory of *lamellar* sources is developed.[4] As was remarked then, Thomson had in turn found the theory in the pages of Ampère and Poisson.[5]

In general following Thomson, Maxwell defines a *simple magnetic shell* as a "thin sheet of magnetic matter" magnetized in a direction everywhere normal to its surface. The product of the intensity of magnetization by the "thickness" of the shell is termed the strength of the shell, and the strength of a simple shell is the same throughout (*Tr* ii/35). (Like Thomson, Maxwell emphasizes that the "magnetic matter" is a convenient fiction.)[6] Shells may be compounded to form complex lamellar distributions, as we have seen, but for Maxwell's purposes in Part iv, the simple shell will suffice.

4. (*Tr* ii/33, *EM*: 382 ff.). Compare pages 96 ff., above.

5. Page 97n, above. At (*EM*: 375), Thomson expresses his debt to Poisson's memoir on magnetism, which was cited at page 54n, above.

6. (*Tr* ii/6–7). Compare page 96, above.

Gauss had shown that the potential of a shell of strength φ is, at any point, equal to the solid angle subtended at that point by the shell, multiplied by the strength of the shell:[7]

$$V = \varphi\omega. \quad [\varphi \leftrightarrow \Phi]$$

Notice that if we apply the "indirect" program to the theory of the interaction of shells, this expression of Gauss's satisfies the requirements of Step I: it tells us the field at any point, due to any given source.

A second fundamental expression, corresponding to Step II, gives the energy of a portion of a shell of area **dS**, when placed in a known field. If V is the given potential of this field at the position of **dS**, and φ is the strength of the shell, then the potential energy of the element of the shell is (*Tr* ii/35):

$$dU = \varphi \nabla V \cdot \mathbf{dS}. \quad [\mathbf{dS} \leftrightarrow dS(l,m,n)]$$

Maxwell, thinking here in terms of free space, does not distinguish magnetic intensity **H** and magnetic induction **B**, so now he is able to write:[8]

$$\mathbf{B} = -\nabla V. \quad [\mathbf{B} \leftrightarrow a,b,c]$$

We might expect Maxwell now to substitute this expression in the preceding one, in order to express the energy of the shell in terms of the flux **B•dS** which he has called the "magnetic induction through the surface" (*Tr* ii/26); but he does not. He shifts instead to the vector potential, using (*Tr* ii/29):

$$\mathbf{B} = \nabla \times \mathbf{A}. \quad [\mathbf{A} \leftrightarrow F,G,H]$$

He finally writes for the energy U (*Tr* ii/45):

$$U = -\varphi \oint_l \mathbf{A} \cdot \mathbf{dl}. \quad [\mathbf{dl} \leftrightarrow ds]$$

Why does he make the shift to the vector potential? One immediate reason is, no doubt, that the theorem is true for a surface of a given boundary, regardless of the shape of the surface itself. Hence the boundary is the determining invariant in the situation, and it is more significant to integrate with respect to this boundary, rather than over an arbitrarily chosen surface. One boundary determines a multiply infinite range of surfaces for which the flux, and hence the energy, is the same.

The ultimate reason, however, is that the boundary of the *shell* is destined to be transformed into a *circuit* when shell theory is transformed into circuit theory in Part iv; and Maxwell wants now to give to shell theory a form which will be appropriate to circuits later. Actually, it must be

7. (*Tr* ii/35). Gauss, "General Theory of Terrestrial Magnetism" (1838), in Taylor, ed., *Scientific Memoirs, 2* (1841), p. 230.

8. The distinction between **B** and **H** was introduced, with something less than full clarity, by Thomson in 1849. (*EM*: 365 ff.) It is the subject of Chapter II of Part iii of the *Treatise.*

acknowledged, he is looking beyond Chapter I of Part iv when he does this; for in Chapter I, as we shall see, he follows Faraday in making the flux (the "number of lines") the working quantity of the theory. But when he goes beyond this stage to enunciate his own "dynamical theory," which he at the same time believes represents Faraday's deepest insight, it will be the vector **A** and its integral *around the circuit* that becomes fundamental. This is a topic to which we shall return in Chapter 6, below.

Once U is known, the force on the shell would be determined by taking the gradient. Maxwell, apparently interested at this point in drawing attention to energies rather than forces, lets this corollary pass without mention. Finally, Maxwell takes Step III of the indirect method in application to shell theory. He eliminates the expression for the field, to write an equation for the mutual potential energy directly. Again, he prefers to express this in terms of integrations around boundaries rather than over surfaces, and obtains[9]

$$U = \varphi_1 \varphi_2 \oint_{l_1} \oint_{l_2} \frac{\mathbf{dl}_1 \boldsymbol{\cdot} \mathbf{dl}_2}{r}. \quad [U \leftrightarrow M]$$

COMMENTARY ON CHAPTER I OF PART IV OF THE *TREATISE*

In Chapter I of Part iv of the *Treatise,* Maxwell recasts Faraday's thoughts in terms of this classical magnetic shell theory. He begins with an account of the fundamental discovery of Hans Christian Oersted (1777–1851), an account that completely bypasses Ampère's renowned interpretation and instead simply assumes that Oersted's own remark that "the electric conflict acts in a revolving manner" may be taken at face value.[10] Maxwell wishes to draw from the Oersted experiment one concept that was central to Faraday's understanding of it: recognition that the field is circular in shape.[11] He does not pause to speculate about the experiment otherwise—to question Oersted's curious term *conflictus* or to suggest other possible interpretations of the same phenomenon.[12]

9. (*Tr* ii/46). Maxwell will later use the symbol U for the quantity which he here calls M, the mutual potential energy of two circuits; I have therefore made the shift to U at this point.

10. Hans Christian Oersted, "Experimenta circa Effectum Conflictus Electrici in Acum Magneticum" (1820), trans. in R. A. R. Tricker, *Early Electrodynamics: the First Law of Circulation* (Oxford: Pergamon Press, 1965), pp. 113 ff. In general, see this book of Tricker's on the work of Oersted and of Ampère.

11. Faraday wrote of his own early observations of the Oersted phenomenon: "It is, indeed, an ascertained fact, that the connecting wire has different powers at its opposite sides; or rather, each power continues all round the wire, the direction being the same...." (1822) (*XR* ii/132).

12. The "circular" character of the force was much disputed, Ampère maintaining that it must be compounded of actions in straight lines joining the magnetic needle and the wire; he showed mathematically that it could in fact be accounted for in this way. Ampère, "Memoir" of 1825, in Tricker, *op. cit.*, p. 184. (See pages 167 ff., below.)

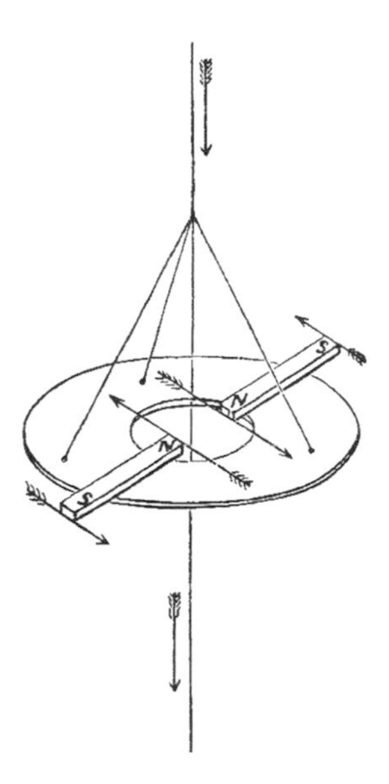

Figure 3. *Treatise*, vol. II, Figure 21

Maxwell instead turns immediately to a demonstration experiment, indeed a superb example of what he had called an *experiment of illustration* in his introductory lecture at Cambridge (*SP* ii/242). The essence of an experiment of illustration is simplicity; a single physical idea must stand out clearly. Here, we shall see how much Maxwell draws from the extremely simple arrangement sketched in Figure 3 (*Tr* ii/139). Current is carried in a long, straight, vertical wire; a horizontal platform is suspended so as to rotate freely about the wire as axis. On this platform, two magnets are placed with their long axes lying along a single line drawn radially through the wire, one magnet on either side of the wire, and equidistant from it so that their weights balance. Both north poles are toward the center, and the south poles outward (note that their tendencies to dip in the earth's field thus balance). Both north poles, one infers, endeavor to revolve in the same sense; both south poles endeavor to revolve in the reverse sense. The result however is null: no net tendency to rotation is observed. This is the best possible outcome for an experiment of illustration. Greatest simplicity is achieved when *nothing at all happens*; no measurements are required, and the mind can instantly grasp the case. Maxwell immediately draws the consequence, from the balancing of the torques, that the field about the wire must be determined by the following rule:

> [T]he electromagnetic force due to a straight current of infinite length is perpendicular to the current, and varies inversely as the distance from it. (*Tr* ii/140)

Maxwell alludes briefly to the fact that we may take this as the basis of a definition of unit current in "the Electromagnetic system of

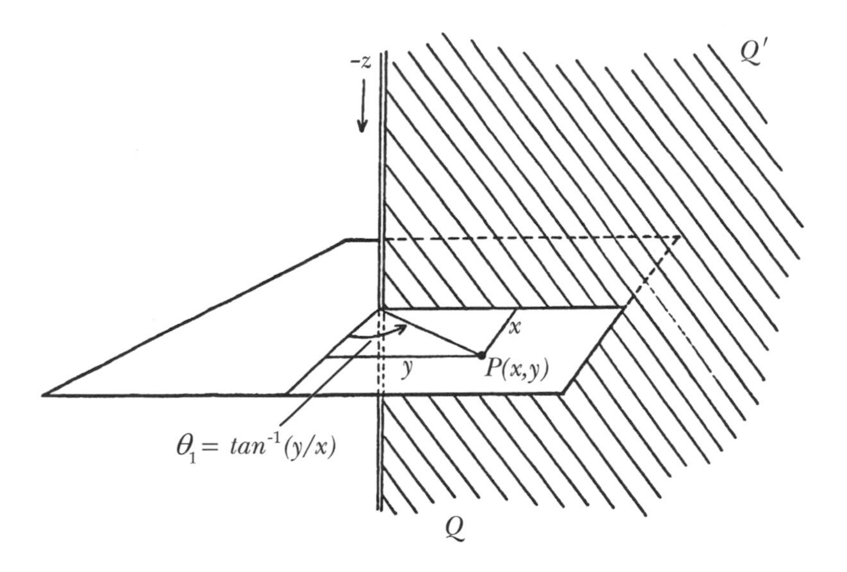

Figure 4. The Oersted demonstration as an experiment of illustration

measurement";[13] in this system, i is defined in such a way that (*Tr* ii/140)

$$H = \frac{2i}{r}. \quad [H \leftrightarrow T]$$

Maxwell moves quickly, however, from this indication of the field of magnetic force H to a reinterpretation of the same experiment of illustration in terms of potentials. He asserts, leaving it to the reader to verify his remark, that the three components of the magnetic force vector **H** are such that **H•dS** is a complete differential—specifically the differential of the potential function V, which is (*Tr* ii/140)

$$V = 2i \tan^{-1}\left(\frac{y}{x}\right) + C .$$

This is a curious case, however, since we see that if angle θ_1 (Figure 4) gives a potential V_1, it must also give potentials $V_1 + n(4\pi i)$, where $n = 1, 2, 3 \ldots$; for all of these potentials give the same value of the tangent, and equally satisfy the equation. The magnetic force in this case is not conservative; work is evidently done by the field in carrying a pole around a path which encloses the wire. Maxwell is intent, nonetheless, on our regarding this as an instance of a potential, and for this purpose introduces a simple device:

13. Maxwell discusses the introduction of absolute systems of units at (*Tr* ii/193–94), where he credits both the electrostatic and the electromagnetic systems to Weber's work in the *Elektrodynamische Massbestimmungen.* Weber refers the electromagnetic definition of unit current to Gauss and Weber, *Resultats aus den Beobachtungen des magnetischen Vereins* (1840); see Weber, *Elektrodynamische Massbestimmungen: insbesondere Widerstandsmessungen* (Leipzig: 1852), p. 219.

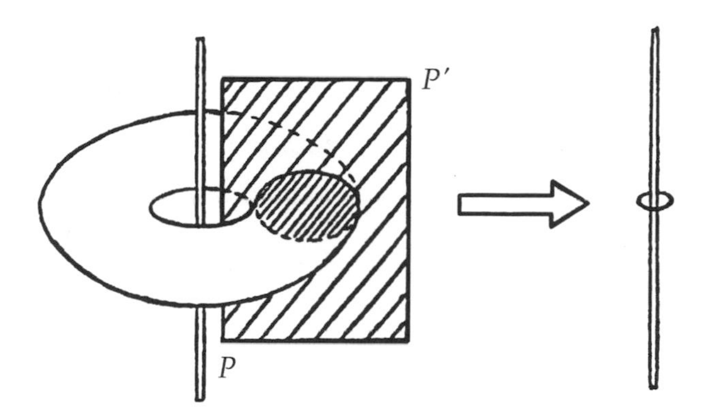

Figure 5. Cyclosis of a ring

> If we consider the space surrounding an infinite straight line we shall see that it is a cyclic space, because it returns into itself. If we now conceive a plane, or any other surface, commencing at the straight line and extending on one side of it to infinity, this surface may be regarded as a diaphragm which reduces the cyclic space to an acyclic one. (*Tr* ii/141)

The "cyclosis" of a region has been explained in Maxwell's "Preliminary" chapter (*Tr* i/17 ff.). Applying the definition which he gives there (*Tr* i/18), we see that this space is *cyclic* in the following sense: it is bounded by an external surface at infinity and by an internal surface which is the surface of the vertical wire. We may think of it by analogy to a ring, which Maxwell offers as the simplest example of a region with cyclosis (*Tr* i/18). If we were to shrink these infinite bounds, we would arrive at a closed path tightly wrapped about the wire (Figure 5).

Theorem II (*Tr* i/19) now applies, telling us in effect that to remove the ambiguity of the potential function, it is necessary to remove the cyclosis of our region. In the case of the ring, we see that we need to pass a diaphragm such as *PP'* in the figure, to remove the cyclosis without destroying the continuity of the space. In the "infinite" region of the Oersted experiment, the analogue to *PP'* is the half-plane *QQ'* of Figure 4.

This solves the problem of applying the potential function to the circular field of the Oersted experiment: our space, now "acyclic," does not admit values of θ greater than 2π. But this seems artificial, and it would indeed be a mere formality were it not for Maxwell's next step:

> The magnetic field is now identical in all respects with that due to a magnetic shell coinciding with this surface, the strength of the shell being *i*. This shell is bounded on one edge by the infinite straight line. The other parts of its boundary are at an infinite distance from the part of the field under consideration. (*Tr* ii/141)

Even today many a reader, if moderately familiar with electromagnetic theory and with the standard examples in the texts, might well be startled by this reconstruction of the Oersted experiment: the infinite straight conductor has become one edge of an half-infinite plane magnetic shell, whose strength is numerically equal to the current in the Oersted conductor. What we are looking at is, in this view, the edge-field of a half-infinite shell. Maxwell has not, indeed, calculated this field for us, but it is a fair challenge to the reader, armed with the techniques of Part iii, to do so for himself.

With strong guidance from Maxwell, the Oersted demonstration has revealed the suggestion that an electric current produces a field equivalent to that of a magnetic shell having the circuit as its boundary. The experiment of illustration has served as a clue, and it is interesting—and quite characteristic—that Maxwell has chosen to begin with an intuitive glimpse, a suggestion, rather than a demonstration. To establish the general principle of the equivalence of circuits and shells, however, we will need more evidence. Maxwell is content to allude to this very briefly:

> It has been shewn by numerous experiments, of which the earliest are those of Ampère, and the most accurate those of Weber, that the magnetic action of a small plane circuit at distances which are great compared with the dimensions of the circuit is the same as that of a magnet whose axis is normal to the plane of the circuit, and whose magnetic moment is equal to the area of the circuit multiplied by the strength of the current. (*Tr* ii/141)

We shall see in a moment how Maxwell intends to draw, from this supposed experimental evidence concerning very small coils, the general principle concerning circuits of any size.

But first, what experiments does Maxwell have in mind? So far as I am aware, Ampère did not report any experiments on very small circuits. In the *Théorie mathématique*, to which W. D. Niven, as editor of the second edition of the *Treatise*, gives a general reference at this point, Ampère does indeed base his arguments on experiments—but very different ones, as Maxwell will himself explain in his Chapter II. Ampère argues from those experiments to his own theory; then from his theory he deduces the forces due to small circular loops. Weber, on the other hand, to whose *Elektrodynamische Maasbestimmungen* Niven also refers, did indeed investigate experimentally the forces between coils, substituting coils for magnets in the procedure which Gauss had developed for the measurement of magnetic intensity, and using an early form of his electrodynamometer.[14] The coils were not as small as Maxwell suggests (the diameters of the coils being approximately 8.8 cm. and 11.1 cm., while the distance ranged from 30.0 to 60.0 cm.); the "accurate" agreement was achieved by the adjustment of three arbitrary

14. Weber (1848), in R. Taylor, *Scientific Memoirs, 5* (1852), pp. 497 ff.

constants,[15] which permit the theory to take into account the departure of actual magnets from the ideal dipole of magnetic moment which Maxwell envisions. Maxwell is oversimplifying considerably in invoking experimental evidence for the principle in the simple form in which he states it, though it is perfectly true that the same theory was well supported empirically at slightly different points.

Maxwell has, it seems, reversed the course of the actual argument for the sake of the exposition. Experiments had been done with larger coils, and conclusions reached for vanishingly small ones (for which alone Maxwell's principle would hold); whereas here Maxwell claims the empirical result for small coils, and will argue to the large circuit. Ampère does, as Maxwell points out, argue in the same way from the vanishingly small circuit to the large; but for Ampère this is a progress from one theoretical result to another.

Let us nonetheless grant Maxwell his wished experimental evidence. We must now move from the principle for vanishingly small circuits to finite circuits of any form. The method which Maxwell uses (*Tr* ii/142) is that of Ampère, which has since been attributed to Stokes.[16] More formally, if ι is current density, then

$$\int_S (\nabla \times \iota)\cdot \mathbf{dS} = \oint_l \iota\cdot \mathbf{dl}\,.$$

Ampère presents it simply, it is true, taking no special care in his phrasing to acknowledge the limiting process; but the core of the argument is present in the *Théorie mathématique* in a form in which it is often given in modern introductory texts. Maxwell derived it analytically in the "Preliminary" chapter of the *Treatise* (*Tr* i/27 ff.). Ampère's argument is based on the physical principle that if two adjacent parallel circuits carry currents that are equal and opposite, the magnetic effects of those currents will cancel everywhere. That this is the case was carefully demonstrated by Ampère in an experiment arranged for the purpose. Maxwell will discuss that experiment in due course (*Tr* ii/160), but here he merely asserts the result as an evident axiom which the reader is expected to grant:

> The effect of two equal and opposite currents in the same place is absolutely zero, in whatever aspect we consider the currents. Hence their magnetic effect is zero. (*Tr* ii/142)

15. *Ibid.*, p. 501.

16. Ampère, *Théorie mathématique*, p. 41. Thomson sent the theorem to Stokes, according to Larmor. (Joseph Larmor, ed., *Memoir and Correspondence of Sir George Gabriel Stokes* (2 vols.; Cambridge University Press, 1907), *2*, p. 31) According to Maxwell, Stokes proposed it as a Smith's Prize Examination question in 1854. Maxwell has good cause to recall the occasion: this was the year in which he and Edward Routh were the Smith's Prizemen. (*Tr* i/29n); Lewis Campbell and William Garnett, *The Life of James Clerk Maxwell* (2nd ed.; London: 1884), p. 124.

It would seem that Maxwell regards the result of Ampère's experiment as evident *a priori*, and the experiment itself as unnecessary.

With this axiom admitted, Ampère's argument, which Maxwell repeats, is that if we fill a surface densely with tiny closed currents, all lying in the surface, having the same sense of rotation and being equal in strength, they will cancel one another everywhere except at the boundary of the surface; thus they will be electrically equivalent to a single current running around that boundary. On the other hand, each such tiny current will, by the principle Maxwell has invoked on empirical grounds, be equivalent to a magnetic dipole with its axis normal to the surface; the collection of currents will then be equivalent to a dipole moment over the whole surface, that is, it will constitute a magnetic shell. Hence such a current around the boundary will be equivalent in its magnetic effects to the magnetic shell which it delimits.

Maxwell now has the general result which he requires to link Parts iii and iv of the *Treatise*: every closed circuit is equivalent in its magnetic effect to a magnetic shell of which it is the boundary, when the strength of the shell and current in the circuit are numerically equal.

The problem of predicting electromagnetic forces between circuits is resolved, for we need only substitute shells for circuits and then refer the problem back to the methods of Part iii. As I mentioned earlier (page 117), Gauss showed how we may represent the potential of any simple shell as the solid angle subtended at the field point. Through the Oersted experiment, Maxwell has led us to think of electromagnetism in terms of potential distributions due to whole circuits. We have achieved a sketch of a complete theory of electromagnetic forces without referring at any point to a force law acting between current elements. This has been Maxwell's objective, in leading us into electromagnetism in a manner appropriate to Faraday's thought. Needless to say, the actual mathematical *route* we have taken would in certain respects be foreign to Faraday's own ways.

The Oersted experiment with which Maxwell began, and which, as an experiment of illustration, has already yielded a surprising sequence of interpretations, finally submits to still another transformation. We have seen the Oersted wire as the boundary of a magnetic shell extending over an infinite half-plane, but we are now asked to consider it anew as an element of a complete electrical circuit. The *infinite shell* is bounded by an *infinite circuit*, a complete loop, of which the infinite straight wire which we have been discussing is merely one side.

There is one difficulty in representing this result as *complete*: it is applicable only insofar as all currents flow in closed circuits. Unclosed circuits, not being reducible to shells, would be altogether omitted from the theory. It appears immediately that there are such unclosed currents; the discharge of a Leyden jar presents a case of current flow, but between the plates of the jar there is apparently no circuit. What are we to say of the electrodynamics of such unclosed currents?

Here Maxwell claims, boldly and unequivocally, that his theory is in fact complete: that, despite appearances, there are no cases of unclosed circuits. Although Maxwell has shaped this chapter throughout according to his own design, this is perhaps the only assertion which is altogether original. He asserts simply:

> In all actual experiments the current forms a closed circuit of finite dimensions. (*Tr* ii/141)

I have taken this sentence slightly out of context; Maxwell is here concerned to show that the "infinite" conductor of the Oersted demonstration falls under the principle—that it is not unclosed current, but forms a loop completed at infinity. Here, therefore, the stress is on the word "finite"; but if there is any doubt that he means what the statement seems to imply, namely that *all* currents form closed circuits, Maxwell clarifies his position a few pages later:

> We may conceive the electric circuit to consist of a voltaic battery, and a wire connecting its extremities, or of a thermoelectric arrangement, *or of a charged Leyden jar with a wire connecting its positive and negative coatings, or of any other arrangement for producing an electric current along a definite path.* (*Tr* ii/155; emphasis added)

And he adds a "Note" shortly thereafter:

> We have reason for believing that even when there is no proper conduction, but merely a variation of electric displacement, as in the glass of the Leyden jar during charge or discharge, the magnetic effect of the electric movement is precisely the same. (*Tr* ii/155)

Thus even the *variation of electric displacement* that necessarily takes place in the "open" portion of a supposedly unclosed circuit constitutes a current, the so-called "displacement current."

The issue is drawn: Maxwell claims completeness for a theory of electrodynamics in which all currents must be treated as occurring in whole circuits. There are, he insists here, no exceptions. All circuits are closed, and all are equivalent in magnetic effect to the shells which they bound.

Much has been written about the origin of the "displacement current" concept in Maxwell's thought.[17] The principle that variations in the electric field, even in the vacuum, have magnetic effects equivalent to those of currents had been well established in Maxwell's work before he began writing

17. Whittaker, *History*, pp. 249 ff.; Mary Hesse, *Forces and Fields* (Nelson and Sons, 1961), pp. 212 ff.; O'Rahilly, *Electromagnetic Theory*, pp. 95 ff.; Alfred Bork, "Maxwell, Displacement Current, and Symmetry," *American Journal of Physics, 31* (1963), pp. 857 ff.; Joan Bromberg, "Maxwell's Displacement Current and his Theory of Light," *Archive for History of Exact Sciences, 4* (1967), pp. 219 ff.

the *Treatise*.[18] Now, on the other hand, he has the opportunity to build a systematic body of concepts in which the displacement current takes its natural place. Thus in his discussion of electrostatics in Part i (to which we shall refer at greater length in Chapter IV of this study), Maxwell has already developed a concept of electric charge that makes inescapable the view that currents are necessarily closed.[19] What we are meeting here, then, is a theory of electromagnetism that is wedded to a concept of electric charge developed earlier. Maxwell's concept of charge as a displacement in the electric field was in turn intimately related to his interpretation of Faraday's crucial Series XI on induction.

Thus, though Faraday does not himself propose the displacement current concept, the electromagnetic theory which Maxwell is now unfolding, and which presupposes a commitment to the displacement current, is fundamentally one appropriate to Faraday's ideas. As Maxwell has interpreted it, Faraday's thought on electrostatic induction and the nature of charge inherently implies an electrodynamics of whole circuits. For, a charge which we associate with any one conducting surface is in Faraday's view only a particular aspect of a state of polarization (Maxwell's "displacement") of the medium, while the initiation of this state is necessarily a shift in the displacement throughout the medium, ultimately forming a closed loop of displacement, and hence a closed current. Although Maxwell is going beyond Faraday in insisting that all currents are closed, we should note that the theory of electromagnetism based on this assumption takes a form in accord with Faraday's intuitions. As we have seen, Faraday's ideas in Volume III of the *Experimental Researches* were based on the interactions of magnets; he deals relatively little with currents. To follow Faraday's lead, then, we must give primacy to magnets, and in turn, of course, to the systems of power with which they are surrounded. This is what Maxwell has done. He has reduced all currents to equivalent magnets, and thus made the theory of magnets the guiding thread through electromagnetism. One need not debate whether "electricity" or "magnetism" is primary. The object of interest in any case is the field; the behavior of the magnet, and hence now of the circuit as well, is merely a reflection of the behavior of the system of the lines of force, if we are to follow Faraday (page 31, above).

Maxwell points to the relationship between the concept of charge which he introduces in Part i, and the electromagnetic theory of Part iv, forewarning the reader at the close of the first chapter of the *Treatise*:

> It follows from this that every electric current must form a closed circuit. The importance of this result will be seen when we investigate the laws of electro-magnetism. (*Tr* i/70)

18. Maxwell had introduced it in his "On Physical Lines of Force" (Part III, 1862; *SP* i/491).

19. (*Tr* i/166–67); compare pages 303 ff., below.

Now, in Chapter I of Part iv, we have begun to see this intimate relation of the two branches of Maxwell's theory. His approach to electromagnetism, in which no cognizance is taken of current elements belonging to unclosed circuits, presupposes his concept of charge. The full significance of this approach will appear only later, however, when the displacement current proves to yield propagating electromagnetic waves.

The whole theory of the forces exerted by one current on another is now implicit, since the corresponding problems have all been solved in magnetic shell theory, although only Step I—the prediction of the field of a given circuit—has been carried through explicitly in Part iv. Even with respect to this first step, however, Maxwell regards his work as incomplete. We have indeed the quantitative expression for the potential field of any circuit at any point; it is the equation of Gauss,

$$V = i\omega \ .$$

But Maxwell is concerned to accomplish a further objective, which he evidently feels the mere possession of a correct mathematical solution has not assured. We have an *equation*, but we may yet lack a *notion*. Having completed the development of the mathematical theory, he says:

> Let us now endeavour to form a notion of the state of the magnetic field near a linear electrical circuit. (*Tr* ii/145)

As we shall discuss further in Chapter V, Maxwell regularly suspects an abstract mathematical result of being barren for the mind: imagination must be invoked to clothe the mathematical result, and so bring into being a notion which the mind can grasp. As Maxwell has shown earlier in the *Treatise* in other contexts, one means for achieving this is a diagram of the configuration of a *field*. Not, it should be noted, primarily a diagram of "lines of force": Maxwell has been directing our attention to the expression for the *potential*, and the lines which diagram Gauss's equation for the potential will be equipotential lines.

The particular case which Maxwell chooses is the field of a circular loop of current, one quadrant of which is diagrammed in Figure 6 (Maxwell's Figure XVIII in the plates appended to Volume II of the *Treatise*). Maxwell says that the diagram was "copied" from Thomson's paper on "Vortex Motion" (1869);[20] it was Thomson who pioneered in diagramming Faraday's lines, as we have seen (page 103, above). Thomson's paper is formulated principally in terms of potentials. In reproducing the diagram, Maxwell has added the lines of force, but his discussion of it centers on the set of equipotential lines, and the lines of force are treated as derivative

20. William Thomson, *Mathematical and Physical Papers, 4*, p. 63. Thomson's paper is based, he says, on Helmholtz' (1858) "On Integrals of the Hydrodynamical Equations, which Express Vortex-Motion," *Philosophical Magazine, 4* (1867), pp. 485 ff. The diagram was constructed, Thomson adds, by a method of Maxwell's; see (*Tr* ii/340–41).

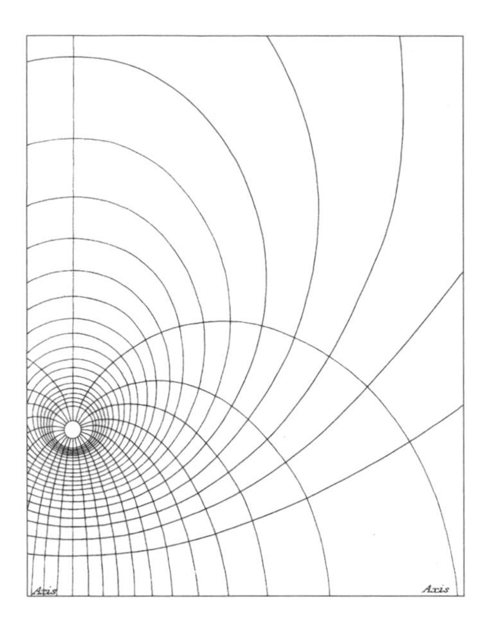

Figure 6. One quadrant of the field of a circular current
(*Treatise*, Vol. II, Figure XVIII)

from the potential configuration (*Tr* ii/146). Maxwell uses the figure essentially to illustrate Gauss's expression; the equipotential surfaces are then "surfaces for which ω is constant" (*Tr* ii/145). Maxwell offers this as the most basic geometrical interpretation of the magnetic field; I have tried in Figure 7 to represent the field as he envisions it. The equipotential lines of the plane figure are merely sections through this surface, and "the force acting on a magnetic pole" is everywhere perpendicular to this surface and "varies inversely as the distance between consecutive equipotential surfaces." The force is therefore to be understood geometrically as the gradient of the potential. If there is no "magnetic matter," and hence no "poles," we can understand that both the "force on a pole" and the "line of force" (insofar as it is derived from this concept) will take a position in Maxwell's theory secondary to that of the potential, from which all observed magnetic relationships can be derived.

Having thus "formed a notion" of the state of the magnetic field in terms of equipotential surfaces and the lines of force, Maxwell now wishes to show how this notion makes it possible in turn to understand the force which acts on a current placed in the field. In so doing, he will complete the second part of the three-part "indirect" theory of the action of a current upon a current. It should be noted that before doing so, Maxwell pauses to discuss a device, highly significant in the history of electromagnetism, by which Faraday succeeded in utilizing the arrangement of the Oersted demonstration to produce continuous rotation of a magnet about a wire (*Tr* ii/144). Maxwell actually discusses Faraday's achievement a second time in this

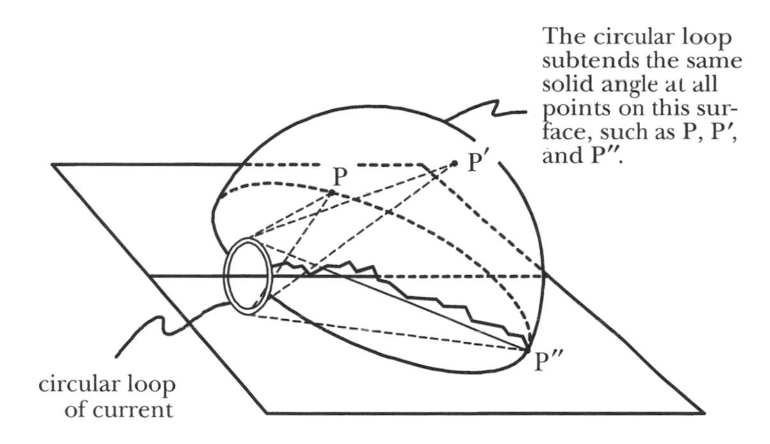

Figure 7. Magnetic field of a circular loop of current

same chapter (*Tr* ii/150). I will examine this topic in the next section of the present chapter.

Since the equivalence of currents and shells as sources of magnetic fields has now been established, the argument to the force on a circuit could follow precisely the corresponding argument for the force on a shell, which we have already reviewed briefly (pages 115 ff.). One first determines the potential energy of a shell placed in a given field, and then takes the gradient of this potential to find the force—in the simplest case treating the shell as a rigid body. Maxwell does in fact apply his theory of force on a shell to find the force on a circuit, with only one significant difference.

We saw (page 117, above) that, having written the potential energy of an element of a shell,

$$dU = \varphi \nabla V \cdot \mathbf{dS},$$

Maxwell rather deliberately shifted to the vector-potential **A**, and then integrated around the boundary rather than over the surface. The result so obtained was

$$U = \varphi \oint_l \mathbf{A} \cdot \mathbf{dl},$$

and the flux through the shell did not appear in the energy expression. Here, for circuits, it does. The vector **A** is not introduced, and the surface integral is taken directly (*Tr* ii/147):

$$U = \varphi N$$

where N, the flux, is

$$N = \int_S \mathbf{B} \cdot \mathbf{dS}.$$

Why is there this difference between the treatment of circuits and the treatment of shells? It is not a minor distinction in Maxwell's view, even though the easy mathematical equivalence of the two expressions would make the choice seem trivial from a purely formal point of view. For Maxwell, it is not symbols that are at stake, but concepts. Maxwell, I believe, is choosing here to follow closely Faraday's own formulation, and to reserve for the later exposition of his own "dynamical theory" the representation which he himself believes is ultimately the best interpretation of Faraday's insight. Maxwell in effect translates Faraday in two stages. In the discussion of shells, he has prepared us for the final interpretation in terms of **A**; but here, in the first stage of the account of electromagnetism, he regresses in order to translate more literally "the language of Faraday." He says of the above surface integral:

> [I]t represents the quantity of magnetic induction through the shell, or, in the language of Faraday, the *number* of lines of magnetic induction, reckoned algebraically, which pass through the shell from the negative to the positive side. (*Tr* ii/147)

We have seen that Faraday takes the *number of lines* intercepted, as indicated by the ballistic galvanometer and the "moving wire," as the measure of the power of a magnetic sphondyloid; the total power of a magnet, or the power of the total sphondyloid, was the number of lines cut by an exploring loop brought from outside the field to a position of tight fit about the equator of the magnet (pages 22–25, above). By writing the integral in the form he has chosen here, Maxwell most directly translates Faraday's thinking, born of the experiments with the "moving wire." In these experiments the "number of lines" swept out by the moving wire, as it was brought into place in the field, became the central concept.

We begin to see one sense in which Faraday's measure (using the exploring loop and galvanometer) is not altogether inappropriate in evaluating a "sphondyloid of power." Flux is not in itself a measure of power or energy; nor, more generally, is it a measure of the "tendency to move" or to act, of which Faraday speaks. But the energy of a circuit carrying a given current in a magnetic field *is* directly proportional to the "number of lines" (or equally, to the "sphondyloid of power") which it intercepts; and by taking the gradient of this expression we measure the "tendency to act" which is latent in this power.

Writing the potential energy U of a circuit carrying current i in a magnetic field whose flux through the circuit is N as given above, we have for the energy:

$$U = iN\,.$$

Taking the gradient, assuming the current i constant, we obtain the force on the circuit moving as a rigid body:

$$\mathbf{f} = -\nabla U = -i\nabla N\,.$$

Maxwell now draws the conclusion in this form:

> We have now *determined the nature of the force* which corresponds to any given displacement of the shell. It aids or resists that displacement accordingly as the displacement increases or diminishes *N*, the number of lines of induction which pass through the shell.
>
> The same is true of the equivalent electric circuit. Any displacement of the circuit will be aided or resisted accordingly as it increases or diminishes the number of lines of induction which pass through the circuit in the positive direction. (*Tr* ii/147; emphasis added)

Maxwell shares, deeply, Faraday's desire to find out the "nature" of the force. He has given here a first account of this nature, corresponding to that which Faraday was able to articulate in the *Experimental Researches.* The reformulation in Maxwell's dynamical theory will yield a new understanding of the force as well, one that Maxwell feels Faraday had intuited but was unable to express—an understanding that corresponds to Faraday's "electrotonic state."

Maxwell later notes the important assumption that the current is maintained constant as the displacement of the circuit occurs (*Tr* ii/149). This is an essential consideration whenever total energies are being reckoned, as work must in general be done to maintain such a rigid current.[21] On the other hand, it corresponds to the assumption of the *perfectly hard magnet*, which, as we have seen, was an important presupposition in Faraday's thinking about the interactions of magnets (page 24, above). The first of Maxwell's two paragraphs just quoted states the force on such a perfect magnet, and explains in Faraday's terms the field equivalent of the notion of "attraction" of opposite poles; it applies equally to the repulsion of like poles, or—on Weber's hypothesis, which Maxwell on the whole endorses (*Tr* ii/475)—diamagnetic repulsion. Faraday's laws of motion for magnetic and diamagnetic substances, quoted on page 28, above, correspond precisely to the interpretation of the "tendency to move" as the *gradient* of the "power," where the "power" is understood as the flux of the magnetic field through the coil. Tyndall, we saw, insisting on action-at-a-distance physics, was unable to make any sense of talk of such a local tendency (pages 83–84, above). Thomson, in an early paper we have discussed, triumphantly derived Faraday's principles from Poisson's theory, and thus showed clearly the possibility of a field physics of local tendencies; but his integral does not admit of such a simple interpretation in terms of an increase or decrease of the number of Faraday's lines intercepted (page 99, above). Maxwell has managed to distill the mathematical theory into a form that makes an easy connection with Faraday's view.

Maxwell has now found the force on a whole circuit and has, it would seem, completed Step II of the "indirect" program. But, as he says, "this is

21. Thomson, "Dynamical Relations of Magnetism" (1860), (*EM*: 447n–448n).

not all" (*Tr* ii/148). He has obtained Faraday's law of motion for magnets: he now goes on to derive Ampère's law of force for current elements *from Faraday's principle.* It is important to take note of a distinction at this point. Ampère assumed an interaction between current elements and established his electromagnetic theory on this basis; he had no empirical evidence for such an interaction, and Maxwell in effect denies its existence—for Maxwell and Faraday, circuits act as wholes. On the other hand, it *is* possible to demonstrate the action of a whole circuit (as source of a magnetic field) upon an element of another circuit. Mechanically, one has only to hold rigid all of the circuit being acted upon except for a very small part (as small as we please) that is allowed to move; one may then observe experimentally the forces on it. Ampère carried out experiments of this form.[22] A complete theory of the electromagnetic force therefore cannot ignore the force upon a current element, although, as we have mentioned, Faraday had very little to say about forces of currents upon currents (page 35, above). Maxwell now fills this gap by showing that Faraday need not have deferred to Ampère in this realm; his own principle leads to all of those results of Ampère's that admit empirical confirmation.

Maxwell works out this force in terms of the change in the flux N through a coil as only one elementary part of it is permitted to move; the gradient of the flux due to the motion of this small part will give the force on it. The procedure is straightforward and leads directly to a result which is equivalent to Ampère's.[23] Maxwell gives it first in geometrical form, and then reformulates it (with an increase in clarity which he no doubt hopes will strike the reader) in the language of quaternions:

> We may express in the language of Quaternions, both the direction and magnitude of this force by saying that it is the vector part of the result of multiplying the vector $i\mathbf{ds}$, the element of the current, by the vector $\mathfrak{B}$, the magnetic induction. (*Tr* ii/149)

In vector notation, this is the familiar law for the force on a current element as given in modern texts:

$$\mathbf{df} = i(\mathbf{ds} \times \mathbf{B}).$$

What Maxwell has achieved at this point is a meeting of two theories, approaching the "motor rule" from opposite directions. In Faraday's view, as Maxwell has translated in terms of the flux integral, this force on a portion of a circuit must be seen not as an effect on the element alone, but as a consequence of the alteration of the flux through the loop as a whole.

22. Ampère discussed this problem in an 1820 memoir reprinted in R. A. R. Tricker, *Early Electrodynamics* (Oxford: Pergamon Press, 1965), pp. 144–46.

23. Ampère, *Théorie mathématique*}, p. 76; (*Tr* ii/148). The result is derived again by Maxwell, this time from Ampère's theory, at (*Tr* ii/169).

"Elements" in Ampère's sense do not, for Faraday, enter into interactions as such—the term $i\mathbf{ds}$ in the equation is in this respect derivative and deceptive. For Faraday, who thought of sphondyloid acting upon sphondyloid (i.e., field upon field), the *whole* has to become the true *element*, and the part is affected only because of its altered contribution to the configuration of the whole.

For Ampère, it is just the reverse. He has *built up* to the same force equation from a theory of the direct interaction of true, vanishingly small geometrical elements, as the fundamental entities of which circuits are merely assemblages. The vector **B** has no physical significance at all for Ampère; as we shall see later in this chapter, he introduces an equivalent vector incidentally, as a convenience of calculation; as though to emphasize its merely formal character, he calls that vector simply the "directrix."[24] Opposites meet, but for Maxwell their very agreement in an equation emphasizes the fundamental disagreement in their premises.

Step III, in which we determine the action of one current upon another by mathematically eliminating the field between them, is taken very readily. Again, it has already been done for magnetic shells in Part iii, where the mutual energy of the two shells was calculated (page 118, above). Maxwell now simply recalls this result. As he does so, however, he institutes a distinction which he had not been careful to make before, between what he calls the *mutual potential* of two closed *curves*, and the mutual *potential energy* of two circuits coincident with these curves, and carrying specified currents. The mutual potential, to which the symbol M is now assigned, though it was in fact used otherwise earlier, is (*Tr* ii/151):

$$M = \oint_{l_1}\oint_{l_2}\frac{\mathbf{dl}_1 \cdot \mathbf{dl}_1}{r}.$$

It is then doubly specific: the mutual potential energy per unit current (of one circuit), per unit current (of the other). The mutual energy, which we have denoted U, is the product of this potential by the two shell strengths or currents:

$$U = \varphi_1\varphi_2 N = i_1 i_2 N.$$

Maxwell finds it impressive that so much of electromagnetic theory is a matter of geometrical intuition, as Faraday's view has so dramatically demonstrated to him—an intuition "of the fundamental forms of space..." (page 45, above). As Maxwell has defined it, M, in which so much of the theory is implicit, is a geometrical quantity, with the dimensions simply of length, and dependent on the two geometrical paths alone. The "power" of Faraday's "system of lines," for unit shells or for unit currents, is measured by this geometrical quantity.

24. Ampère, *op. cit.*, p. 31.

Figure 8. Uniform field disturbed by straight current-carrying conductor (*Treatise*, Vol. II, Figure XVII)

It is perhaps perplexing to find that, in Step III of the "indirect method," the mediating field cancels out and we are left with an expression which would be equally appropriate in an action-at-a-distance theory: it is a distance equation pertaining to points separated by the distance r. So far is the equation, in itself, from representing the theory; one has to remember the steps by which it was derived. When we do so, we see the equation as measuring the "power" of a field configuration—which, for unit currents, is the number of lines that link the circuits. Its symmetry reflects the symmetry Faraday has seen in the interaction of magnets: it is not that one magnet, as source, acts upon another, as object, but rather that two systems of power interact. (In one of Faraday's formulations, the two sphondyloids "coalesce.")

In invoking the above result for the mutual energy from shell theory, Maxwell has not pointed out to the reader that he has quietly set aside the commitment, temporarily adopted during this chapter, to "Faraday's language." The representation in terms of integration around two paths, rather than over two surfaces, depended upon the introduction of the vector **A**, which he has avoided in this translation of Faraday. This anticipates the form Maxwell finally will give to Faraday's theory; but it seems an unfortunate move for Maxwell to have made at this point, where it is already difficult to keep in view the significance of this action-at-a-distance equation as the representation of a theory of the field.

By taking the gradient of this mutual energy, we can obtain once again the force of one circuit on another (*Tr* ii/151). Maxwell remarks that the expression for the force on a "portion of an electric circuit due to the action of another electric circuit" may be obtained from it as well; he does not carry this out, but it would yield the expression familiar in modern texts:

$$\mathbf{df} = i_1 i_2 \mathbf{dl}_2 \times \oint \frac{\mathbf{dl}_1 \times \hat{\mathbf{e}}_r}{r^2},$$

where $\hat{\mathbf{e}}_r$ denotes a unit vector in the direction of $\mathbf{r}$.

Again Maxwell is unwilling to leave the final result in the form of a symbolic expression alone; once more he takes us to a diagram of the configuration which this equation measures (Figure 8). But where his earlier diagram in Figure 6 represented the field of a source alone (corresponding to Step I), the diagram in Figure 8 represents a dynamic interaction (corresponding to Step III) between a uniform field and the field of the long straight conductor with which Maxwell's chapter opened (see Figures 3 and 4, pages 119–120, above); the two fields "coalesce," to use Faraday's term. Both the potential and the force fields are drawn, as two orthogonal sets.[25] In Maxwell's figure, the uniform field is directed to the left (regarded as "north") and the current is into the page, so that its field runs clockwise.

We are faced with a problem discussed in Chapter 1, in connection with Faraday's claim to be able to "read" the iron filing patterns. We there reviewed certain principles of Faraday's concerning the interpretation of a combined field pattern (pages 33–35, above). Maxwell enunciates a principle of his own:

> [T]he force acting on the wire will be from the side on which the forces strengthen each other to the side on which they oppose each other. (*Tr* ii/153)

This is an immediate corollary of a principle Faraday asserted when thinking of the systems of lines in separation: that parallel lines repel, and anti-parallel lines attract (page 32, above). For where the two systems of lines run parallel, the combined field is strong (here, to the "west" of the wire), while the anti-parallel lines, tending to cancel, leave the field weak (to the "east" of the wire). There is then repulsion from the strengthened side, and motion toward the weakened side.

Faraday, when thinking of the resultant pattern, offered a different principle, which we quoted on page 33. It included the rule that the lines tend "to contract in length," suggesting an analogy between the whole system and a strained elastic medium. This is certainly the model which Maxwell's diagram most immediately urges upon us—a medium appears to be

25. Maxwell published a crude version of Figure 8, showing only the lines of force, as Figure 5 of "On Physical Lines of Force" (1861), (*SP* i/484–85, oppos. 488). It was apparently first presented to the British Association in 1856, as an example of a simple graphical method for constructing the combined field when two separate fields are known; the method is analogous to that by which optical interference patterns had been constructed by Young. The text at (*SP* i/241), presumably that of the British Association reporter, appears to be confused; the method must be that of the addition of potentials described (for the electric field) at (*Tr* i/183).

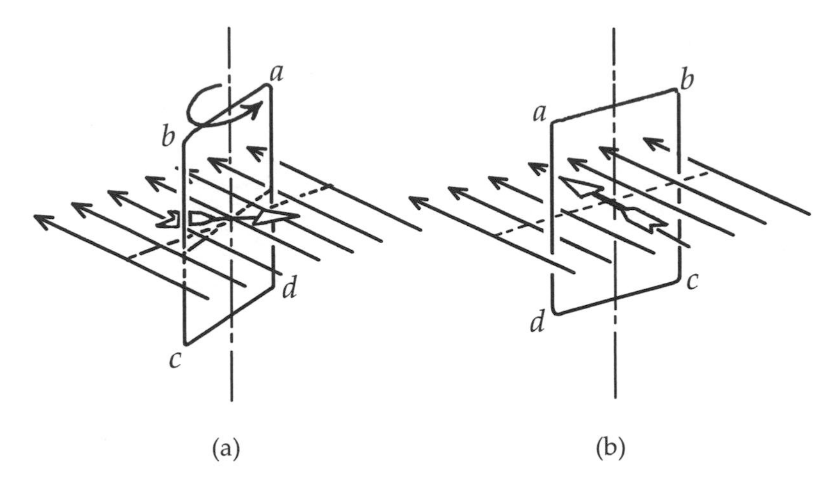

Figure 9. Straight conductor in uniform field, represented as part of a rectangular loop

compressed to the west of the wire, and rarefied to the east, so that the tendency of lines in it to contract would carry the wire from west to east. This whole analogy to an elastic solid, as we have already seen in an early letter of Maxwell's to Thomson (page 109), is most interesting to Maxwell, and he devotes a section of the *Treatise* to a discussion of it, following his presentation of the dynamical theory (*Tr* ii/278–283). The dynamical theory itself is of course independent of such images or hypotheses.

Maxwell was often criticized as a teacher. Although this first chapter in Part iv of the *Treatise* proves, on careful reading, to build a most effective translation of Faraday's ideas into a systematic theory of "indirect" action, the devices which Maxwell has chosen place every sort of pitfall in the way of the student. This is done, I suspect, out of his delight in taking an unexpected turn of mind (transforming the Oersted demonstration, for example, out of all recognition, several times over). For a theory which will reduce all interactions to those of complete circuits, an infinite straight conductor is hardly a helpful beginning! Again, at the end of the chapter, Maxwell has given us an example which is particularly difficult to relate to the very equation he wishes to illustrate. To relate the straight wire in a uniform field to the equation for the potential of the circuit, we must of course see the wire as part of a circuit (as it became necessary to do with the Oersted wire). In Figure 9, for example, the wire is part of a rectangular loop that is free to turn about the vertical axis .

If the loop is initially oriented east-west, it experiences no net tendency to move, but the equilibrium is unstable. For if it is slightly deflected, as shown in the figure, it experiences a torque which will carry it around until it reaches stable equilibrium 180° from its original position (Figure 9b). In Figure 9a, the flux through the coil, in terms of the principle of page 131 is a minimum, as the sense of the coil and the field are reversed. In

Figure 9b, the flux is a maximum, and there is no further tendency to motion. If we think of the uniform field as established, for example, by a large coil standing perpendicular to the paper, the mutual potential of the two coils will be minimized by the motion.

We may now, therefore, reinterpret Faraday's principle of the tendency to "set to each other" as a tendency to minimize the mutual energy U. Better yet, we can interpret Faraday's further principles in the same terms. "[T]he lines from all the sources tend to coalesce," he says, and "to pass through the best conductors" (page 33, above). *Coalescence*, or mingling the systems of lines of forces of two magnets (or now, equally, of two coils) will, in the same way, mean precisely minimizing the mutual energy. And similarly, if a good "conductor," a bar of soft paramagnetic material, becomes magnetized by induction in the field of another magnet, it too will turn and move to increase the flux through it and hence make the quantity U a minimum. This illustrates, more fully perhaps than Maxwell had shown us, how well a translation in terms of the mutual energy of two systems fits Faraday's thoughts—how closely Faraday's ideas of "systems of power," and "tendencies to motion" derived therefrom, accord with a formulation in terms of mutual potential energy. This is true despite the fact that the latter is a concept drawn from the mathematical theory of action-at-a-distance, the tradition of Laplace and Neumann, which would seem utterly remote from Faraday's concepts of nature and science. Maxwell, living intellectually in both worlds, is uniquely able to reveal the appropriateness of one set of thoughts as the translation of the other.

Maxwell summarizes the new approach to a theory of electromagnetic forces in two sweeping principles. He prefaces the summary with his assertion of the universality of his theory, which we have discussed above (page 125). He states what we may identify as his First Principle of Electromagnetism:

> If any closed curve be drawn, and the line integral of the magnetic force be taken completely round it, then, if the closed curve is not linked with the circuit, the line-integral is zero, but if it is linked with the circuit, so that the current i flows through the closed curve, the line-integral is $4\pi i$... (*Tr* ii/155)

This is the integral which first confronted us in the attempt to understand the experiment of illustration in terms of potential. To establish a unique potential in the region about the linear current, it proved necessary to make the space acyclic by dividing it with a semi-infinite diaphragm. Each penetration of the diaphragm meant an increment of potential of amount $4\pi i$, so that there was no longer a unique assignment of potentials in the space unless the diaphram were regarded as impenetrable. All of this is most dramatically illustrated by the experiment of the Faraday rotator, the discussion of which will be reserved for the next section of this chapter.

Maxwell did not need to repeat the demonstration of this principle in Chapter I, since it is a part of the magnetic shell theory of Part iii. There, Maxwell demonstrated:

> Hence, if a closed curve be drawn so as to pass once through the shell, or in other words, if it be linked once with the edge of the shell, the value of the integral ... extended round both curves will be $4\pi i$. (*Tr* ii/43)

Thus this principle, which we now recognize as the finite form of one of "Maxwell's equations," is grounded in shell theory.[26] The fact that Maxwell had thoroughly discussed the mathematical theory in Part iii enables him to draw upon illustrations rather than derivations when re-introducing shell theory in Part iv.

This first principle stands, in the summary, for Step I of the indirect method, the characterization of the field in terms of its source. He now states a second principle, asserting the effect of a field upon a circuit. This, which we may call the "Second Principle of Electromagnetism," states:

> If by any given motion of the circuit, or of a part of it, this surface-integral ["of the magnetic induction through any surface bounded by that circuit"] can be *increased*, there will be a mechanical force tending to move the conductor or the portion of the conductor in the given manner. (*Tr* ii/156)

From this, the Ampère law of force on a current is drawn as an immediate consequence.

Finally, Maxwell, like Faraday, finds that this view of electromagnetism reacts on the understanding of the magnetic line of force (compare page 22, above). It had originally been defined in terms of the force on a magnetic pole, but in electromagnetism poles do not appear. In due course, Maxwell will assert that on the theories whose probability he certainly accepts, there are no poles:

> If we adopt the theories of Ampère and Weber as to the nature of magnetic and diamagnetic bodies, and assume that magnetic and diamagnetic polarity are due to molecular currents, we get rid of imaginary magnetic matter, and find that everywhere $m = 0$ and [$\text{div}\,\mathbf{B} = 0$]. (Compare *Tr* ii/282)

Looking ahead to this principle (which is another of the equations now known as "Maxwell's"), Maxwell now proposes alternative definitions for the "line of magnetic induction":

26. Maxwell's "curl **H**" was given in Part II of his first paper on electromagnetism, though at that time it did not include the displacement current (1856) (*SP* i/194). In finite form, as a principle for the potential of shells, it had been given by Thomson (1850) (*EM*: 387); Thomson had worked it out in differential form for currents, in 1849, but had not published it (*EM*: 428–30). Note that it is included under a general principle for line integrals of total differentials in cyclic spaces, where the quantity $4\pi i$ is termed the cyclic constant" (*Tr* i/19).

> It is that line to which the force on the conductor is always perpendicular. It may also be defined as a line along which, if an electric current be transmitted, the conductor carrying it will experience no force. (*Tr* ii/157)

This definition formally erases the distinction between "magnetism" and "electromagnetism." Having met it, we understand the structure of the *Treatise* anew: the magnetic theory of Part iii has proved so useful to the electromagnetic theory of Part iv, precisely because magnets are aggregates of circuits. Part iii of the *Treatise*, on "Magnetism," has proved to be Part iv, on "Electromagnetism," in disguise. The structure of the book incorporates a revolution in the concepts of the theory. Originally, this was an historical discovery which followed upon Ampère's theory of electrodynamics; but for Maxwell it has become a deliberate device in guiding the student to successive levels of understanding of electromagnetic theory.

Maxwell has to pay a price for his reformulation of electromagnetism in Faraday's field terms; nothing in it points to a possible physics of particle-particle interactions, whereas Weber had shown, long before Maxwell began his work in electricity, that Ampère's view of interacting current elements transforms readily into equations for forces between moving charged particles. It is not easy to stretch Maxwell's concept of displacement current in the field to include the case of the moving charged particle, although Maxwell's principle admits no exceptions; this task necessarily fell to his interpreters.[27] But the *Treatise* as well as Maxwell's other writings are silent about what came to be called the "Lorentz force"—the force upon a charged body moving in a magnetic field or, more fundamentally, the force of one charged particle upon another moving with respect to it. It may be easier, in other words, to build circuits out of Ampèrian elements, and thereby to construct a physics of *both* circuit-circuit and particle-particle interactions, than to begin with principles such as those two of Maxwell's, and derive from them a law of particle-particle forces. The two principles which Maxwell has presented here lead beautifully into equations of electromagnetic propagation, but they turn the mind away from the equally rich areas of charged-particle physics.[28]

27. Whittaker, *op. cit.*, pp. 306 ff. William Thomson's exasperation with any efforts to trace displacement currents in a simple electrostatic discharge is well known (S. P. Thomson, *Life*, p. 1039).

28. O'Rahilly's thesis, developed in the two volumes of *Electromagnetic Theory* (*cit.*), is that the only satisfactory basis of modern electromagnetic theory is an equation of particle-particle action in the tradition of Weber. "Whether we like or dislike the admission, we are logically back again in the pre-Maxwellian epoch" (p. 211). For non-relativistic electromagnetism he recommends the Liénard-Schwarzchild equation (p. 218). In his view, a theory of fields was an unnecessary digression in presentations of electromagnetic theory. For other authors suggesting a particle electrodynamics as an alternative to Maxwell's field theory, see page 8n, above.

In a final paragraph, elaborated by Niven as editor, Maxwell makes a distinction which is related to the above consideration of the true elements of the science, namely, a distinction between two categories of force. One is the *mechanical* force on conductors, which has been the subject of this chapter; the other is the *electromotive* force, which will be the subject of Maxwell's Chapter III in Part iv. The first is a force which acts, Maxwell says,

> not on the electric current, but on the conductor which carries it. (*Tr* ii/157)

The second Maxwell avoids calling a force "on electricity," not wishing to suggest a commitment to any notion of the electric current as a fluid. It is, nonetheless, that force which leads to the phenomena we call those of the "electric current."

We are now so accustomed, I believe, to assume a rigid bond between "charge" and "mass," schooled to this by the discovery of the elementary "charged particles" (a discovery which Maxwell almost certainly does not envision), that Maxwell may seem to be insisting here on a distinction which has little effect. But for all the evidence before him, no such primitive coupling of charge and mass, locked in the bond of an "elementary particle," exists. He therefore quite properly assumes an independence of electricity, whatever it is, and mass; and he accordingly distinguishes forces which act on mass as such from those which do not. He does entertain the question whether charge is intrinsically inertial, but his experiments on this question will lead to negative results, as we shall see in Part II.

Assuming, then, a fundamental independence of mass and charge, and yet observing that the electromagnetic forces discussed in this chapter act in an invariable way upon the ponderable masses of the conductors, he assumes that these electromagnetic forces act intrinsically upon the masses, even though the action is dependent upon the presence of the currents. He rejects at the outset, without discussion, the modern view that the electromagnetic forces act intrinsically upon the charges, and only incidentally upon the masses, through the coupling of mass and charge in the atomic and molecular structures. In other words, we have found a far tighter linking of mass and charge in the structure of matter than Maxwell had considered; we therefore can account in these terms for an invariable relation of mechanical force on conductors to currents in them, even while assuming that the electromagnetic force acts primarily on the charges. Maxwell would have expected that if the force thus acted upon the charge, the coupling between current and conductor would have been looser than he observed, and that the mechanical forces on conductors, for the same currents, would have depended upon details of cross-section, material, etc., in ways which do not in fact complicate the electromagnetic equations.

Convinced of this essential independence of the two types of force, the one acting on the mass of the conductor in the presence of currents, and the other on what we call "electricity" itself, Maxwell regularly distinguishes them by name as, respectively, the *mechanical* and the *electromotive* forces, both exerted by the electromagnetic field. One well-known consequence of this insistence of Maxwell's was that Oliver Lodge was, by his own account, discouraged from discovery of the Hall effect.[29] A note on the Hall effect (discovered in 1880, the year after Maxwell's death), was added at this point to the third edition of the *Treatise* by J. J. Thomson. In the Hall effect, a potential is produced in a solid conductor precisely by the action which Maxwell denied, that is, by the action of the forces discussed in this chapter on the current itself rather than on the mass of the conductor.

A further discussion of these two supposed types of forces will arise in connection with our discussion of the general dynamical theory of electromagnetism, in Chapter 6, below.

AMPÈRE'S THEORY IN THE *TREATISE*

Having given us an account of the mechanical force between currents, shaped to Faraday's concepts, Maxwell now, in the second chapter of Part iv, gives an entirely different theory of the same thing. It is more than likely that he felt obliged to include Ampère's theory in the *Treatise*; it had been a standard Cambridge topic since the publication of Murphy's treatise for Cambridge students in 1833.[30] Beyond that, it was a complete and elegant theory, virtually universally accepted, and capable of accounting for all of the known phenomena. It had furnished the foundation for such extensions as Neumann's and Weber's, through which Faraday's discovery of electromagnetic induction and the phenomena of diamagnetism had been successively integrated into a single connected account. Maxwell no doubt felt a responsibility to insure the literacy of Cambridge students in this action-at-a-distance theory. It is significant, at the same time, that he placed the account in Faraday's terms first, to introduce the ideas of field theory firmly in the reader's mind at the outset.

My own impression is that Maxwell would have included his chapter on Ampère's theory even if he had not been required by circumstances to do so. As I have suggested, he was eager to place alternative types of theory before the reader in order to make the significance of Faraday's method clearer, and no doubt also to give the student freedom of choice in an area

29. Whittaker, *op. cit.*, p. 289.

30. Robert Murphy, *Elementary Principles of Electricity, Heat, and Molecular Actions: Part I, On Electricity* (Cambridge: 1833). Maxwell began his studies with "a little antipathy to Murphy's Electricity" (Larmor, *Origins*, p. 3).

of science whose future was very much in doubt. Furthermore, he was in possession of a joke which he probably could not resist sharing. It is at once a joke on Ampère, and a telling consequence of the use of excessively formal methods in mathematical physics. Ampère had claimed, in the very title of the treatise which constitutes the ultimate statement of his theory, a special *methodological* victory. Not only had he achieved a completely successful mathematical theory of electrodynamics, but he proudly claimed to have *deduced* it "uniquely" [*uniquement*] from the phenomena.[31] In this, as in other respects, Ampère is following the strict tradition of the *Principia* as closely as possible: Newton likewise claims, not to *induce*, but to *deduce* the law of gravity from the phenomena. Given the laws of force and the propositions which follow from them in Book I of Newton's *Principia*, together with the experimental evidence concerning the lines of apsides of the planetary orbits summarized briefly at the beginning of "The System of the World," the inverse-square law of gravity indeed follows as the necessary consequence of a deductive argument. In the same way, Ampère was confident that his law of electrodynamics followed inevitably from an elegant set of experimental results, together with certain undeniable principles.

Maxwell's presentation of Ampère's theory in the *Treatise* is not the same as Ampère's, and the difference reveals the joke: not one unique law, but an infinite variety of possible force laws emerge rather ridiculously from Ampère's argument as Maxwell reconstructs it. Maxwell by no means feels that he has introduced this difficulty through his reformulation, but rather that he has revealed a problem that Ampère had not acknowledged. Maxwell certainly takes satisfaction in this perplexity, which results, he is convinced, from the artificiality of Ampère's methods.

In order to follow Maxwell properly, it will be necessary to review the theory very briefly as Ampère himself presents it, and then to outline the argument which Maxwell gives. To proceed as economically as possible, I shall put Ampère's argument into the terms of Maxwell's figures and symbols from the start, and I shall give the argument only in schematic outline.[32]

As we have noted earlier (page 118n, above), an issue was drawn between Ampère and others, among them Faraday, on the question of electromagnetic rotations. Ampère's theory, true to its Newtonian paradigm, allows only forces which act in direct lines between the elements of circuits; the circular form of the Oersted phenomenon, on the other hand, did not appear to be reducible to such linear forces. Ampère soon proved himself able to derive these circularly-directed forces from his own

31. See page 18n, above.

32. For a more complete outline of Ampère's theory, see Tricker, *Early Electrodynamics*, pp. 42 ff.

principles, but the question of rotational effects nonetheless took on special interest partly because of the challenge it seemed to offer to a theory which took linear forces as fundamental. Faraday's first triumph in the science of electromagnetism was in the demonstration of continuous electromagnetic rotations, and for him, as for Maxwell, they are highly important as clues to the circular configuration of the field. Maxwell's analysis of the Faraday rotator, in terms of the concepts of Chapter I which we have just reviewed, is particularly interesting. Therefore, after we have derived Ampère's force law both by his own methods and by Maxwell's, we will turn to this question of electromagnetic rotations, reviewing its history briefly, and then examining the account of it which Maxwell gave in Chapter I of Part iv of the *Treatise.*

We have said that the Oersted phenomenon seemed to resist inclusion in the framework of Newton's *Principia.* But there was no other mathematical physics than that of Newton, as perfected by Laplace. The issue for Ampère was thus very simple: either to find some way to include Oersted's discovery within the program of Newton and Laplace, or to abandon hope of dealing with it through strict mathematical reasoning. The extension of Newton's work to include the phenomena of electricity known up to this time had already been carried out brilliantly by Coulomb and Poisson; it fell to Ampère to carry out this seemingly impossible task for the new electromagnetic phenomenon. In the opening paragraph of his *Théorie mathématique,* Ampère places his effort squarely in relation to Newton:

> L'époque que les travaux de Newton ont marquée dans l'histoire des sciences n'est pas seulement celle de la plus importante des découvertes que l'homme ait faites sur les causes des grands phénomènes de la nature, c'est aussi l'époque où l'esprit humain s'est ouvert une nouvelle route dans les sciences qui ont pour objet l'étude de des phénomènes.[33]

This is the "road" which Ampère is resolved to follow. Specifically, this is the demand upon the Newtonian theorist:

> Newton nous a appris que cette sorte de mouvement doit, comme tous ceux que nous offre la nature, être ramenée par le calcul à des forces agissant toujours entre deux particules matérielles suivant la droite qui les joint, de manière que l'action exercée par l'une d'elles sur l'autre soit égale et opposée à celle que cette derniere exerce en même temps sur la première. ... Mais il ne suffisait pas de s'être élevé à cette haute conception, il faillait trouver suivant quelle loi ces forces varient avec la situation ... en exprimer la valeur par une formule.[34]

33. Ampère, *Théorie mathématique,* p. 1.

34. *Ibid.,* pp. 1–2.

Oersted, looking at the action of his wire upon the magnetic needle, was full of visions of vortices in a medium.[35] Ampère, speaking of the time before Newton, and hence literally referring to Descartes and the Cartesians, says:

> [P]artout où l'on voyait un mouvement révolutif, on imaginait un tourbillon dans le même sens.[36]

The remark applies, however, to Oersted as well as to Descartes. Newton had written the *Principia* to refute the Cartesian hypothesis of vortices; Ampère now sees himself called upon to perform the same task with respect to electricity and magnetism: once again to perform the Newtonian magic, to show that a rotational motion can be reduced to the operations of a rectilinear force—to dispel the vortices, and thereby to preserve mathematical intelligibility of nature.

The first step, which Ampère performed with legendary swiftness after hearing Oersted's experiment reported to the French academy, was to produce a further phenomenon.[37] In terms of a "current" (a concept which was not clear in Oersted's account), what Ampère showed was that one current exerts a force on another, quite independently of any magnets or poles. The program which was to dispel vortices was this: first, to show that the forces with which currents act on currents can in every case by analyzed as the sum of forces acting between differential elements of the currents, each of these elementary forces being strictly Newtonian in the sense Ampère had specified in the quotation above, and second, to show that magnets are equivalent to aggregates of currents flowing in small closed paths, so that the Oersted phenomenon in which currents act on magnets, as well as all other magnetic forces, is reducible to these Newtonian actions of the first type.

The interaction of currents was at first shown by Ampère with movable frames carrying currents, an apparatus that has become familiar in school laboratories ever since. But when putting his argument into rigorous form in the *Théorie mathématique*, he chose to base it rather on the smallest possible number of independent experiments of a somewhat different sort, and, he insists, on no other empirical evidence. Ampère, in listing five principal accomplishments upon his application for appointment to the Collège de France, states as the fifth:

35. Oersted, *op. cit.*, pp. 116–17. What is translated "circular motion" in this reference was rendered "*mouvements tourbillonnaires*" in the French translation of 1820; Jules Joubert, ed., *Collection de memoires relatifs a la physique vols. 2 & 3: Memoires sur l'electrodynamique* (Paris: 1885, 1887), *2*, p. 6. This is a very useful collection of papers of Oersted, Ampère, Biot and Savart, et al., with valuable notes by Joubert.

36. Ampère, *op. cit.*, p. 1.

37. Ampère (1820), "The Mutual Action of Two Electric Currents," in Tricker, *op. cit.*, pp. 140 ff.

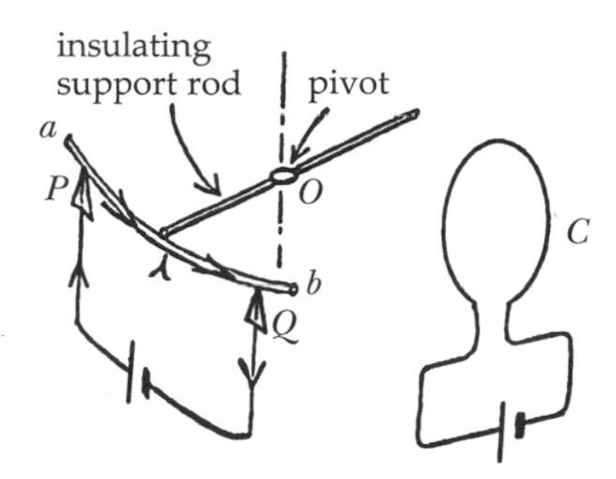

Figure 10. Ampère's Experiment III, to determine whether segment *ab* will move along its own length under the influence of current in coil C.

> 5. La marche qui m'a conduit à cette formule sera toujours un modèle de celle qu'on doit suivre pour arriver à de telles formules par l'expérience seulement et sans aucune supposition.[38]

Interestingly, one of those who held up Ampère's work as a model of scientific method was James David Forbes, Maxwell's teacher at the University of Edinburgh.[39]

Ampere's four basic experiments were these:[40]

Experiment I: If a wire is doubled on itself, so that the same current flows in opposite directions in adjacent conductors, the net force on a nearby current-carrying conductor is zero.[41]

Experiment II: If in the above experiment, one of the wires is given a series of tight twists or bends, which do not however take it far at any point from the original straight line, the results are as before.[42]

Experiment III: A current-carrying conductor experiences no force along its length as the result of the action of a second conductor carrying current in a closed loop (Figure 10).[43]

38. Launay, *Le grand Ampère: d'après des documents inédits* (Paris: 1925), p. 211.

39. James David Forbes, *Dissertation Sixth ... of the Progress of Mathematical and Physical Science}* (Edinburgh: [1856?]), p. 975: "He is at least as well entitled as any other philosopher who has yet appeared to be called 'the Newton of Electricity'." Forbes compares Ampère's reduction of magnets to circulating currents, as an artifice, with Newton's "fits of easy reflection and transmission" of light. Maxwell echoes Forbes' characterization of Ampère as the "Newton of electricity" at (*Tr* ii/175).

40. The four experiments are diagrammed in the *Treatise* in very poor, sometimes indeed unintelligible, reproductions of Ampère's figures from Plate I of the *Théorie mathématique*. They are adequately reproduced in Tricker, *op. cit.*, pp. 164, 168, and 171.

41. Ampère, *op. cit.*, pp. 9–10; (*Tr* ii/160).

42. *Ibid.*, pp. 10–14; (*Tr* ii/160–61).

43. *Ibid.*, pp. 14–17; (*Tr* ii/161–62).

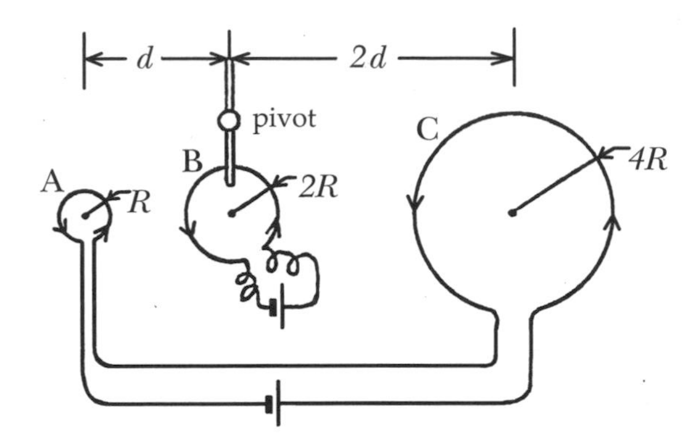

Figure 11. Ampère's Experiment IV

Experiment IV: If two current-carrying conductors lie in the same plane, the force between them will be the same whatever the scale of the apparatus, provided the currents are held constant as the scale is changed, and the geometry is kept similar to itself during the scale change (Figure 11).[44]

The null results of these four experiments are so elegant and have such an inherent plausibility that it may hardly seem necessary to perform the experiments at all; indeed, although Ampère makes no such admission, the *Théorie mathématique* closes with what must be one of the great anticlimaxes of scientific history:

> Je crois devoir observer, en finissant ce Mémoire, que je n'ai pas encore eu le temps de faire construire les instruments répresentés dans la figure [that of Experiment IV]. Les expériences auxquelles ils sont destinés n'ont donc pas encore été faites....[45]

He explains that the result has been assured by other means.

No less an experimenter than Weber, whom Maxwell very much admired, later criticized Ampère's experiments in detail and rather severely, pointing out that a null experiment is nonetheless a measurement which yields the value zero only to within certain limits of accuracy which Ampère ought to have reported; on the whole, it seems likely that Weber was right in regarding Ampère as a thinker rather than an experimenter.[46] Nevertheless, there

44. *Ibid.*, pp. 17–18; (*Tr* ii/162–63).

45. Ampère, *op. cit.*, p. 151. Tricker (*op. cit.*, pp. 46–48) calls attention to this, and describes his own version of the fourth experiment.

46. Weber terms Ampère "mehr Theoretiker als Experimentator" (*Wilhelm Weber's Werke* (6 vols.; Berlin: 1892–94), *3*, p. 213). He makes a number of comments to the same effect in the *Elektrodynamische Maasbestimmungen* of 1846. Carl Neumann, citing Weber's criticism, recommends regarding "the results of Ampère's so-called fundamental experiments not as experimental facts, but as hypotheses" and he offers a list of six such hypotheses. Carl Neumann, ed., *Franz Neumann's Gesammelte Werke* (3 vols.; Leipzig: 1912), *3*, pp. 340–41.

can be no objection to null-experiments in principle (provided they are performed!), and Ampère's experiments have as precedent the Cavendish determination of the force law for electricity, an experiment which was repeated with great precision at the Cavendish Laboratory under Maxwell.[47] Newton's Proposition 70 of Book I of the *Principia*, on which the Cavendish experiment was based, is the ultimate prototype for Ampère's reasoning in interpreting his own Experiment IV.[48] Whether it is in fact owing to experimental care on Ampère's part, or to an insight into the ways in which nature *ought* to work, the four null results are empirically quite correct, as verified by a much later repetition.[49]

Experiments I and II yield postulates of Ampère's theory directly. From I we conclude that currents can be added algebraically, and from II, that the force exerted by one current on another is equal to the vector sum of components into which the first may be resolved in any way. But Experiments III and IV are best interpreted in the context of the theory as it develops.

Ampère assumes initially that the elementary force must be proportional to the length of the element of the circuit through which the current is flowing, and that, with other factors constant, the magnitude of the force may be assumed proportional to the quantity of the current as well:

$$df \propto ii'ds\,ds',$$

where i and i' are the currents in the two elements of conductors $\mathbf{ds}$ and $\mathbf{ds}'$.[50] The distance between ds and ds' is taken as unity in the above expression (thereby implying a definition of unit current).[51] Ampère takes a

47. See Maxwell's note as editor of Cavendish's electrical papers: Henry Cavendish, *The Scientific Paper of the Honourable Henry Cavendish, R.R.S.: Volume I, The Electrical Researches*, ed. James Clerk Maxwell (revised by Joseph Larmor) (Cambridge: Cambridge University Press, 1921), pp. 404–9; cf. (*Tr* i/81–86).

48. Maxwell points out that Laplace had improved upon Newton's demonstration of the basic theorem (*Tr* i/85n); it was Laplace who deduced an inverse-square law from Biot's electromagnetic experiments and it was this demonstration in turn which suggested to Ampère the direct argument from his Experiment IV which he appends to the *Théorie mathématique*, pp. 152 ff., and which Maxwell in effect repeats (*Tr* ii/162–63).

49. Von Ettinghausen, "Ueber Ampère's elektrodynamische Fundamentalversuche," *Königliche Akademie der Wissenschaften, Wien: Mathematische-Naturwissenschaftliche Klasse, Sitzungsberichte* (11) 77 (1878), p. 109. Cf. J. J. Thomson, "Report on Electrical Theories," *British Association Reports, 1885*, p. 98.

50. Ampère, *op. cit.*, p. 19. Maxwell follows Ampère unquestioningly in this same assumption (*Tr* ii/166).

51. Ampère's definition falls in naturally with his force law, but it differs from the definition of unit current in the cgs-emu system which Maxwell and the subsequent tradition adopted. Maxwell points out that "the unit current adopted in electromagnetic measure is greater than that adopted in electrodynamic measure [Ampère's] in the ratio of $\sqrt{2}$ to 1" (*Tr* ii/173).

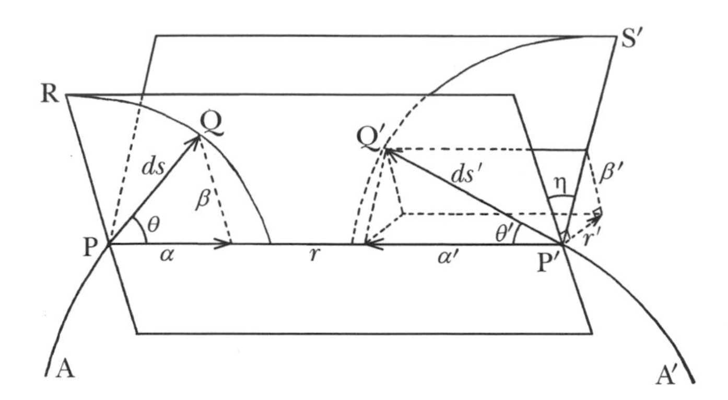

Figure 12. The geometry of Ampère's theory

positive sign as denoting attraction. It must be emphasized that this first equation is not the result of an experiment, and cannot in principle be confirmed by any experimental test at Ampère's command. Electrodynamic analysis of a circuit, perhaps unlike chemical analysis of a compound, is a purely intellectual act. One is certainly impressed by the extent to which the *Théorie mathématique* is, from this first step, *made* by Ampère, and not found in experiment.

There is a major difficulty lurking in this first assertion of an elementary force: in what direction does it act? There is certainly no doubt in Ampère's mind on this point, and yet it is precisely the question which Maxwell insists on raising. Ampère asserts:

> [O]n ne peut pas concevoir cette force autrement que comme une tendance de ces deux points à se rapprocher ou à s'éloigner l'un de l'autre suivant la droite qui les joint....[52]

But there is no direct evidence in Ampère's favor; indeed, in electromagnetic experiments, we meet forces which act in very strange directions, as we have already seen.

Ampère's next step is to consider the effect of the *positions* of the current elements. For Newton, dealing with planets, this was only a question of their separation r, but here Ampère must consider as well their spatial attitudes with respect to each other—hence the far greater complexity of Ampère's mathematical problem. In Figure 12 (corresponding to Maxwell's Figure 29 (*Tr* ii/164)), the two elements **ds** and **ds′** are resolved into components in the following manner: pass the plane RP' through **ds′** and **r**. Let α and β denote the components of **ds** parallel to **r**, and perpendicular to **r**, respectively, in this plane. Then resolve **ds′** into components of α', β', and γ', respectively in the direction of **r**, perpendicular to **r** in the plane RP', and perpendicular to that plane. Finally, pass a second plane PS'

52. Ampère, *op. cit.*, p. 86. Contrast Maxwell: "[W]e shall not at first assume that their mutual action is necessarily in the line joining them" (*Tr* ii/165).

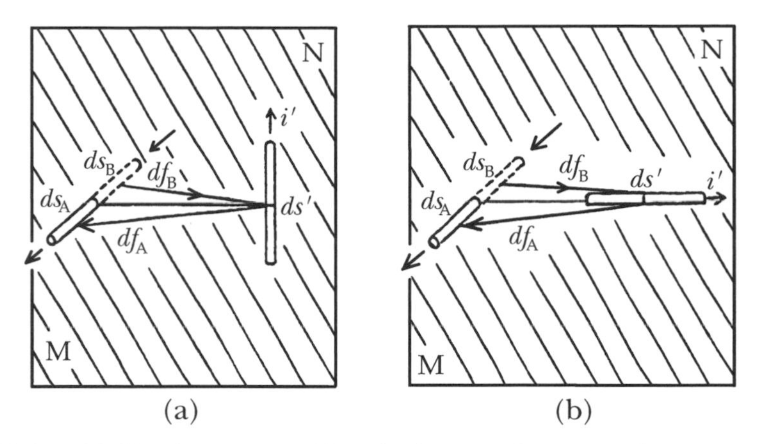

Figure 13. Two instances of Ampère's crossed current elements.

through **ds′** and **r**, and let η denote the angle between the two planes; let $\theta = \measuredangle(\mathbf{ds},\mathbf{r})$ and $\theta' = \measuredangle(\mathbf{ds}',\mathbf{r})$, as shown in Figure 12.

If we assume, as both Ampère and Maxwell do without question, that the dependence on distance of separation **r** and the dependence on the spatial attitude characterized by θ, θ', and η are *independent*, then the force equation will be separable. If we represent the dependence on *attitude* by the unknown function $\rho(\theta, \theta', \eta)$, and if we assume, in the manner of Newton, that the force will be inversely proportional to some power n of the distance, we may write with Ampère: [53]

$$df = \frac{\rho(\theta,\theta',\eta)}{r^n} i\, i' ds\, ds'.$$

Ampère's next step is to use the right afforded by Experiment II to consider the interactions of the components of **ds** and **ds′** separately. In general, then, each component of **ds** will interact with each component of **ds′**, but Ampère argues that not all of these interactions could in fact occur in nature. He asserts:

> [U]ne portion infiniment petite de courant électrique n'exerce aucune action sur une autre portion infiniment petite d'un courant situé dans un plan qui passe par son mileur et que est perpendiculaire a sa direction.[54]

Ampère's "proof" is this (see Figure 13): let **ds′** lie in the plane perpendicular to **ds** through the latter's center. Then current i flowing through $\mathbf{ds}_B$ and through $\mathbf{ds}_A$ must exert opposite effects on **ds′**, one being attractive and the other repulsive, *since one flows toward the plane MN and the other flows away from it*. In the limit, the two components cancel; hence there can be no net force of this kind in nature.

53. Ampère, *op. cit.*, p. 19.

54. *Ibid.*, p. 20.

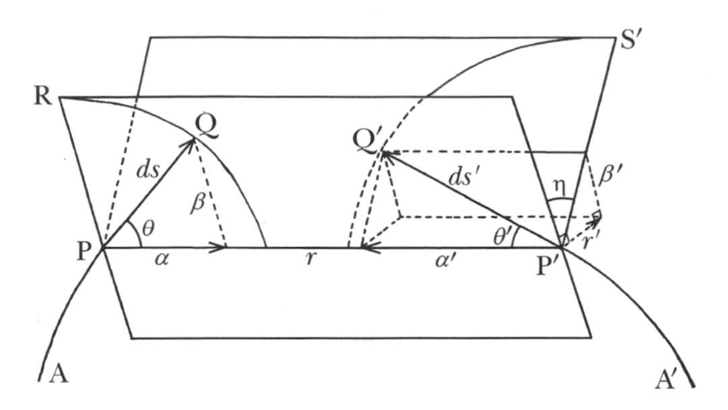

Figure 12 (repeated)

This theorem is crucial, since it very much simplifies the force equation by allowing us to disregard all possible terms due to such crossed currents. What Ampère is assuming is in effect that the world must behave the same way as its mirror-image. Suppose the current in $\mathbf{ds}_A$ is as shown in Figure 13a and that the force on it is repulsive. Now suppose that the plane *MN* is a mirror; the mirror-image of $\mathbf{ds}_A$ with its current away from the mirror would be an element $\mathbf{ds}_B$ similarly carrying current *away from the mirror.* Since $\mathbf{ds}_A$ experiences a repulsion, so would the image in the mirror. Therefore if (disregarding mirrors) we set up an experiment with a current element $\mathbf{ds}_B$ carrying current away from the plane *MN*, it (like the mirror image of $\mathbf{ds}_A$) would necessarily be observed to be repelled. If the current in it were then reversed, it would be attracted, as Ampère affirms. Maxwell endorses Ampère's assumption; he says:

> The only action possible between elements so related is a couple whose axis is parallel to *r*. (*Tr* ii/166)

The "couple" is evidently that formed by $\mathbf{df}_A$ and $\mathbf{df}_B$ as assumed by Ampère and drawn in Figure 13; Maxwell apparently allows the possibility that both forces might act, yielding a couple but no net force. But his theory does not take couples into account, so he, like Ampère, omits this term. J. J. Thomson, commenting as editor of the third edition of the *Treatise,* questions the assumption by proposing an alternative in terms of a vector cross-product relation, and then answers his own question in this way:

> The reason for assuming that such a force does not exist, is that the direction of the force would be determined merely by the direction of the currents, and not by their relative position. (*Tr* ii/166n)

By "relative position" he evidently means, their position relative to such a reference plane as *MN* in Figure 13. It was Carl Neumann who, in listing

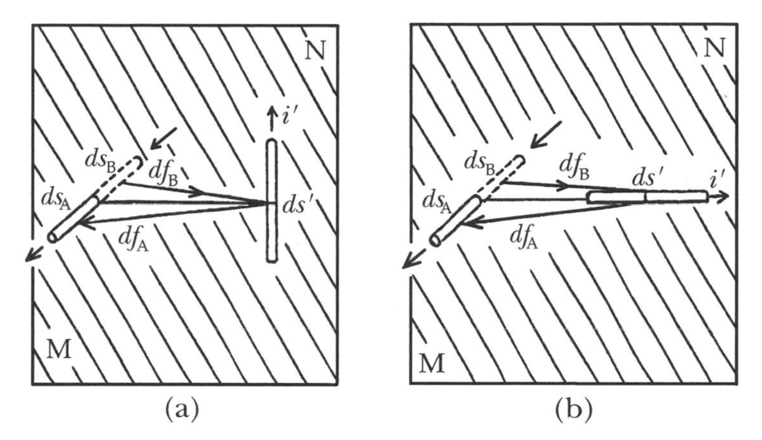

(a) (b)

Figure 13 (repeated)

Ampère's axioms, made the parity-assumption explicit.[55]

Only two pairs of components now remain which might exert forces on one another: the collinear components α and α', and the parallel components β and β' (Figure 12). Let us denote these respectively df_1 (due to α, α'), and df_2 (due to β, β'). We have no information at all about their relative strengths, or even whether both in fact exist, although we know they both *might* act. This information is the rabbit in the hat in the third experiment.

For the present, Ampère merely introduces a constant k to measure the ratio of the force df_1 to the force df_2. He defines unit current in terms of parallel current elements, so that he writes first

$$\text{PARALLEL ELEMENTS:} \quad df_2 = \frac{ii'\beta\beta'}{r^n},$$

then

$$\text{COLLINEAR ELEMENTS:} \quad df_1 = k\frac{ii'\alpha\alpha'}{r^n}.$$

55. Carl Neumann, ed., *Franz Neumann's Werke, 3*, pp. 340–41. Note that the development of mathematical representations of spatial relations suggested alternative force laws, and was closely related to the problems of electrodynamics. This is true of Hermann Grassmann's "Ausdehnungslehre" (see Grassmann, "Neue Theorie der Elektrodynamik," *Annalen der Physik, 64* (1845), pp. 1 ff.); Grassmann remarks with some surprise that he had invented his new mathematics "... zwar ehe ich von dieser neuen Theorie eine Ahnung hatte," but it proved very appropriate (*Ibid.*, p. 11). P. G. Tait took electrodynamics as a problem upon which to exercise the art of quaternions: "Quaternion Investigations Connected with Electrodynamics and Magnetism," *Quarterly Journal of Pure and Applied Mathematics, 3* (1860), pp. 331 ff.

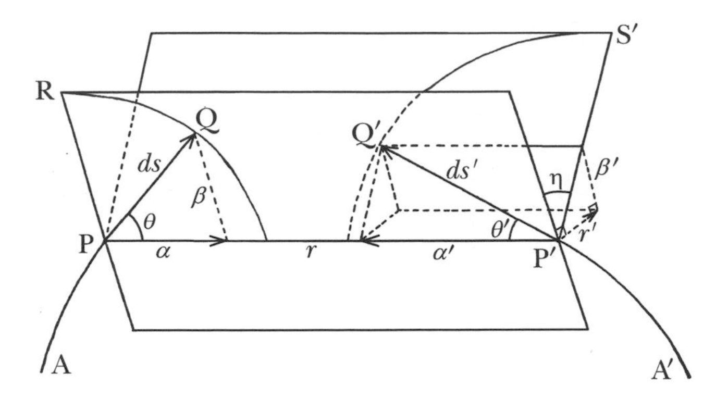

Figure 12 (repeated)

Since both forces act along the direction of **r**, they add algebraically, and we might write for the total force:

$$\mathbf{df} = \left(df_1 + df_2\right)\hat{\mathbf{e}}_r\,.$$

Substituting, and expressing the elements in terms of the angles of Figure 12 in a way which is readily confirmed:

$$\mathbf{df} = \frac{ii'}{r^2}\left(\sin\theta\cos\theta'\cos\eta - \frac{1}{2}\cos\theta\cos\theta'\right)ds\,ds'\,\mathbf{e}_r$$

Note that k and n are both unknown; these two elements of the force equation are to be determined from the information of Experiments III and IV.

Let us look only at the result of the argument by which Ampère brings Experiment III to bear on the evaluation of k and n. He transforms the force equation into an expression in terms of partial derivatives (substituting these derivatives for the corresponding trigonometric functions of the angles), and he then uses the following strategy: he evaluates the force *along the length* of a current element ds' due to a complete circuit of elements ds (let us call this df'). He gets this expression:[56]

$$df' = -\frac{1}{2}ii'ds'(1-n-2k)\oint\frac{\cos^2\theta'}{r^n}ds\,.$$

The analysis has been cut to fit the experiment, for in Experiment II a length (ds') of a conductor is pivoted so as to be free to move only along its length, and it is submitted to the action of a current in an adjacent loop. The fact that the experimental result is nil permits Ampère to set the above expression equal to zero, and since the integral is not in general zero (in the integrand, r and θ' are independent), he obtains this relation between the two coefficients:

56. Ampère, *op. cit.*, p. 26.

$$k = \frac{1-n}{2},$$

leaving only one further relation to be found.

Through an argument analogous (as has been mentioned) to that of Proposition 70, Book I, of the *Principia*, Experiment IV yields the exponent $n = 2$ in the law of force.[57] This in turn tells us that $k = -\frac{1}{2}$. We recall that k determines the relative direction and magnitude of the forces due to parallel and collinear components: the analysis has therefore revealed that collinear components contribute half as much as parallel components of the same length and carrying the same current, and collinear components repel whereas parallel attract (compare pages 32 ff., above). The force law may now be written explicitly:[58]

$$\mathbf{df} = \frac{ii'}{r^2}\left(\sin\theta\cos\theta'\cos\eta - \frac{1}{2}\cos\theta\cos\theta'\right)ds\,ds'\,\mathbf{e}_r\,.$$

It is convenient to use a relationship from spherical trigonometry which helps us to condense and interpret the above. If ε denotes the angle between **ds** and **ds′**, (compare Figure 12), then:[59]

$$\cos\varepsilon = \cos\theta\,\cos\theta' + \sin\theta\,\sin\theta'\,\cos\eta,$$

and finally

$$\mathbf{df} = \frac{ii'}{r^2}\left(\cos\varepsilon - \tfrac{3}{2}\cos\theta\,\cos\theta'\right)ds\,ds'\,\mathbf{e}_r\,.$$

The fact that $k < 0$ shows that Ampère's analysis has reported a force of repulsion between two collinear elements, α and α'. Does this force really exist in nature? Our earlier discussion indicates that experiment cannot yield an answer: we cannot experiment with two current elements. But strangely, in this case Ampère and de la Rive felt that they had verified this conclusion empirically.[60] This could only appear to be a verification of the theory if we forget that, as Ampère is usually the first to point out, it is not the interaction of two current elements which we observe in an experiment,

57. Note that in the body of Ampère's text, a long argument intervenes before the consequence of Experiment IV can be drawn. The simplified argument is appended (pp. 152 ff. of Ampère's text).

58. Ampère gives his result first in another form (*op. cit.*, p. 44; but see p. 116, where the result above is given explicitly).

59. For any spherical triangle, if a, b, and c are the sides and A the angle opposite side a, then $\cos a = \cos b \cos c + \sin b \sin c \cos A$ (the spherical "law of cosines"). Here if **ds**, **ds′**, and **r** are laid out from a single point along three radii of a sphere, the arcs $(\pi - \theta')$, θ, and $(\pi - \varepsilon)$ constitute three sides of a spherical triangle, with $(\pi - \eta)$ as the angle opposite $(\pi - \varepsilon)$, and the theorem applies.

60. Ampère, *op. cit.*, p. 28; Joubert, *op. cit.*, pp. 33–34.

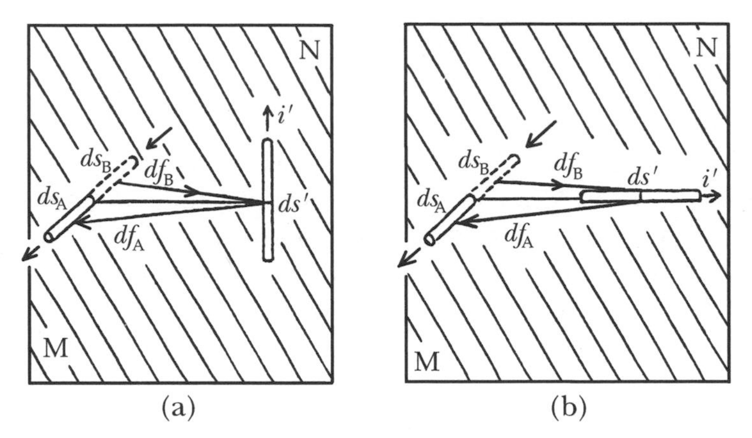

Figure 13 (repeated)

but the effect of *one whole circuit* on an element. It is difficult to understand how Ampère allowed this misunderstanding to arise. I suspect the error arises through the uncritical inclusion of an earlier paper in a later work.

The fact is that no experiments in which currents flow in closed circuits can confirm this fundamental law directly.[61] It must stand or fall by the strength of Ampère's argument, and Maxwell was not the first to question this. Hermann Grassmann (1809–1877) pointed to a peculiarity in the result when the law is applied to parallel elements which make an angle θ with the line joining them: he showed (as is easily seen) that when that angle takes the value $35°30'$, the force implausibly goes to zero, and for smaller angles changes from attraction to repulsion. He remarked:

> Already the tangled form of this equation must raise a doubt about it. This doubt must be enhanced when we try to apply it.[62]

For the sake of obtaining a law with less arbitrary behavior, Grassmann was willing to abandon Ampère's first requirement of equal and opposite forces acting along the line **r**.

Ampère's law, we should note in passing, is not the same as the "law" usually given in modern texts, which is in open contradiction to Newton as well as to Ampère.[63] The modern law, which we may write

$$\mathbf{df} = ii'\left(\mathbf{ds} \times \frac{\hat{\mathbf{e}}_r \times \mathbf{ds}'}{r^2}\right),$$

61. The possibility of experimenting with the electrodynamic effects of unclosed circuits, and thereby obtaining empirical evidence which would decide between alternative theories, interested several investigators, among them Helmholtz. "Ueber die Bewegungsgleichungen der Elektrodynamik" (1870), in Hermann Helmholtz, *Wissenschaftliche Abhandlungen* (3 vols.; Leipzig: 1882–95), *1*, pp. 547, 563. Cf. J. J. Thomson, *op. cit.*, pp. 144 ff., where a number of contributions are reviewed.

62. Grassmann, *op. cit.*, p. 4 (compare page 151n, above).

63. Tricker, *op. cit.*, p. 43.

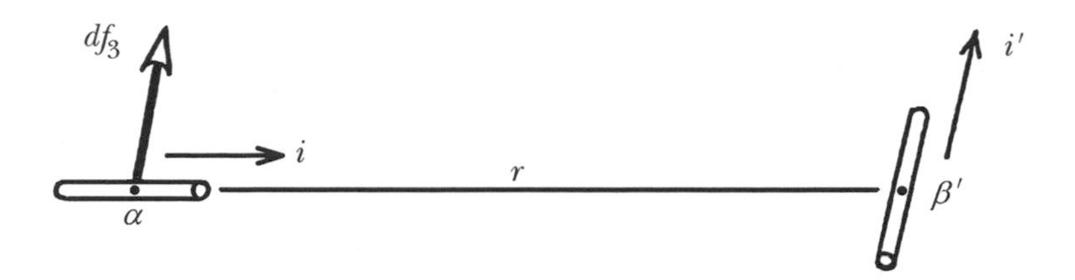

Figure 14. Maxwell's force $\mathbf{df}_3$

appears fleetingly in Ampère's text, but by no means as a law of nature; it is *Reynard's law*, apparently first enunciated in 1870.[64]

Maxwell's analysis in Chapter II of Part iv of the *Treatise* agrees with Ampère's with respect to the forces we have called $\mathbf{df}_1$ and $\mathbf{df}_2$; he reformulates the expressions for them slightly (*Tr* ii/166):

$$\mathbf{df}_1 = \mathbf{A}\alpha\alpha' ii'\hat{\mathbf{e}}_r$$
$$\mathbf{df}_2 = B\beta\beta' ii'\hat{\mathbf{e}}_r .$$

The only respect in which Maxwell diverges from Ampère's argument is in the admission of one form of crossed-element force. Like Ampère, he denies a force due to elements perpendicular both to each other and to the line joining them (the case of Figure 13a, discussed earlier), but he includes a force due to elements perpendicular to each other, one of which however lies along the direction of **r** (Figure 13b). Why is this not subject to the stricture against the first case? It would be, if it were supposed to take the form of an attraction or a repulsion. Maxwell, though, rejects this possibility, assuming instead that, if it exists, it must be a force $\mathbf{df}_3$, parallel to the current element which is perpendicular to **r** (Figure 14).

Three different pairs of components can give rise to such a force, namely $\alpha\beta'$, $\beta\alpha'$, and $\alpha\gamma'$. A typical expression for the force will be

$$\mathbf{df}_3 = C\alpha\beta' ii'\hat{\mathbf{e}}_{\beta'} .$$

Arguing for this choice of direction, Maxwell says:

> The sign of this expression is reversed if we reverse the direction in which we measure β'. It must therefore represent either a force in the direction of β' or a couple. (*Tr* ii/166)

In saying this Maxwell rejects the possibility of a cross-product relation such as $\alpha \times \beta'$ determining the direction of the force; such a relation would formally reverse $\mathbf{df}_3$ with a reversal of either α or β'—as it should—but it would assume that "handedness" could be distinguished in a fundamental law of nature, an assumption that Maxwell tacitly rejects.

In admitting a force which acts perpendicularly to the line joining the elements **ds** and **ds′**, Maxwell has denied Ampère's interpretation of the

64. Joubert, *op. cit.*, *3*, p. 123; F. Reynard, "Nouvelle théorie des actions électrodynamiques," *Annales de chimie et physique, 19* (1870), p. 272.

Newtonian program, according to which "one cannot conceive" (on ne peut pas concevoir) the force except as acting along the line of centers (page 148, above). Maxwell does say, at the end of the chapter, when surveying the array of possibilities which his new analysis has produced,

> [T]hat of Ampère is undoubtedly the best, since it is the only one which makes the forces on the two elements not only equal and opposite but in the straight line which joins them. (*Tr* ii/174)

But by putting the question this way, Maxwell has made it a matter of taste rather than a scientific necessity. Throughout the analysis which follows, he continues to deal with a component of the force which does not obey Ampère's principle; Maxwell must therefore regard it as conceivable.

Having introduced components of the force in directions other than along **r** itself, Maxwell has to carry out his analysis as a vector problem. Taking coordinates in the directions of α', β', and γ', we have now:

$$\mathbf{df}_{\alpha'} = A\alpha\alpha' + B\beta\beta'$$
$$\mathbf{df}_{\beta'} = C(\alpha\beta' - \alpha'\beta)$$
$$\mathbf{df}_{\gamma'} = C\alpha\gamma'.$$

To avoid carrying the currents i and i' through the equation, Maxwell takes **df** as an intensive measure of the force between circuits carrying unit currents (force per unit current per unit current).

I shall abridge the complexities of Maxwell's analysis as much as possible, in order to focus upon the relation of his results to Ampère's. Of a number of quantities which he introduces, we will need to use the following (*Tr* ii/168, 170):

$$P = \int_r^\infty \frac{A+B}{r^2} dr$$
$$Q = \int_r^\infty C\,dr$$
$$\rho = \tfrac{1}{2}\int_r^\infty (B-C)dr$$

It will be useful to define two further quantities as well which are not separately identified by Maxwell, namely potentials of the forces $\mathbf{df}_1$ and $\mathbf{df}_2$:

$$W = \int_r^\infty A dr$$
$$V = \int_r^\infty B dr.$$

Maxwell's quantity Q presents a peculiar problem of interpretation. The force $\mathbf{df}_3$, of which C is the measure, is not a central force, but acts in every case normally to **r**. If we carried the element **ds′** along **r** at right angles to the current element **ds** from $\mathbf{r}_0$ to infinity, the component $\mathbf{df}_3$ would contribute nothing to the work done. Q then seems to be a fictitious potential: the potential which $\mathbf{df}_3$ *would* have if it were central. Since $C = C(r)$, this works out perfectly well mathematically. Since Q enters into the other apparent potential expressions in Maxwell's chapter, they all share this fictitious character. For example, ρ may be written

$$\rho = \frac{1}{2}\left(\int B\,dr - \int C\,dr\right) = \frac{V-Q}{2},$$

and later Maxwell will work with $(Q+\rho)$, which is

$$(Q+\rho) = \frac{V-Q}{2} + Q = \frac{V+Q}{2}.$$

It should be made clear that Maxwell does not apply the term "potential" to these; but he is arranging his theory to yield the true scalar and vector potentials M and $\mathbf{A}$ later in the argument, as we shall see.

The general force expression with A, B, and C undetermined in relation to one another is too complex to be of interest, but it simplifies in the case of *that component* of the force of $\mathbf{ds}'$ on $\mathbf{ds}$ which acts in the direction of $\mathbf{ds}$—that is, the component which is pertinent to Ampère's Experiment III. Writing this expression and integrating around a closed circuit of elements $\mathbf{ds}'$, Maxwell obtains for the net force along the length of $\mathbf{ds}$:

$$df_s = -\oint (2Pr - B - C)\,l'\lambda\,ds'.$$

Here l' is $\cos\varepsilon$, and λ is $\cos(x, r)$. From the independence of these two quantities, we conclude that (*Tr* ii/169)

$$(2Pr - B - C) = 0,$$

corresponding to Ampère's evaluation of k in terms of n. That is, Experiment III has this time told us that

$$P = \frac{1}{2r}(B+C).$$

Using this result to eliminate P, and with it any reference to force $\mathbf{df}_1$, Maxwell now obtains a general expression for the total force of a closed circuit upon a current element $\mathbf{ds}$. Notice that we are now approaching a point of convergence with the Faraday theory of Maxwell's preceding chapter. Maxwell defines an auxiliary quantity which we may write

$$\mathbf{D} = \oint \frac{B-C}{2}\,\hat{\mathbf{e}}_r \times \mathbf{ds}' \quad [\mathbf{D} \leftrightarrow \alpha', \beta', \gamma'],$$ [65]

where the integration is around the circuit of the elements $\mathbf{ds}'$. In terms of this quantity, the force $\mathbf{df}$ on $\mathbf{ds}$ is (*Tr* ii/169)

65. The symbol **D** is is not related to Maxwell's displacement vector **D**, to be discussed in Chapter 6. It is in fact the symbol Ampère uses to denote his "directrix" (Ampère, *op. cit.*, p. 31).

$$\mathbf{df} = (\mathbf{ds} \times \mathbf{D})ii'.$$

In a passage in the *Théorie mathématique* which we did not cite, Ampère derived a result that will be seen to agree exactly with the above expression once B and C have been determined.[66] Maxwell's enlargement of Ampère's theory does not affect any testable case such as this.

Maxwell says of the quantity we have written **D**, but which he writes in terms of the components (α', β', γ'—no relation to his earlier quantities of the same name!):

> The quantities α', β', γ' are sometimes called the determinants of the circuit s' referred to the point *P*. Their resultant is called by Ampère the directrix of the electrodynamic action. (*Tr* ii/169)

The reader has only to recall a result from Maxwell's preceding chapter to recognize what has happened: Faraday's line of magnetic force has emerged *out of the formalism of analysis*. Maxwell goes on to add:

> Since we already know that the directrix is the same thing as the magnetic force due to a unit current in the circuit s', we shall henceforth speak of the directrix as the magnetic force due to the circuit. (*Tr* ii/169–70)

Strictly, as Maxwell has defined the quantity here, it is equal to the magnetic force per unit current in the source loop. This coincidence of the formal theory and Faraday's view was certainly unknown to Ampère and Faraday; I do not know that it was remarked by anyone before Maxwell pointed it out.[67]

Maxwell postpones, seemingly until the last moment, the evaluation of the dependence of the force on the distance r; he accomplishes this very simply, interpreting Experiment IV essentially as Ampère did. The factor $(B-C)/2$ in the equation previously written for the directrix is now shown to be $1/r^2$ (*Tr* ii/173), so that

$$\mathbf{D} = \oint \frac{\hat{\mathbf{e}}_r \times \mathbf{ds}}{r^2}.$$

To compare Maxwell's result with Ampère's, let us set aside for a moment a section of Maxwell's chapter which has no counterpart in Ampère (*Tr* ii/170–71), and go directly to his final statement of the force law (*Tr* ii/173). He formulates it in terms of a set of oblique coordinates which we may represent by the non-orthogonal unit vectors $\hat{\mathbf{e}}_s$, $\hat{\mathbf{e}}_{s'}$, and $\hat{\mathbf{e}}_r$,

66. Ampère, *op. cit.*, p. 31.

67. Tait, for example, in discussing it, does not point out any relation to Faraday: "This vector ... which is of great importance in the whole theory of the effects of closed or indefinitely extended circuits, corresponds to the line which is called by Ampère 'directrice de l'action électrodynamique'. It has a definite value at each point of space..." (*op. cit.*, p. 334); cf. p. 151n, above.

parallel to **ds**, **ds′**, and **r**.[68] Then

$$\mathbf{df} = \left(R\hat{\mathbf{e}}_r + S\hat{\mathbf{e}}_s + S'\hat{\mathbf{e}}_{s'}\right)\mathbf{ds}\,\mathbf{ds}'.$$

Maxwell finds that[69]

$$R = \frac{1}{r^2}\left(\cos\varepsilon - \frac{3}{2}\cos\theta\cos\theta'\right) + r\frac{\partial^2 Q}{\partial s\,\partial s'}$$

$$S = -\frac{\partial Q}{\partial s'}$$

$$S' = \frac{\partial Q}{\partial s}.$$

We have traced schematically how Ampère's four experiments serve to eliminate the unknown coefficients in the force equation, P having been eliminated by Experiment III, and $(B-C)/2$ by Experiment IV. What about Q, an unknown function of r, which remains? The answer is that Q, appearing always as a perfect differential, is immune to experiment; in integration around the closed loop of the source circuit, each of the terms involving Q goes to zero, so Q is left to our free choice:

> Since the form and value of Q have no effect on any of the experiments hitherto made, in which the active current at least is always a closed one, we may, if we please, adopt any value of Q which appears to us to simplify the formulae. (*Tr* ii/174)

Maxwell mentions three interesting options in addition to Ampère's; this last arises if $Q = 0$, thereby suppressing the components other than $\hat{\mathbf{e}}_r$. The others are (1) that there be no force between collinear elements (that is, $A = 0$); (2) that the attraction R be proportional to $\cos\varepsilon$; (3) that the attraction and the oblique force depend on θ and θ' only. Each yields a new law of force—but others might be invented without limit.[70]

Maxwell's opinion of the process is perhaps reflected in this comment in a Royal Institution lecture:

> The formula of Ampère, however, is of extreme complexity, as compared with Newton's law of gravitation, and many attempts have been made to resolve it into something of greater apparent simplicity.

68. Maxwell had before him a paradigm of the use of oblique coordinates, in Newton's analysis of the three-body problem (*Principia*, Book I, Proposition 66.) He taught the *Principia* and knew it well, though I do not know that he had this example in mind here.

69. (*Tr* ii/173). I have transcribed Maxwell's result by substituting the geometrical equivalents of the partial derivatives as given at (*Tr* ii/165), taking into account the fact that Maxwell's θ' is the supplement of the angle θ' shown in Figure 12 (page 148 above).

70. On the generalization of Ampère's law of force—by others as well as by Maxwell—see J. J. Thomson, *op. cit.*, p. 115; Whittaker, *History*, pp. 233–36; O'Rahilly, *Electromagnetic Theory*, pp. 102–123.

> I have no wish to lead you into a discussion of any of these *attempts to improve a mathematical formula*. Let us turn to the independent method of investigation employed by Faraday in those researches ... which have made this Institution one of the most venerable shrines of science. (*SP* ii/318; emphasis added)

This suggests, I think, the point of the chapter for Maxwell: there is a moral in the elusive function Q. This kind of investigation with which Ampère has endowed science is an attempt "to improve a mathematical formula"; that it is not an investigation of nature is proven by the fact that the mathematics proves absolutely deaf to anything nature might say. Maxwell is impatient to return to Faraday, and no doubt hopes that the reader shares this feeling.

Before returning to Faraday, however, we must consider the section of Maxwell's Ampère chapter referred to above, a section that does not correspond to anything in Ampère. This is a section (*Tr* ii/170–71) in which the result for the force on a current element is reformulated in terms of potentials. To turn immediately to the result, Maxwell shows that the force on **ds** due to **ds′** can be written in this way (*Tr* ii/171):

$$\mathbf{df} = ii'\frac{\partial}{\partial s\,\partial s'}\left(\nabla M - \nabla L + \mathbf{A} - \mathbf{A}'\right),$$

in which

$$M = \int_0^s\int_0^{s'}\frac{\mathbf{ds}\cdot\mathbf{ds}'}{r},$$

$$L = \int_0^r (Qr+1)dr,$$

$$\mathbf{A} = \int_0^s\frac{\mathbf{ds}}{r}, \qquad \mathbf{A}' = \int_0^{s'}\frac{\mathbf{ds}'}{r}.$$

We see that here, Maxwell is looking at Ampère's "elements" from the point of view that was naturally adapted, in Maxwell's Chapter I of Part iv, to Faraday's concepts involving a view of systems as wholes. This is perhaps a little like looking the wrong way through a telescope. Maxwell has rewritten Ampère's force law in terms of an incremental mutual potential, an incremental vector potential, and one quantity, L, that will disappear on the first integration.

The force between two finite currents, neither of which is closed, would be

$$\mathbf{f} = ii'\left(\nabla M - \nabla L + \Delta\mathbf{A} - \Delta\mathbf{A}'\right),$$

where $\Delta\mathbf{A}$ is the increment in the vector-potential of conductor s between the end-points of conductor s', *etc*., and M and L are as given above. Since Maxwell has denied the existence of such open currents, this is for him purely an exercise of thought, but not necessarily for that reason altogether idle, as it may lead to a clearer understanding both of the vector potential and of the significance of the earlier insistence on closed currents. We see that **A** would have a role in the force between currents, if currents did not always form closed circuits.

Maxwell investigates whether the force between unclosed currents would have a potential, and shows that because of the vector-potential terms, it would not. Forming the differential of work $\mathbf{f}\cdot\mathbf{dl}$, the terms $\Delta\mathbf{A}\cdot\mathbf{dl}$ and $\Delta\mathbf{A}'\cdot\mathbf{dl}$ are not perfect differentials.[71]

If we integrate around one closed loop only, say of circuit s', then $\Delta\mathbf{A}$ goes out, and we may form an empirically meaningful expression for the force on an element **ds**. This force will be **dF**, having components such as:

$$dF_x = \frac{\partial}{\partial x}M + \mathbf{ds}\bullet\nabla A_x\,.$$

This is an alternative form for the law of force on a current element in a magnetic field.[72] Again, this force does not have a potential, since $\Delta\mathbf{A}\cdot\mathbf{dl}$ is not a perfect differential. We see that so long as either circuit is unclosed, we do not have to do with the mutual potential alone; instead it is modified, and the force is made nonconservative, by a contribution from the vector potential.

It is only when we deal with two complete circuits that the remaining term in the vector potential disappears, and the force is conservative:

$$\mathbf{F} = \nabla M\,.$$

If Maxwell feels that Faraday's method involves the interactions of "systems of power" which should be expressible as potentials, this analysis has shown that Faraday's insights are relevant only to whole circuits; we must have displacement currents if we are to have a potential theory of electromagnetism. We might say Maxwell has found that incomplete circuits do not have sphondyloids.

To review for a moment, we have followed two derivations of a law of force between current elements, one Ampère's, and the other what is almost Maxwell's parody of Ampère's. Both start from the same set of four null experiments, and use the evidence of the experiments in essentially the same way. They differ in that Ampère denies from the outset the possibility of a force which acts otherwise than along the line connecting the two elements, while Maxwell is willing to admit a force perpendicular to this line; as a result, Maxwell is able to include a force arising from one case of crossed elements, while Ampère rejects all such cases. Maxwell is thus led to develop a much more complex theory of a force in which three vector

71. O'Rahilly similarly demonstrates that it is not in fact possible to sustain a theory of the mutual potential of two current elements, or of a current element and a complete circuit; O'Rahilly, *op. cit.*, p. 117.

72. The equation can be obtained from the vector triple-product for **df** of our contemporary texts (page 154, above), by expanding the triple product as the sum of two terms.

components must be considered, instead of Ampère's one, along the line of centers.

Maxwell's analysis employs functions analogous to potentials, but is on the whole parallel to Ampère's. One unknown function $Q(r)$ runs through Maxwell's analysis and remains, untouched by any possible evidence based on experiments with circuits, in the final result; it represents the irremovable uncertainty about the forces introduced by the crossed current elements. The two results are these:

AMPÈRE:

$$\mathbf{df} = \frac{ii'}{r^2}\left(\cos\varepsilon - \frac{3}{2}\cos\theta\cos\theta'\right) ds\, ds'\,\hat{\mathbf{e}}_r$$

MAXWELL:

$$\mathbf{df} = df_r\,\hat{\mathbf{e}}_r + df_s\,\hat{\mathbf{e}}_s + df_{s'}\,\hat{\mathbf{e}}_{s'}$$

$$df_r = ii'\left[\frac{1}{r^2}\left(\cos\varepsilon - \frac{3}{2}\cos\theta\,\cos\theta'\right) + r\frac{\partial^2 Q}{\partial s\,\partial s'}\right] ds\, ds'$$

$$df_s = -ii'\frac{\partial Q}{\partial s'} ds\, ds'$$

$$df_{s'} = -ii'\frac{\partial Q}{\partial s} ds\, ds'$$

Finally, Maxwell throws his force equation into the form of an incipient equation of potential, as it were. His effort here, I believe, is to find an intelligible relation between Ampère's view and Faraday's:[73]

$$\mathbf{df} = ii'\frac{\partial^2}{\partial s\,\partial s'}\left(\nabla M - \nabla L + \mathbf{A} - \mathbf{A}'\right).$$

ELECTROMAGNETIC ROTATIONS

Let us turn now to the question of electromagnetic rotations. As has been mentioned, the possibility of producing a continuous rotation, either of a current about a magnet, or a magnet about a current, led to a confrontation of two modes of analysis—Ampère's, in which the rotation was a complex phenomenon arising out of couples produced by simultaneous attractions and repulsions; and Faraday's, in which the rotation followed naturally from a circular configuration of the field. The rotations were

73. Compare this early remark in a letter to William Thomson (September 13, 1855): "[Y]ou are acquainted with Faraday's theory of lines of force & with Ampère's laws of currents and of course you must have wished to understand Ampère in Faraday's sense." (Larmor, *Origins*, p. 18)

therefore at the center of the early discussion of Ampère's theory. Furthermore, Maxwell takes the electromagnetic rotation as an illustrative experiment which reveals the power of Faraday's mode of interpreting the phenomena. We first review the early history of the rotations very briefly, and then turn to Maxwell's analysis of them, which we have postponed from our discussion of Maxwell's Chapter I.

We have seen (page 118, above) that Oersted in reporting his discovery announced that the effect could not be attributed to attractions. Though Ampère subsequently showed that it is possible to do so through a mathematical *tour de force*, the question remained, is this sound physics? Oersted clearly had a predisposition toward vortices, but Faraday, who studied the phenomena more open-mindedly, reached a similar conclusion. Trying Oersted's arrangement of the wire and the magnetic needle for himself, Faraday first sees the effects in terms of positions in which a pole is attracted or repelled by the wire. Yet after a struggle, prolific in marginal sketches—indeed an impressive instance of the restructuring of a concept—he is ready to speculate:

> Perhaps the settling of the needle perpendicular to the wire across it and then its approach on that side and its passing off on the other is the most instructive instance of attraction and repulsion. It is shown that the attraction and repulsion, as they seem to be, are only the combined action of the two circles in which the poles endeavour to move round the wire.[74]

A week later he writes to de la Rive:

> I find all the usual attractions and repulsions of the magnetic needle by the conjunctive wire are deceptions, the motions not being attractions or repulsions, nor the result of any attractive or repulsive forces, but the result of a force in the wire, which ... endeavours to make it move round in a never-ending circle. ... The law of revolution, and to which all the other motions of the needle and wire are reducible, is simple and beautiful.[75]

Faraday thus reversed Ampère's understanding of the situation: the observed phenomena may indeed be complex and call for interpretation, but instead of analyzing circular motions into the combined effects of attractions and repulsions, he resolves apparent attractions and repulsions into combinations of circular motions.

Faraday published his claim that motions "which resemble attractions and repulsion" should be resolved into a more fundamental motion of rotation, and a French translation of this provoked a reply in the form of a commentary of which Ampère himself was co-author. The issue is thereby drawn very clearly:

74. Faraday, *Diary*, *1*, pp. 51–52 (September 4, 1831).

75. Bence Jones, *Life and Letters of Faraday*, *1*, p. 356.

> [Faraday] prend l'action révolutive pour le fait primitif, et montre très bien que ces attractions et répulsions peuvent y être ramenées; mais nous venons de faire voir qu'en considérant, au contraire, comme fait primitif les attractions et répulsions entre les petites portions de courants électriques ... on en déduit immédiatement les mouvements circulaires des fils conducteurs et des aimants les uns autour des autres. ... les faits ... s'expliquent également bien des deux manières.[76]

This does not mean that the two theories weigh equally in the balance:

> [E]n adoptant la théorie de M. Ampère, ses faits rentrent dans les lois générales de la Physique, et ... on n'est pas obligé d'admettre comme fait simple et primitif une action révolutive dont la nature n'offre aucun autre example, et qu'il nous paraît difficile de considerer comme tel.[77]

Furthermore, Ampère's theory has powers which Faraday's was presumed to lack:

> L'action révolutive du fil conducteur et d'un aimant ... que M. Faraday considère comme fait primitif ... ne suffirait pas pour soumettre les phénomènes au calcul.[78]

Ampère was hardly able to consider the objections which were raised by Faraday. Since Faraday did not answer arguments, but only complained afterwards of wanting "something more on which to steady the conclusions," there would be no way of reasoning with him. The critic with whom Ampère was willing to debate—though only with signs of great exasperation—was Jean Baptiste Biot (1774–1862). Biot had, like Faraday, begun by examining the phenomena of currents in relation to magnets, rather than Ampère's interaction of two currents, and had in fact preceded Ampère in a mathematical theory of the phenomena.[79] Though the actual investigation concerned the force of a long straight conductor upon a magnetic dipole, the announced law asserted an inverse-square attraction between current elements and magnetic molecules. The force clearly caused the magnetic pole to tend to revolve about the wire, so the elementary action was described as a mechanical *moment*, and the discussion turned about the hypothesis of "*un couple primitif*" between a current element and a magnetic pole. The possibility of such a rotational moment as an elementary fact in nature was a point long at issue between Ampère and Biot. Biot was no match for Ampère in analysis. If Ampère represents

76. *Annales de chimie et de physique, 18* (1827), p. 337; Joubert, *op. cit., 2*, p. 187.

77. *loc. cit.*

78. Joubert, *op. cit., 2*, p. 184.

79. Biot and Savart, "Sur l'aimantation imprimée aux metaux par l'électricité en mouvement," *Annales de chimie et physique, 15* (1820), pp. 222–23. Tricker, *op. cit.*, pp. 23 ff., pp. 118 ff.

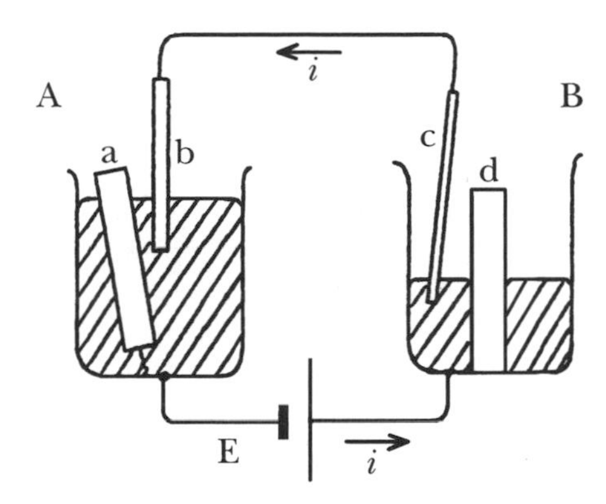

Figure 15. Faraday's electromagnetic rotations

him correctly, Biot was still vainly attempting to counter Ampère's analysis of magnets into currents with an alternative analysis of currents into closed rings of magnets—long after it had been decisively demonstrated both theoretically and empirically that closed rings of magnets have no effects on one another at all.[80]

The issue had already been formulated between Ampère and Biot when Faraday announced his successful production of the *continuous rotation* of a magnetic pole about a wire.[81] It was this announcement of Faraday's, in November, 1821, we are told, which led Ampère to resume his work in electricity after having abandoned it for metaphysical investigations.[82] Faraday's apparatus is diagrammed schematically in Figure 15; the two jars *A* and *B* contain mercury, *a* in the left-hand jar being a permanent magnet floating vertically and tied by a light thread at the bottom of the jar, while *b* is a conductor, fixed in position. In the right-hand jar, *d* is a permanent magnet fixed vertically in the center, while *c* is a light conducting rod dipping into the mercury at the bottom, and rotating freely about a suspension-point at the top. Current flows through the circuit completed by the conductors *c* and *b* and the mercury in the two jars. The top of the magnet *a* is observed to rotate about the fixed conductor *b*, while the bottom of

80. Ampère, *Théorie mathématique*, p. 140.

81. Joubert, *op. cit.*, *2*, p. 125.

82. Ampère, letter to Bredin, November 30, 1821, in Launay, *Correspondance*, *3*, p. 909: "[L]a metaphysique remplissait ma tête, mais depuis que la memoire de M. Faraday a paru, je ne rève plus que courants électriques." Faraday's announcement was made in the journal of the Royal Institution: *Quarterly Journal of Science*, *12* (October, 1821), p. 186. Figure 15 of the present text is sketched from Figure 1, plate IV, of the *Experimental Researches*, in which Faraday's original paper is reproduced (*XR* ii/147 ff.). Ampère repeated the rotation experiment, which he called "cette belle éxperience," together with Hachette and Savary, and reported it to the Académie on November 12, 1821, (Launay, *op. cit.*, *2*, p. 805, quotes the *procès-verbaux*).

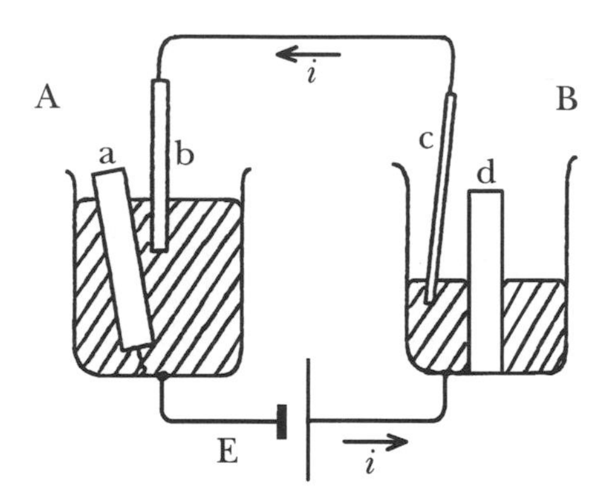

Figure 15 (repeated)

the wand *c* rotates about the magnet *d*. The rotations gradually accelerate, until the torque is balanced by the friction of motion through the mercury.

The apparatus alluded to by Maxwell in Chapter I of Part iv is essentially the same. He discusses first the rotation of the magnet about the wire (the arrangement at the left-hand of Figure 15) (*Tr* ii/150). In terms of the discussion of Chapter I, the first arrangement, which directly reveals the circular shape of the field about the straight wire, illustrates Step I of the "indirect method"; the second arrangement Maxwell interprets by means of the principle which epitomized Step II, the force on a circuit placed in a given magnetic field (page 138, above). We shall return to Maxwell's analysis of these rotations in a moment, but it will be important first to examine briefly the discussion which ensued upon Faraday's discovery.

One might wonder whether there could be a clearer demonstration of a *rotational* effect in nature—a more decisive phenomenon to support Biot against Ampère. Recall that Biot had himself observed no more than the *orientation* of a magnetic dipole in the vicinity of a current; he had interpreted this as a tendency of each pole of the needle to revolve about the central wire, and he had made the purely theoretical assertion that a constant *couple* acts on each pole, accelerating it in rotation. Now Faraday had shown how to free one pole from the other, and he had thereby exhibited precisely the effect predicted by the Biot theory: a continuously accelerating rotation about the conductor.

On the strength of such evidence, I think, it would have been quite possible for science to have abandoned the Ampère theory once and for all, concentrating future attention on the cultivation of rotational principles, whether in the form of Biot's *couple primitif* or Faraday's circular lines of force. Only a mind of the remarkable tenacity of Ampère's could have resisted this conclusion. For Ampère it seems to have been only a still more stimulating challenge. He admits freely the difficulty Faraday's experiment poses for him:

> [J]'ai expliqué en détail cette expérience, parce qu'elle semble, plus qu'aucune autre, appuyer l'hypothèse du couple primitif, quand on ne l'analyse pas comme je viens de la faire.[83]

We have compared Ampère's problem to Newton's, but it is important to recognize that the present difficulty is considerably greater than Newton's: Ampère must generate, not merely a rotational motion, but a force always tangential to a circular path, through the action of linear forces, each always directed to the center of the circle. Newton introduces such tangential forces either through the action of a third body or when the center of force is not at the center of the orbit; moreover those forces never act continuously, but always tend to balance out over the orbit. Ampère is thus entering upon new ground.

Experimentally Ampère demonstrates, as Faraday had not succeeded in doing, the possibility of setting a magnet spinning about its own axis.[84] Theoretically, he proves (1) that the observed rotations can all be deduced from his own theory; (2) that continuously accelerating rotations cannot arise from the interaction of rigid closed circuits (and hence, not from the interactions of magnets); and finally (3) that Biot's theory leads to the same predictions as his own in all of the experimental cases. What might have seemed a defeat to his theory was converted to a dramatic triumph—a triumph of purely mathematical analysis.

I shall not delay here to examine Ampère's analysis,[85] but it is important to notice the challenge posed by his second theorem: that continuous rotations cannot arise from the interaction of two rigid circuits. (This is an immediate consequence of the equivalence of circuits to magnetic shells; since shells have potentials, their interactions must be conservative: the work done by one upon the other must sum to zero over any closed cycle.) How, then, can Ampère account for the rotation Faraday observed? In the case of the rotating wire, a magnet would seem to give rise to the continued rotation of a circuit. The answer is that it is not the *whole* circuit which rotates, but only a part of it corresponding to the rotating wand *c* of Figure 15.[86] To prove the point, Ampère constructed a modification of the apparatus, in which the central magnet *d* was free to turn on its axis.[87] It nonetheless remained at rest, demonstrating that it was subject to no net torque from another closed circuit. Furthermore, that part of the circuit which does rotate can do so only because an equal and opposite moment is exerted on

83. Ampère, *Théorie mathématique*, p. 133.

84. Communicated to the Académie January 7, 1822 (Launay, *Correspondance*, 2, p. 806.

85. Ampère, *op. cit.*, pp. 131 ff.

86. *Ibid.*, pp. 131–32.

87. *Ibid.*, Figure 13, Plate 1.

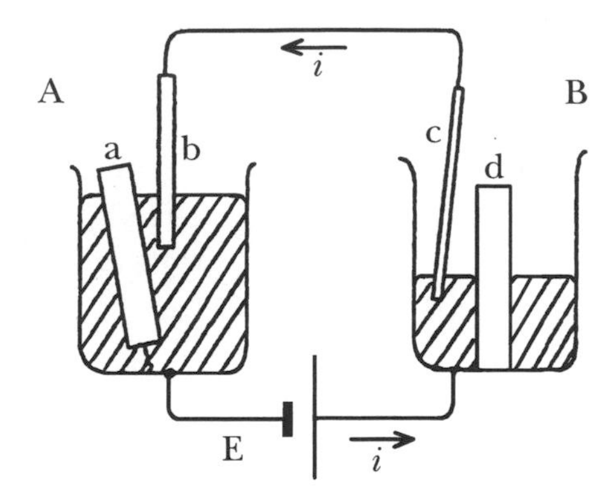

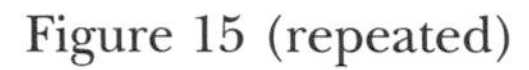
Figure 15 (repeated)

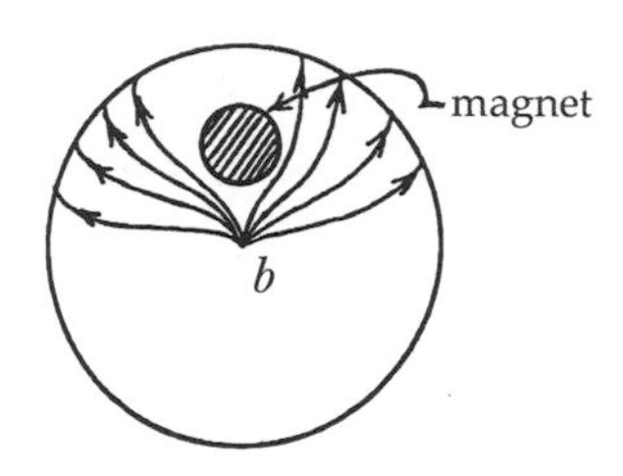

Figure 16. Ampère's analysis of the rotation of a magnet about a conductor.

the remainder of the circuit; Davy had, in a similar piece of apparatus, demonstrated the counter-rotation of the mercury.[88]

Every motor involving steady currents (as distinguished from alternating-current motors) must in some way take Ampère's theorem into account, and provide for relative motion of one section of the current with respect to another. Maxwell was deeply impressed by Faraday's ingenuity in discovering, through the device of Figure 15, a practical resolution of this difficulty. Referring to Faraday's apparatus of the form of the right-hand arrangement in Figure 15, Maxwell wrote:

> The thing cannot be done unless we adopt in some form Faraday's ingenious solution, by causing the current, in some part of its course, to divide ... so that the middle of the magnet can pass across without stopping it, just as Cyrus caused his army to pass dryshod over the Gyndes by diverting the river into a channel cut for it in his rear. (*SP* ii/788)

Similarly, in analyzing the other form of the experiment, in which the magnet is carried around the conductor, Ampère supposed that it is *not* the vertical conductor *b* (Figure 15) that is acting on the magnet, but rather the currents fanning out radially through the mercury from *b* to the conducting cup (seen from above in Figure 16). Thus it is not the question of the interaction of one circuit with a magnet, but rather, of a *sequence* of circuits distributed through the mercury, acting upon the magnet successively as it move through the mercury in its revolution.[89]

Let us turn now to Maxwell's analysis of the two forms of Faraday's continuous rotation apparatus of Figure 15. Happily, they correspond

88. Humphry Davy, "On a New Phenomenon of Electromagnetism," *Philosophical Transactions, 113* (1823), pp. 153 ff.

89. Ampère, *op. cit.*, pp. 125–26.

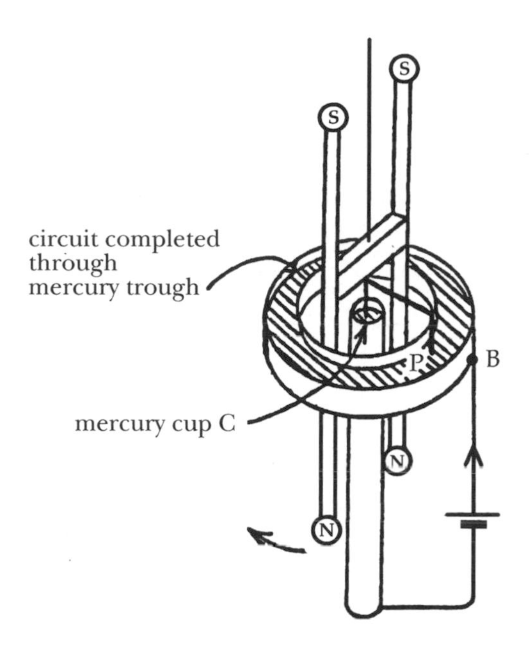

Figure 17. Modified form of apparatus for continuous rotation of a magnetic pole about a conductor.

respectively to Step I and Step II of the "indirect method" of Maxwell's Chapter I, and he discusses them as illustrations of the mathematical expression he has given to this method. The first, the motion of the pole about the magnet, is therefore discussed—as Step I was expressed—in terms of the magnetic field about a circuit, as characterized by its potential. The actual apparatus Maxwell has in mind, diagrammed in his own Figure 23 (*Tr* ii/150), is a slight modification of Faraday's original form; it is drawn in Figure 17.[90]

First, Maxwell points out the nature of the problem: if we had a long, flexible magnetic solenoid (or magnetized wire) placed parallel to a long straight conductor, the magnet would wind itself tightly about the conductor. If the current flowed vertically downward (as in the Oersted demonstration, Figure 3, page 119 above), and the solenoid were oriented with its south pole on top, the south pole would rotate counterclockwise and the north pole clockwise, the result being a winding with a single sense throughout.[91] This would not constitute continuous rotation; to achieve that, one pole and not the other must pass, he says, "from one side of the current to the other" during the revolution (*Tr* ii/144). If we are thinking

90. The trough was used by Ampère: Gustav Wiedemann, *Die Lehre vom Galvanismus* (2 vols.; Braunschweig: 1874), *2*, Part 1, p. 13: cf. his Figure 103, p. 161, with Figure 17 above.

91. Maxwell points out the analogy to the twisting of a vine in the case of such a screw motion; the present case would be an example of "the system of the vine," that of a right-handed thread, as opposed to the "system of the hop," that of a left-handed thread (*Tr* i/25n).

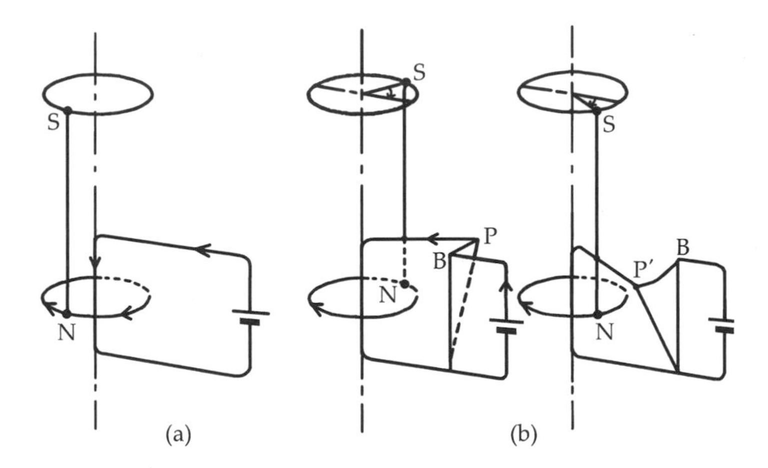

Figure 18. Analysis of Figure 17 in terms of an equivalent magnetic shell.

of the single vertical conductor as the "current," this remark does not seem to apply. Maxwell is asking us to make again the shift of viewpoint he required in the Oersted demonstration, from the thought of the current as flowing in an isolated straight conductor, to that of flow along the edge of a closed loop—the loop in turn being equivalent to a shell, which does indeed have "sides." Figure 18 indicates the way in which we are to see this, first (Figure 18a) diagrammatically, and then (Figure 18b) in terms of Maxwell's division of the current, like Cyrus's division of the river, into two branches. If B is the point of fixed connection to the mercury trough, and P is the gliding connection, the clockwise sense of rotation will carry P to P′, in effect reshaping the vertical shell so that the north pole is first on one side of it (the south side, to which it is attracted), and then on the other (the north, by which it is repelled), to continue another cycle of the rotation. It is as if an isolated north pole, let us say of strength m, were to pass through the magnetic shell equivalent to the current in the Oersted demonstration: each passage, corresponding to each cycle of rotation, would impart $4\pi im$ units of energy. The south pole of the rotator does not, of course, pass through the shell: if it did, it would *lose* the amount of energy the north pole gained, and there would be no net increase.

In its context of Chapter I, Maxwell is asking us to see the motion of the pole as taking place through the field of potential energy—Faraday's "system of power"—about a closed circuit; the experiment reveals a striking and universal feature of this field. We see at once that each cycle about a wire will mean one penetration of the equivalent shell, or the withdrawal of energy of amount $4\pi im$ from its magnetic field. It is, then, a dramatic illustration of the idea expressed by what we have identified as Maxwell's "First Principle," that which expressed Step I of the indirect method, or the nature of the magnetic field about a circuit (page 137, above).

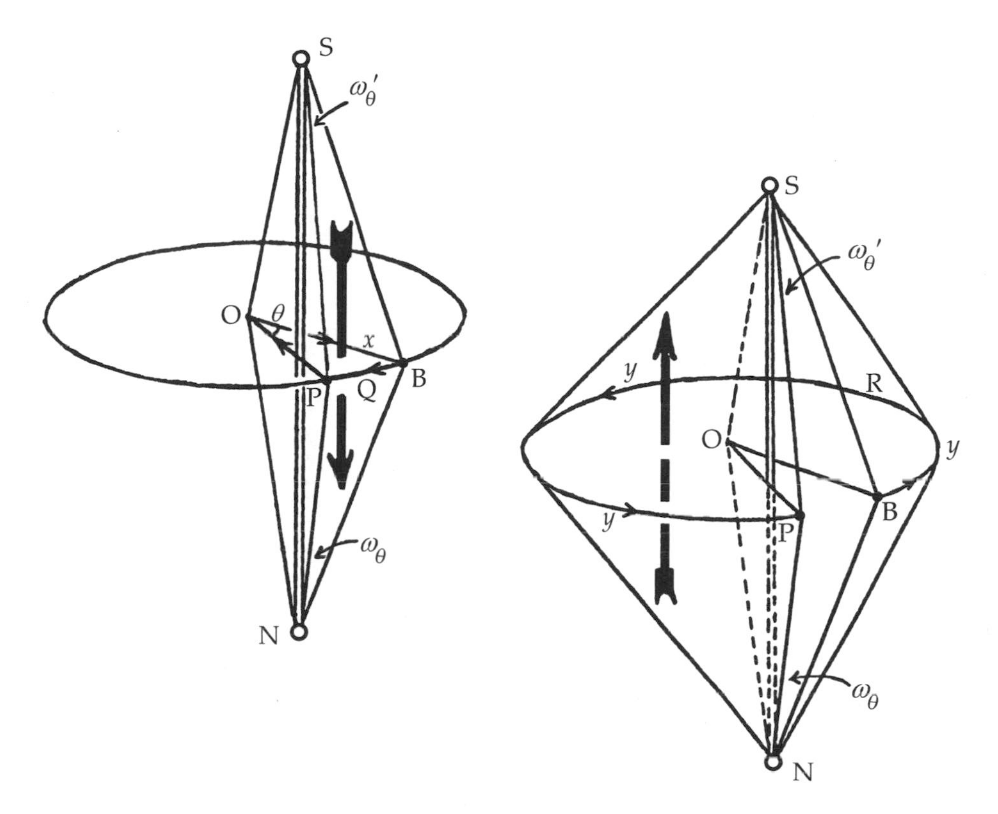

Figure 19. Potentials measured by solid angles subtended at the poles in the rotator of Figure 18.

Actually, it is not the case that the rotor of Figure 18 gains energy of amount $4\pi im$ with each revolution, since there is in fact another circuit to take into account. This is the circuit, actually best understood as a pair of circuits, in the plane of the trough itself. These too contribute to the field in which the poles move. As we shall see, although the north pole in effect falls to constantly decreasing energies in the field of the shell, *both* poles constantly climb to higher energies in the field of the currents in the trough, so that this energy must be deducted from the amount gained in a cycle. Maxwell calculates this by the use of Gauss's principle, the energy at any point being known by the solid angle the circuit subtends. In Figure 19 the circuits of Figure 18 have been projected into the horizontal plane, and the solid angles ω and ω' subtended at the north and south poles drawn—angles which we must see directly as measures of energies. The current entering through the wire at B divides at the trough; if P again denotes the point of gliding contact, let x represent the current flowing to it by the path BQP through the mercury, and y that by the path BRP. As the north pole N, and with it the point P, move clockwise, the path PQB and the angle θ will increase; the current x will diminish and the current y will increase, as the "river" shifts from one side of N to the other.

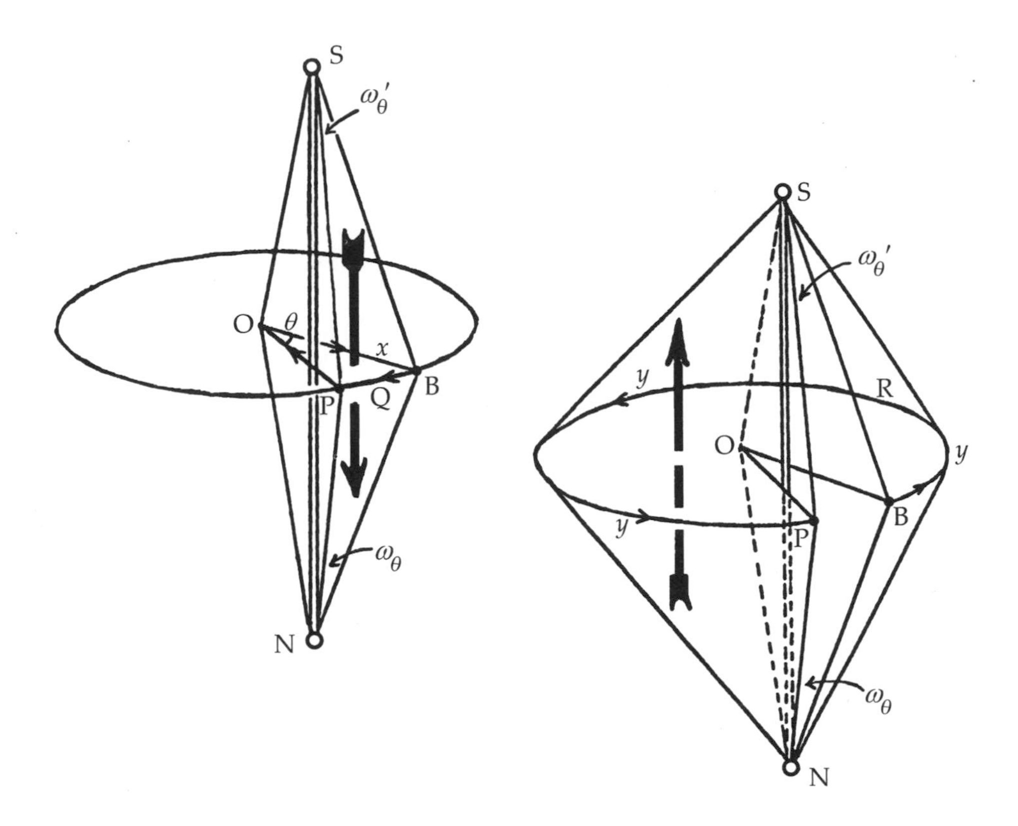

Figure 19 (repeated)

Niven, commenting on this paragraph of the *Treatise*, details Maxwell's argument in the following way: write the energies of the north and south poles at any moment for which the angle BOP is θ, such as that illustrated in Figure 19, calling ω_θ and ω'_θ the solid angles subtended at that moment by the sector BQPOB respectively at the north and south poles. By the sense of the current in that sector, we see that both energies are positive (both poles are present against a repelling force):

$$U_x = mx\omega_\theta + mx\omega'.$$

Now write the energies due to the other sector, BRPOB; these will be negative:

$$U_y = -\left[my(\omega - \omega_\theta) + my(\omega' - \omega'_\theta)\right].$$

Then the total energy of both poles, at the instant in question, will be

$$U_i = U_x + U_y$$

$$U_i = -my(\omega + \omega') + m(x + y)(\omega_\theta + \omega'_\theta).$$

This energy, U_i, is the initial energy at the beginning of a cycle of rotation. During a cycle, ω_θ and ω'_θ will increase by the amounts ω and ω', so that the final energy will be

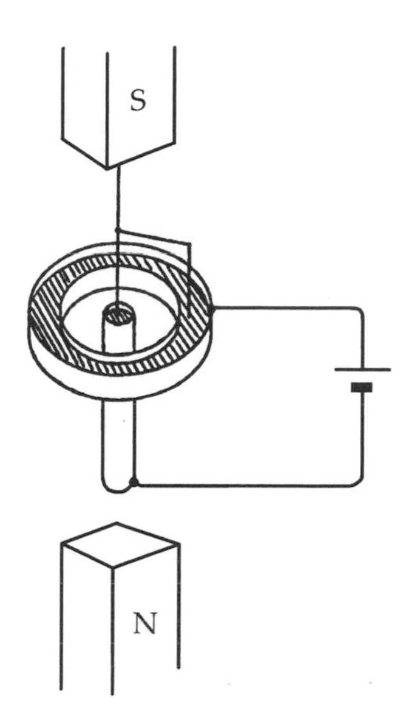

Figure 20. Alternate form of the rotator (compare *Treatise*, Vol. II, Figure 23).

$$U_f = -my(\omega + \omega') + m(x + y)(\omega_\theta + \omega + \omega'_\theta + \omega'),$$

and the increase in energy will be

$$\Delta U = -im(\omega + \omega').$$

This is indeed an *increase* in energy, since work is done against the field in the process. Thus the energy supplied to the pole during a cycle, which would otherwise have been $4\pi im$, is less by the above amount, or

$$W = im[4\pi - (\omega + \omega')].$$

The two limiting cases are interesting: if the dipole is very short, ω and ω'each approach 2π, and the rotator will not run. On the other hand, if the dipole is very long, with both poles very far from the trough, we approach as nearly as we please the limiting case of the isolated north pole traveling about a long straight conductor, both solid angles approaching zero, and the energy expression yielding $4\pi im$, as expressed in the First Principle.

Virtually the same apparatus serves for the right-hand form of Faraday's rotator (Figure 15, page 168 above), the revolving dipole being replaced by a fixed vertical magnetic field—either that of a permanent magnet at the center, as in Faraday's device, or that of the earth, which is what Maxwell assumes in his discussion. This case is illustrated in Figure 20, based on Maxwell's Figure 23; it is unfortunate that Maxwell uses the same diagram to represent *both* forms of the Faraday rotator, thereby making the essential distinction between them obscure to the reader.

The analysis is now completely different: we are concerned, not with the force on a magnetic pole, but with the force on a current in a magnetic

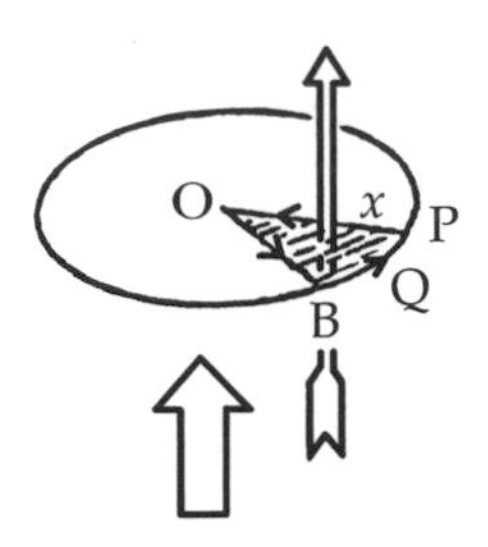

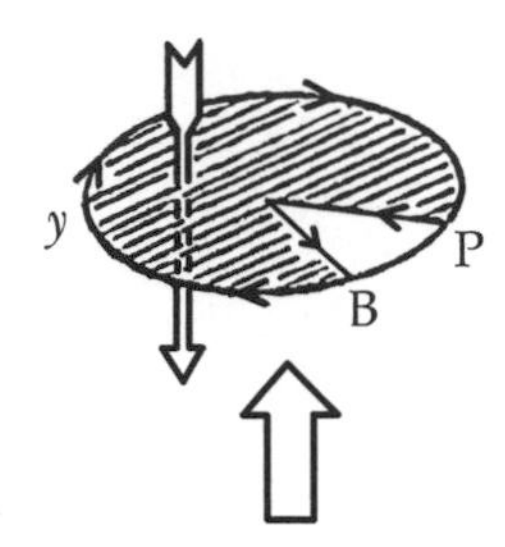

Figure 21. Flux through the working circuit of the apparatus of Figure 20.

field. Again, however, Maxwell immediately requires that we think of the current in terms of a closed circuit. Since the vertical flux will cut only the horizontal projection of the circuit, this time we need consider only that equivalent circuit in the plane of the trough—again, divided into the two currents x and y. The analysis, which is extremely simple, is an immediate application of the Second Principle, which was expressed in terms of the magnetic induction through the circuit:

> If by any given motion of the circuit, or of part of it, this surface-integral can be *increased*, there will be a mechanical force tending to move the conductor or the portion of the conductor in the given manner. (*Tr* ii/156)

In Figure 21, we see that if the magnetic field is assumed to run vertically upward, the field of the sector carrying current x will have the same sense as that of this fixed field. If the flux through this sector is N_x, the energy will be (compare *Tr* ii/147)

$$U_x = -xN_x .$$

The negative sign reflects Faraday's principle of coalescence: the fields concur, and the system will run (field energy decreasing) to increase the coalescence continually. The field of the other sector is reversed with respect to the fixed field, and the system will run to diminish this anti-coalescent field. The energy available per cycle is immediately given by the product of the whole current and the flux through the trough (*Tr* ii/150).

In saying this it is important not to lose sight of the principle of the reversal of the circuit in the course of the cycle: if the operation were simply a question of the increase in area of a single sector with the corresponding increase of flux, the device would come to a halt at the end of a single cycle. As the point P moves, and the sector BQPOB enlarges, the resistance to the passage of current x increases, and x diminishes, while y correspondingly increases. In the second half of the cycle, current y predominates, and the net polarity of the circuit has reversed: once again, the magnet has in effect "transferred from one side of the current to the other" (here both poles have transferred, passing through in opposite directions). We might well regard Faraday's trough as a mercury-commutator, and recognize in it the principle of every subsequent form of electric motor: in the conventional direct-current motor, a split-ring commutator accomplishes at a single moment of switching the reversal which is accomplished gradually over the cycle in the mercury; while the alternating-current motor employs a gradual reversal of polarity of the field quite analogous to the commutation achieved by the split path in Faraday's mercury trough.

What Maxwell has done through his use of the two forms of the Faraday rotator as illustrations of Chapter I, is to show how his own mathematical interpretation of Faraday's concepts could be brought to bear upon this much disputed apparatus, and how—a half-century after Ampère's challenge had been issued—Faraday's method could be used to give not only an intuitive understanding of the rotational motion, but a strict quantitative account. There has been little talk of force on poles, or on elements of conductors—everything has been described either in terms of potentials of whole circuits, or of fluxes through whole circuits, and the corresponding energies. One can *derive* from this, as Maxwell has shown us elsewhere in his Chapter I, conclusions concerning forces exerted at particular points, but he has scrupulously kept whole circuits and their energies before us in his own analysis. At the same time, he has put his finger on the theoretical principle of the ingenious device by which Faraday accomplished the first continuous electromagnetic rotation, and has shown that this, too, can be clearly expressed in terms of whole circuits, their lines of force and fields of potential energy—concepts which Maxwell sees as giving precise form to Faraday's thought on electromagnetism.

COMMENTARY ON CHAPTER III OF PART IV OF THE *TREATISE*

Maxwell's project is now fully launched: he has shown us with great care the division of the ways—the way of Faraday and the way of Ampère—and it is already quite clear which road we shall take. Until the last chapter, which is a survey of action-at-a-distance theories beyond the horizon of the work itself, the remainder of the *Treatise* will be devoted in one way or another to the exploration of the realm of field theory. The two chapters which we have just discussed concerned a topic which was common to Faraday and to Ampère, that of the mechanical forces between conductors. We now go on to a topic which was more particularly Faraday's own: electromagnetic induction, which he discovered, and with which his *Experimental Researches* opened. The connected sequence of reasoning of the *Experimental Researches* evolved from this topic, and the crucial notion of the "electrotonic state" was first articulated in connection with it. Yet it must be added, as has been mentioned before, that although electromagnetic induction is particularly "Faraday's topic" in this sense, and Ampère himself did not work it into his theory, others—above all Franz Ernst

Neumann—had done so, with the result that both Faraday's concepts and the methods of Ampère's theory ultimately extend over the whole range of electromagnetic phenomena. In following Faraday's method from this point on, we are thus deliberately turning our backs upon an alternative that is completely satisfactory according to what would ordinarily be regarded as scientific criteria. Maxwell does not withhold any of the credit that Ampère's theory deserves:

> The whole, theory and experiment, seems as if it had leaped, full grown and full armed, from the brain of the 'Newton of electricity.' It is perfect in form, and unassailable in accuracy, and it is summed up in a formula from which all the phenomena may be deduced, and which must always remain the cardinal formula of electro-dynamics. (*Tr* ii/175)

The purpose of the *Treatise*, however, is to pursue Faraday's ideas (with deference, it has been acknowledged, to the necessities for coverage of other topics in a Cambridge text); and we now return to that chase, making very few further compromises with action-at-a-distance theories.

Maxwell's Chapter III is prefaced by an essay, three pages in length, in which he takes a long look at the two models he has set before us in Chapters I and II. It is extremely interesting to observe the terms in which Maxwell makes this comparison of Faraday and Ampère, and these pages are among the finest anywhere in Maxwell's writings. Many of the themes with which Maxwell is concerned separately in other places converge in such a way that it is not easy to isolate them, nor appropriate to do so.

We see how deeply Maxwell is concerned with this difference in theoretical structures. We have at once a contrast of two modes in mathematics, which can be illustrated out of Euclid, and two *styles*, which go to the very characters of the men who use them. Again, it is not simply a question of two kinds of science, but of the "scientific" value—that is, the value for science-*making*—of two approaches to science. The question is that of power to generate new thoughts. One excellence of Faraday's style is his willingness and ability to draw the reader candidly into the vicissitudes of a *nascent* science (*Tr* ii/175). The record of the actual birth of Faraday's concepts gives rise in turn to nascent ideas in the reader's own mind (*Tr* ii/176).

Without attempting to report in detail Maxwell's thoughts in these pages, which are too compressed and interwoven to be epitomized, and without collecting references to parallel passages in other writings of Maxwell's, where resonances of these thoughts abound, let us turn directly to a single aspect of the contrast of Ampère and Faraday which is particularly revealing of Maxwell's view of the enterprise of the *Treatise*.

A reader of Maxwell's first two chapters might not have anticipated one of the points of comparison which he most emphasizes in this preface to the third. We are likely to think of Faraday's field view as characterized primarily by its emphasis on *local* action, for which Faraday's term is "contiguity," as opposed to action-at-a-distance. It was this local "tendency" to move, as contrasted to an attraction to a remote center of force, which proved to be the crux of Faraday's point of view in the dialogue with Tyndall (page 84, above), and it was the same "tendency" which Thomson showed how to formulate in a differential equation (page 99). We might expect, then, that Maxwell in contrasting Ampère's view and Faraday's would emphasize that the former introduces remote, unmediated actions, while the latter ascribes every motion to physical agents immediately present at the site of the body moved. This is true, but it is not the aspect of the contrast which Maxwell draws attention to here; presumably, it is not for Maxwell the really essential point in the field concept. Rather, he makes the distinction between a theory which begins with parts and builds a whole by summation (Ampère's), and one which begins with wholes and derives particulars from them (Faraday's). Maxwell writes:

> We are accustomed to consider the universe as made up of parts, and mathematicians usually begin by considering a single particle, and then conceiving its relation to another particle, and so on. This has generally been supposed the most natural method. To conceive of a particle, however, requires a process of abstraction, since all our perceptions are related to extended bodies, so that the idea of the *all* that is in our consciousness at a given instant is perhaps as primitive an idea as that of any individual thing. Hence there may be a mathematical method in which we proceed from the whole to the parts instead of from the parts to the whole. (*Tr* ii/176 77)

From this point Maxwell's thought will move quickly to sketch the new mathematical physics to which he sees Faraday's methods leading. But we must interrupt to observe that the question which Maxwell is raising—that of the order as to priority of whole and part—is concerned with more than mathematical procedure. It is a question as large as the universe and is, moreover, interwoven with a problem of the nature of human perception. I shall argue in Chapter 5 that Maxwell founds his physics on a metaphysics that is of constant importance to him, and that has a continuing influence on the shape the physical theory takes. Without initiating that investigation here, let me simply remark that the principle of the priority of the *whole* is articulated in the metaphysical system of William Hamilton (1788–1856), Maxwell's teacher at Edinburgh, as a corrective to the tradition of Locke, Dugald Stewart, Thomas Brown, and ultimately Mill; and I believe Maxwell is bringing this philosophical principle to bear in finding the essence of the difference between Ampère's method and Faraday's. Hamilton wrote:

> I formerly explained why I view the doctrine held by Mr. Stewart and others in regard to perception in general and vision in particular, as

> erroneous; inasmuch as they conceive that our sensible cognitions are formed by the addition of an almost infinite number of separate and consecutive acts of attentive perception, each being cognizant of a certain *minimum sensible*. On the contrary, I showed that, instead of commencing with minima, perception commences with masses. ... Having first acquired a comprehensive knowledge of it as a whole, we can descend to its several parts. ... We decompose, and then we recompose.[92]

Ampère, I believe Maxwell feels, shut out this wholeness of the act of perception by a deliberate effort to construct a system on the concept of least parts. Faraday, proceeding (as Maxwell points out) *naturally*, accepts what he observes in its immediate unity. He sees magnets as wholes, he sees the patterns of lines of force in their unity as systems, and he names that unity, the "sphondyloid." Our senses are aware of wholes first, according to Maxwell; Faraday, willing to be instructed by his senses, proceeds by what Maxwell earlier in the *Treatise* called "eye-knowledge."[93] A nascent science, in this view, will by necessity of our mode of perception be a science of such wholes; Ampère is able to produce a science of another sort only by working from the beginning within a rigid tradition, and by erasing the record of the actual investigation to impose the desired atomic theory upon his results. Field physics is ultimately a question of style, but style in turn relates to truth. Lockean atomism and Newtonian physics are brilliant artifices, worthy of praise and admiration, but forced upon nature. Newton himself, Faraday and Maxwell emphasize, was discontent with the result in the form in which he left it in the *Principia*.[94] Faraday is emancipated from this; he can set his own style, and shape his own language. The result is his ability "to coordinate his ideas with his facts," a "cultivation of ideas under the direct influence of experiment," and their expression "in natural, untechnical language" (*Tr* ii/175–76). Style is crucial to truth, because only a "natural, untechnical" style will articulate a science of nature as we perceive it. Elsewhere, too, Maxwell puts the distinction between Ampère and Faraday in terms of the rhetoric they used. Ampère speaks the language of traditional mathematics, the "tongue of the learned"; his successors, instead of learning from nature, are trapped in "attempts to improve a mathematical formula" (*SP* ii/318). Faraday is forced to leave this completely aside, and "to explain the phenomena to himself by means of a symbolism which he could understand," the "new symbolism" of the lines of force:

> This new symbolism consisted of those lines of force extending themselves in every direction from electrified and magnetic bodies, which

92. William Hamilton, *Lectures on Metaphysics and Logic*: *Volume I, Metaphysics*, ed. H. Mansel and J. Veitch (Boston: 1875), p. 498.

93. (*Tr* i/178), cf. Hamilton, *op. cit.*, pp. 168–170.

94. (*XR* iii/532, 571), (*SP* ii/316).

> Faraday in his mind's eye saw as distinctly as the solid bodies from which they emanated. (*SP* ii/318)

In this context, we see how much the metaphor of "the mind's eye" carries with it: as the eye immediately grasps the *whole* in perception, so the mind's eye intuits the concept of the field as an extended entity, the *system* of the lines of force "extending themselves in every direction." Field physics must be the product of such a candid reporting of nature, if Maxwell is right about the character of perception.

In its essence, then, the field view for Maxwell is not primarily a question of contact-actions, but a principle of unity in nature: the field is to be thought of essentially as a *whole*, as an entity. Describing the field of force, Maxwell is really characterizing the entity Faraday had called the "sphondyloid":

> He conceives all space as a field of force, the lines of force being in general curved, and those due to any body extending from it on all sides, their directions being modified by the presence of other bodies. (*Tr* ii/177)

Maxwell weighs carefully the extreme formulation of this view, which had tempted Faraday:

> He even speaks ... of the lines of force belonging to a body as in some sense part of itself, so that in its action on distant bodies it cannot be said to act where it is not. (*loc. cit.*)

He gives two references to Faraday, both to speculative papers outside the regular series of the *Experimental Researches*; one is the letter to Taylor to which we referred in connection with the discussion of Mossotti (page 72, above),[95] and the other is a passage in a later letter to Phillips, which refers back to the first. This was the pole of thought in which Mossotti's hypothesis becomes linked with Faraday's interpretation of Boscovich, and it becomes conceivable to him that "each atom extends throughout the solar system" (*XR* ii/293). Maxwell discounts this extreme field-view, and proposes a middle position as the truest account of Faraday's own conviction.

> This, however, is not a dominant idea with Faraday. I think he would rather have said that the field of space is full of lines of force, whose arrangement depends on that of the bodies in the field, and that the mechanical and electrical action on each body is determined by the lines which abut on it. (*loc. cit.*)

Here the system of lines of force is one entity, the body another, and the phenomena of electromagnetism are the product of the interaction of

95. "A Speculation Touching Electric Conduction and the Nature of Matter" (1844) (*XR* ii/290–91).

these two substances. Note, by the way, that, like Faraday, Maxwell uses the term "field" to denote the *container*, rather than the entity. The theory is thus not so much a "field" theory as a theory of the entities in the field—a theory of the system of the physical lines of force or, as we suggested earlier, "sphondyloid theory" (page 25, above).

We must understand, then, that Maxwell is seriously embarking on a theory in which the primary emphasis is on entities filling all space, interacting with the bodies directly observable in our experiments. The insistence that they be considered in their integrity as whole suggests an almost organic image, and it is striking that when Maxwell does propose a metaphor for the system of lines of force, he chooses it from the realm of life:

> This is quite a new conception of action at a distance, reducing it to a phenomenon of the same kind as that action at a distance which is exerted by means of the tension of ropes and the pressure of rods. When the muscles of our bodies are excited ... the fibres tend to shorten themselves and at the same time to expand laterally. A state of stress is produced in the muscle, and the limb moves. (*SP* ii/320)

The system of lines of force is a whole which, if it is not more than the sum of its parts, is at least best understood as a single substance. This is an understanding of field theory that bears strongly upon a view of the universe: at least some of the things which go to make it up may be infinitely extended, mutually interpenetrating substances. It must be noted in passing that this aspect of Maxwell's theory was of importance to Whitehead, where it indeed becomes part of a fundamentally revised view of the universe, an "organic theory of nature."[96] We recognize the consequence with which Faraday toyed, and from which Maxwell drew back, in Whitehead's characterization of the electron: "The electron is its whole field of force."[97]

Corresponding to Maxwell's emphasis on the electromagnetic system as a single entity is his characterization of the mathematics of the field. There are two aspects of field mathematics: the partial differential equation which characterizes the local configuration, and the integral of this equation over all space, under boundary conditions. Maxwell, thinking of the unity of the system, stresses the fact that the differential equation speaks, not about one point, but about *every* point at once, and he thinks of field mathematics primarily in terms of the integral over space. The distinction is elusive,

96. Alfred North Whitehead, *Science in the Modern World* (New York: Mentor Books, 1948), p. 106. Victor Lowe, *Understanding Whitehead* (Baltimore: The Johns Hopkins Press, 1962), p. 166. In relation to the present study, Lowe's note 13 on the same page, which has recently come to my attention, is particularly intriguing.

97. Whitehead, *The Concept of Nature* (Ann Arbor: University of Michigan Press, 1957), p. 159.

because both Ampère's method and Faraday's are expressed mathematically by integrals; but in the former, the integral is over the sources and is no more than a summation of their elementary effects, while in the latter the integral is itself the elementary expression. It is helpful to keep in mind the example of Euclid, with which Maxwell continues his discussion of this topic:

> For example, Euclid, in his first book, conceives a line as traced out by a point, a surface as swept out by a line, and a solid as generated by a surface. But he also defines a surface as the boundary of a solid, a line as the edge of a surface, and a point as the extremity of a line.
>
> In like manner we may conceive the potential of a material system as a function found by a certain process of integration with respect to the masses of the bodies in the field, or we may suppose these masses themselves to have no other mathematical meaning than the volume-integrals of $\frac{1}{4\pi}\nabla^2\Psi$, where Ψ is the potential. (*Tr* ii/177)

Whether the science be geometry or mechanics, the question is, What are the "elements?" Euclid, Maxwell says, recognizes that he has an alternative; the elements might be points, or they might be solids—one can work synthetically from points, or analytically from solids. The same alternatives confront mechanics, where the question takes the form: what is the "mathematical meaning" of the term *mass*? (When the question is put this way, we are reminded of the root meaning of the term "mass"—a *measure* of matter, not a small hard piece of it.) The first approach named in the passage just quoted conceives "the potential of a material system as a function found by a certain process of integration with respect to the masses of the bodies in the field." In the second approach, the mass is itself the volume-integral of the Laplacian of the gravitational potential over all space. The primitive fact is the configuration of the space-filling function $\nabla^2 V$; the elementary measure is the definite integral over the whole system of the Laplacian of it. If one sticks to the notion of the integral as a summation of elements, this does not seem to make much sense; but Maxwell is asking us to consider an alternative view of the integral, corresponding to a reformed concept of perception. The integral is then a synoptic view of the extended function, a single measure of it; we recognize at once the parallel to Hamilton's insistence that the eye sees figure primarily, or the ear hears the orchestra rather than the instruments—and to Faraday's intuitive grasp of the reading of his ballistic galvanometer as recording in a single impulse of the needle a quantity, the "power," which belonged invariably to the whole sphondyloid extended in its natural figure through all space.

Maxwell's second candidate for the "mathematical meaning" of mass accords broadly with the extreme position of Boscovich and with Faraday's speculations, in which matter *is* the field; its measure, mass, is the measure of a property of this field.

Maxwell had characterized the mathematical alternatives earlier in the *Treatise*, in connection with the discussion of the electric field. The distinction there was between the *direct* and the *inverse* methods, terms which are certainly echoed in the distinction between the "direct" and the "indirect" methods in electromagnetism.[98] The direct method, he says, assumes action-at-a-distance, and determines the potential by integrating over given sources. He then explains the inverse method in this way:

> If we call this the direct method of investigation, the inverse method will consist in assuming that the potential is a function characterised by properties the same as those which we have already established, and investigating the form of the function.
>
> In the direct method the potential is calculated from the distribution of electricity by a process of integration, and is found to satisfy certain partial differential equations. In the inverse method the partial differential equations are supposed given, and we have to find the potential and the distribution of electricity. (*Tr* i/123)

These might seem to represent no more than the exigencies of two different kinds of practical problem, calling for different techniques within a single intellectual framework. For Maxwell, however, the inverse method brings with it a new idea of the world. He writes two equations for the potential; Laplace's differential equation, and the volume integral over the sources, and says of them, speaking first of Laplace's equation:

> The mathematical ideas expressed by this equation are of a different kind from those expressed by the definite integral. ... In the differential equation we express that the sum of the second derivatives of V in the neighbourhood of any point is related to the density at that point in a certain manner, and no relation is expressed between the value of V at that point and the value of ρ at any point at a finite distance from it.
>
> In the definite integral, on the other hand, the distance ... is denoted by r, and is distinctly recognised in the expression to be integrated.
>
> The integral, therefore, is the appropriate mathematical expression for a theory of action between particles at a distance, whereas the differential equation is the appropriate expression for a theory of action exerted between contiguous parts of a medium. (*Tr* i/123–24)

We notice a difference between this statement, from Part i of the *Treatise*, and the other statements we have discussed in Part iv. In electrostatics the emphasis is on contiguity, as we would, I think, be inclined to expect of an account of field theory. In electromagnetism, as we have seen, the emphasis is on the integral over all space and on the system of lines as a space-filling entity. This, I believe, corresponds directly to the difference

98. Thomson had divided methods as "synthetical" and "analytical," the latter having to do with "inverse" problems—i.e., with boundary-value problems. (*EM*: 50).

in Faraday's own interests between Series XI and the serieses in Volume Three of the *Experimental Researches*: briefly, there is no counterpart in electrostatics to the *moving wire* of the electromagnetic researches—there is no opportunity for observation of the mechanical relations of charged systems, nor an equally convenient equivalent of iron filings to present patterns synoptically and thereby suggest electric sphondyloids. Therefore field theory changes, between Volume One and Volume Three of the *Experimental Researches*, and Maxwell's account of even the mathematical expression of field theory changes emphasis in the same way, between Part i and Part iv of the *Treatise*. In Part iv he says:

> The mathematical process employed in the first method is integration along lines, over surfaces, and throughout finite spaces, those employed in the second method are partial differential equations *and interactions throughout all space.* (*Tr* ii/177, emphasis added)

Sphondyloid theory may seem to have tempted Maxwell into a retrograde step in the characterization of field physics. Faraday's vision of space-filling entities corresponding to individual magnets may work well for the equilibrium relations of permanent magnets, out of which it is sprung; but it is harder to apply to the interactions of non-rigid currents, and it hardly seems applicable at all to the propagation of waves. One wonders whether Maxwell had the electromagnetic theory of light in the forefront of his mind as he wrote the passage about "the *all*." Yet we see a hint that behind this passage there lies a larger conviction about the universe and our perceptions of it, to which field theory as Maxwell is here characterizing it is particularly appropriate. The "dynamical theory" of the field, toward which Maxwell is certainly building at this point, will be an effort to characterize as precisely as the evidence permits that single connected system which remains for him the essential object of the new physics which Faraday has introduced. It will mean a new view of the world, in which the whole is prior to the part.

We turn now to Maxwell's discussion of electromagnetic induction, which occupies the balance of Chapter III of Part iv of the *Treatise*. In §530, the reader is first commanded to read the first two series of Faraday's *Experimental Researches*, in which the discovery of electromagnetic induction was announced, and Faraday's ideas about it given their initial form. Then the principal phenomena of induction are summarized, whether due to variation of the primary current or to relative motion of the primary and secondary conductors. Here we are concerned, not with the mechanical force on the conductor, which has been the subject of Maxwell's Chapter I, but with the *electromotive* force—the "force," whatever it may be, which gives rise to the phenomena we call the flow of the electric current. Perhaps in order to emphasize a correspondence between this account and the discussion of Chapter I, Maxwell again directs attention to the straight wire as the object for consideration, so that although he begins by considering two complete circuits, a primary and a secondary, he quickly shifts attention to a pair of parallel straight segments of these circuits. Once again, however, these are not "elements" of the circuits in Ampère's sense, and he is not suggesting here, any more than in the case of the electromagnetic force, that *part* of one circuit in fact acts on *part* of another. This discussion invites comparison with that of the force between parallel conductors, page 132 above.

Maxwell closes a review of the phenomena of induction (which we need not repeat) by taking a step precisely like that with which he reformulated the Oersted experiment at the outset of Chapter I: he substitutes, as the source of the effect, a magnetic shell. But where he introduced the shell at that time as a theoretical consequence of the experiment of illustration—i.e., as the insight which the experiment was to induce—he here introduces the equivalence of the primary circuit to a magnetic shell as an additional experimental fact:

> If we substitute for the primary circuit a magnetic shell, whose edge coincides with the circuit, whose strength is numerically equal to that of the current in the circuit, and whose austral [north] face corresponds to the positive face of the circuit, then the phenomena produced by the relative motion of this shell and the secondary circuit are the same as those observed in the case of the primary circuit. (*Tr* ii/179)

The status of this assertion is like that of the "numerous experiments" to which he alluded earlier to establish the equivalence of the electromagnetic force of a small plane circuit and a magnetic shell (page 122, above): the assertion has plenty of indirect empirical support, though probably not in quite the form which Maxwell describes.

Assuming now the principles of the "indirect method" which he labored to explain in Chapter I, Maxwell goes directly to a summary of the phenomena in a law cast in terms of the field, in a form indeed parallel to that which summarized Step II of that chapter. He concluded there (page 131, above):

> Any displacement of the circuit will be aided or resisted according as it increases or diminishes the number of lines of induction which pass through the circuit in the positive direction. (*Tr* ii/147)

The analogous law for the electromotive force is this:

> When the number of lines of magnetic induction which pass through the secondary circuit in the positive direction is altered, an electromotive force acts round the circuit, which is measured by the rate of decrease of the magnetic induction through the circuit. (*Tr* ii/179)

This is the law often referred to in the literature as "Faraday's law of induction." The question, raised by Williams, whether Faraday in fact understood the relation as a "law" in the sense in which it is formulated above has been discussed earlier (page 13, above).

Maxwell has not really given the evidence here for his quantitative assertion ("...which is measured by..."). He prefers, it seems, to assert this result abruptly, much as he produced his successive interpretations of the Oersted experiment. Only after having illustrated the proposed principle with examples, and thereby having established the idea in question in the mind of the reader, does he wish to move on to the serious examination of its experimental credentials. This, I suspect, is part of Maxwell's idea of the role of *illustration*: one must first induce the idea as a whole, and then, having established the notion, move on to more precise argument and rigorous experiment. In general, then, the plan of this chapter, as of Chapter I, is to move from "illustration" in broad strokes—a bold sketch which may perhaps leave the reader staggered—to a later refined and quantitative analysis. In the present case, the corresponding quantitative phase, which begins at (*Tr* ii/182), is in terms of the precise measurements possible with the Felici induction balance.

If I am right about this movement as characteristic of Maxwell's style in the *Treatise*—and there is support for it, I believe, in the general plan as enunciated in Maxwell's Preface (*Tr* i/xii)—it is impossible not to notice that we have here a theory of learning which corresponds to the theory of perception discussed earlier. Maxwell teaches us by first confronting us with the whole theory, presented in its unity in the case of an illustrative instance—thereby proposing the largest ideas and general principles at the outset, without building to them by reasoning step-by-step from elementary principles. Later, he considers details, and supplies the missing links in the argument. This is the precise opposite of Ampère's method in the *Théorie mathématique*, in which the final, general result emerges only at the end of a long sequence of argument, pursued with impeccable care.

If this is the case, then the *Treatise* meets a certain literary test: for better or for worse, its style mirrors its content—form corresponds to function. In the terms in which Maxwell has now characterized field theory for us, as grasping the whole as primary, and moving from it to consideration of the part, the *Treatise* is written to conform to a *field theory of learning*. I do not think this is an extravagant statement, but rather an observation which follows quite naturally from the very deep sympathy which Maxwell evidently feels for Faraday's comprehensive insights, and for the view of learning and the world which was expressed in Hamilton's lectures. Evidence for the latter will be discussed in Chapter 5, below.

Maxwell gives two examples to illustrate the principle he has just put forth. The first is a moving railway train. The effect of such a choice is perhaps to remind the reader of the ubiquity of induction phenomena in the earth's magnetic field. Faraday had made just this point:

> [S]carcely any piece of metal can be moved in contact with others ... without an electric current existing within them. It is probable that amongst arrangements of steam-engines and metal machinery, some curious accidental magnetoelectric combinations may be found, producing effects which have never been observed, or, if noticed, have never as yet been understood. (*XR* i/52)

The second illustrative example is still more specifically inspired by his reading of the prescribed series of the *Experimental Researches*. Faraday had speculated there about the possibility of currents being induced under various circumstances as a result of the earth's motion. Maxwell's account here is, in effect, a gentle corrective to Faraday's. Faraday envisions that since currents have been shown to be induced by the motion of a conductor in a magnetic field, then when a conductor is carried with the rotation of the earth, a current ought similarly to be observed to pass through it. (Note that Faraday is assuming here, as he does regularly, that when a magnet is rotated on its axis, it passes through its own magnetic lines of force—compare page 26, above.)

Faraday does not necessarily expect a positive result if a loop of wire is placed upon the earth's surface and a current is sought in it, but he understands a negative result, *not* by saying that there is in fact no motion of the earth relative to its own field, but rather by supposing that, the induction being the same in opposite parts of the loop, no *net* current is observed to flow. Possibly, he thinks, there would be current flow if the induction were a function of the material of which the conductor is composed. Then if a rectangular loop were stretched out in which the two north-south segments were made of different materials, the experiment might yet succeed. He actually carries out experiments on this hypothesis, perversely understanding them as crucial tests of the question whether induction is a function of material. A fascinating example is the loop composed, one half of copper wire and the other half of the water of Kensington Palace lake. The scale is appropriately large, and the materials, copper and water, are as diverse as possible. Faraday concludes from the negative result that induction is *not* a function of the substance of the conductor (*XR* i/54).

The illustration which Maxwell now proposes simply translates this situation into terms correspondent with the assumption that the earth's field in fact turns with the earth, so that there was no relative motion between conductor and field in Faraday's Kensington lake experiment. Faraday, recognizing for reasons of his own that no current will be induced in any loop at rest on the earth's surface, had suggested:

> [I]f another bar in the same direction be connected with the first ... [and if it] be carried from east to west, which is equivalent to a

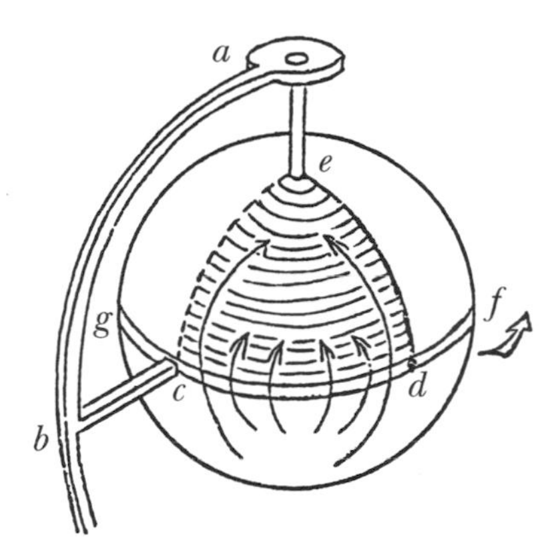

Figure 22. Maxwell's "quadrantal arch"
(compare *Treatise*, Vol. II, Figure 31).

> diminution of the motion communicated to it from the earth ... then the electric current communicated to it from south to north is rendered evident. ... [I]f collectors could be applied at the equator and at the poles of the globe, as has been done [in earlier experiments in this First Series] with the revolving copper plate ... and also with magnets ... then negative electricity would be collected at both poles. But without the conductors, or something equivalent to them, it is evident these currents could not exist, as they could not be discharged. (*XR* i/53)

Maxwell proposes this interpretation:

> Suppose a metal girdle laid round the earth at the equator, and a metal wire laid along the meridian of Greenwich from the equator to the north pole.
>
> Let a great quadrantal arch of metal be constructed, of which one extremity is pivoted on the north pole, while the other is carried round the equator, sliding on the great girdle of the earth, and following the sun in his daily course. There will then be an electromotive force along the moving quadrant, acting from the pole towards the equator. (*Tr* ii/180–81)

The arrangement which Maxwell describes as the "quadrantal arch" is illustrated in Figure 22, which corresponds to Figure 31 of the *Treatise*. The arch is *ab*, the meridian wire *ed*, and *gcdf* is the metal girdle around the equator. The points *c* and *e* suggest Faraday's "collectors" at the equator and the poles. According to Faraday's view, if *ab* and *ed* both turned with the earth, *both* would have tensions induced in them, but that induced in *ed* could not be discharged through *ab*, being met by an equal and opposite tension in it. (This would be the case, as the Kensington lake experiment had proved, even if *ab* were of a different material.) On the other hand, according to Faraday, if *ab* had a motion from west to east, against that of the earth, there would be a net current because of a resulting difference in tensions in *ab* and *ed*.

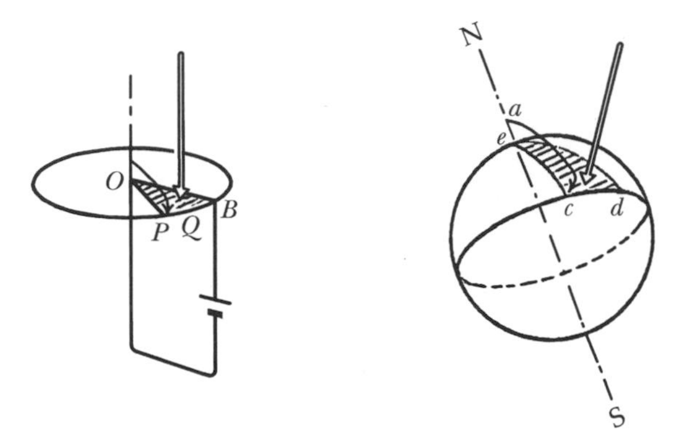

Figure 23. The Faraday rotator and the quadrantal arch.

Maxwell, on the other hand, sees that only in *ab* will there be any induced electromotive force, due to its motion relative to the earth's field. Both Maxwell and Faraday agree on the result, however, and the formulation in terms of the rule Maxwell has given is neutral with respect to the two points of view—precisely because the rule does not regard inductions in *parts* of the circuit, but only changes in the total induction through the circuit *as a whole*. Either approach yields the same conclusion with regard to the change in flux through the arch.

Applying the rule (page 184, above) to this case, the "circuit" will be *abcde*; this is perhaps confusing in that this circuit does not lie in one plane, but it is at the same time illustrative of the fact that the rule requires no such limitation. We need only reckon the total number of lines cutting the circuit, however it is shaped—in this case, the flux will be that of the dip component of the earth's field through the spherical triangle *ecd*. This changes only if *ec* moves relatively to *ed*—that is, if the area of the triangle is changing. This is the function of the "quadrantal arch."

It is not difficult to recognize in the "quadrantal arch" the Faraday rotator which we have discussed in the preceding section. Figure 23 is designed to clarify the parallelism of the two cases, one of which is the paradigm of all motors, and the other of all generators. Maxwell had encouraged us to think of the rotator in its second form (in which a conductor revolved in a fixed magnetic field) as taking place in the dip field of the earth:

> If, as in northern latitudes, $\mathfrak{B}$ acts downwards, and if the current is inwards, the rotation will be in the negative direction. (*Tr* ii/151; compare Figure 20 on page 173 above)

The parallelism of the two pieces of apparatus—the one a practical demonstration experiment, the other a work of imagination—is so close that the reader is undoubtedly intended to discover it. Possibly Maxwell's propensity to bury such jewels in the reader's path is in part explained by the comment he made in praise of Faraday's style: the reader, he said, "is tempted to believe that, if he had the opportunity, he too would be a discoverer" (*Tr* ii/176). The analogous forms of apparatus derive immediately from the strict correspondence of the two fundamental laws, which we set out for comparison on page 184, above. Both the motor and the generator principle depend strictly on the change of flux through a closed circuit; with a fixed field, this means in either case a loop with a sliding connection. The principle and the apparatus made to illustrate are is the same in either case: if the quadrantal arch were energized by a battery, it would run around the earth as a motor; if Faraday's rotator were driven and a galvanometer substituted for the battery, it would be found to be a generator.

Maxwell now returns to a distinction he asserted at the close of Chapter I, between the mechanical force of electromagnetism, which he said was exerted on the body of the conductor, and the electromotive force, which was in some sense exerted on the current itself. The electromotive force, Maxwell now asserts, acts both on the electric current, and on a statically electrified body:

> If the body is electrified, however, the electromotive force will move the body, as we have described in Electrostatics. (*Tr* ii/182)

This suggests a sweeping unification of the present discussion of electromagnetic induction and electrostatics: the force on a pith ball is a force of the same kind as the "force" which leads to current flow in induction. He asserts this, really, without either evidence to support the statement, or signs of doubt that it is true. It is important for us to keep in mind that Maxwell is not thereby committing himself to a naive assumption about either "electric charge" or "electric current." He does not assume that the former is a fluid on the surface of a charged body, or that the latter is a fluid flowing inside a conductor. One may be a state of a physical system in the field, and the other a time-variation of that state. But he is committed here to the statement that the electromotive force, which otherwise would not manifest itself directly in any mechanical form, is indeed the force which exerts a perfectly measurable mechanical action on a charged body. One piece of evidence for the relation of electrostatic and electromotive forces is the Rowland "convection current" experiment, but this—though it is imagined in the *Treatise*—was done several years after Maxwell's assertion was published.

In fact, Maxwell had always presupposed this identity of the electrostatic and electromotive forces. In "Faraday's Lines" he had written an equation which we may transcribe as follows (*SP* i/190–91):

$$\mathscr{E} = \mathscr{E}_{ex} + \mathscr{E}_{in}\,.$$

In the terminology of Maxwell's paper, these terms denote the following quantities:

$[\mathscr{E} \leftrightarrow (\alpha,\beta,\gamma)]$ "The effective electro-motive forces" [i.e., total, or net, electromotive force.]

$[\mathscr{E}_{ex} \leftrightarrow (X,Y,Z)]$ "External electro-motive forces ... either from the relative motion of currents and magnets, or from other causes acting at a distance." [*i.e.*, all the inductive electromotive force under discussion in this section of the *Treatise*.]

$\left[\mathscr{E}_{in} \leftrightarrow \left(\frac{\partial p}{\partial x}, \frac{\partial p}{\partial y}, \frac{\partial p}{\partial z}\right)\right]$ "Internal electro-motive forces ... principally from difference of electric tension at points of the conductor in the immediate neighbourhood of the point in question. The other causes are variations of chemical composition or of temperature in *contiguous parts* of the conductor." (emphasis added)

It would be foreign to the method of the present study to undertake to trace Maxwell's thought on this question (the identity of the electrostatic and electromotive forces) from "Faraday's Lines" to the *Treatise*; nor, of course, would I wish to argue that a belief which Maxwell held in 1855 was necessarily a factor in the composition of the *Treatise* itself. But I believe we see here in the earlier paper a clue to a presupposition which indeed still underlies the *Treatise*, and which will be introduced explicitly, somewhat abruptly, at the close of the exposition of the dynamical theory. The ability to write Coulomb's law and Faraday's law of induction in a single connected set of equations is an important aspect of the unity achieved by "Maxwell's equations." It is in itself not an original contribution of Maxwell's—Weber had encompassed both laws in a single equation in his problematic *Grundgesetz* (page 56, above). It is sufficient at this point to note that Maxwell has at least suggested that he will interpret the electromotive force as a vector additive with the electrostatic force.

Thus far, Maxwell has sketched a theory of electromagnetic induction, but he has postponed giving the evidence for it. He now fills this gap by introducing a further experiment of illustration: the induction balance. It is indeed an elegant instrument, a null device having the essential simplicity and precision of a weight balance—indeed, highly suggestive of the null experiments of Ampère, and serving in the theory of induction much the same function as the Ampère experiments do in the theory of electromagnetic force. That Maxwell himself sees them as counterparts is indicated by this statement:

> [T]he experiments are much more easily performed than those relating to electromagnetic inductions [i.e., the Ampère experiments], where the conductor itself has to be delicately suspended. (*Tr* ii/183)

The instrument, which is diagrammed in Figure 24, was fully described

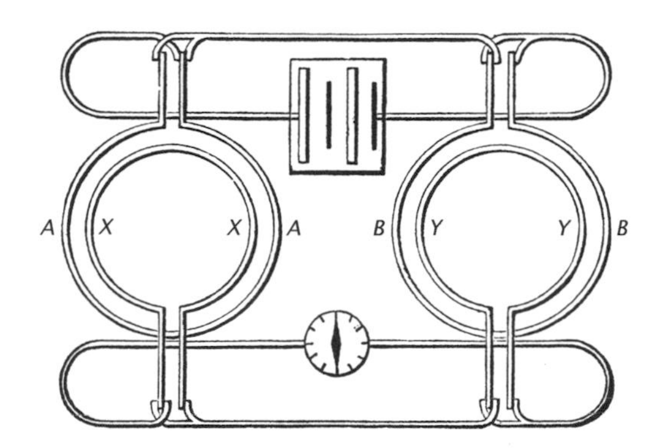

Figure 24. The induction balance
(*Treatise*, Vol. II, Figure 32)

by Riccardo Felici (1819–1902), a student of Matteucci's at Pisa.[99] Maxwell details the ways in which this instrument can be used to give precise verification of the principles of induction he has already set out (*Tr* ii/183).

Felici's second series of experiments, reported in 1859, served to demonstrate that the induction is a function of the initial and final states of the two circuits, and is independent of the intermediate states.[100] Maxwell reports this evidence with care, for it leads us back to the potential point of view which he has already cultivated in the case of the mechanical force on a conductor. The result is formulated in terms of what Maxwell calls the "total induction current," meaning the time-integral of the pulse of induced current in the secondary circuit.[101] The "states" of the two circuits are defined by their position coordinates and the currents in them; these data then constitute the independent variables of state. The law which Felici obtained empirically, following the theoretical work of Franz Neumann, is simply that the "total current" is independent of the *path* the intervening sequence, both of positions and currents. If this is the case, this total current will be equal to the difference in values of some scalar function of the state variables.

One of Faraday's Series I experiments demonstrated that, other things being equal, the induction is proportional to the primary current; and Felici's experiments of course confirmed this. We infer that the function,

99. Riccardo Felici, *Annales de chimie* (3), *34* (1852), pp. 64 ff. Felici describes his work as parallel to Ampère's at p. 74. Three of Felici's memoirs (not including the two that Maxwell cites) are collected by E. Wiedemann, ed., *Mathematische Theorie der elektrodynamischen Induktion: Ostwalds Klassiker, Nr. 109* (Leipzig: 1899).

100. Felici, *Nuovo cimento* (1), *9* (1859), pp. 345 ff.

101. F. E. Neumann uses the term "inducierten Integralstrom," which I suspect Maxwell is translating here. "Die mathematische Gesetze der inducierten elektrischen Ströme" (1845), *Gesammelte Werke* (3 vols.; Leipzig: 1912), *3*, p. 261. Also in *Ostwalds Klassiker*, Nr. 10, and translated in *Journal de mathématiques, 13* (1848), pp. 113 ff.

which Maxwell names F, factors as the product of this current and a function of the position variables of the primary and secondary coils, which he epitomizes respectively as A and B. Maxwell therefore writes, in effect, (*Tr* ii/186):

$$\int i_2\, dt \;=\; C\Delta F \;=\; C(i)\Delta(Mi_1), \quad [i_1 \leftrightarrow \Upsilon]$$

where C is the conductivity of the secondary circuit, and M denotes some function of the positions A and B alone.

When Maxwell writes, innocently, of this position factor: "Let M be this function," he is of course anticipating a result which it will be a major task yet to establish, that this position function is in fact identical with the earlier function (also called M) that has already made its appearance. There, in a quite different context, it was the mutual potential of two circuits with respect to the mechanical force between them. Maxwell knows very well that he has no right as yet to identify this new function with the old one. The device, however, of anticipating a highly interesting result by means of an advance clue is a tongue-in-cheek characteristic of Maxwell's style that is designed the *suggest* an insight before the time has come actually to demonstrate it, as we discussed earlier (page 185, above). The quantity M, as the thread which runs through Maxwell's theory of electromagnetism, manifests itself under different guises. As was just mentioned, we have already met it as the "mutual potential" of circuits. In the present context, it is to be termed the "coefficient of mutual induction"—of the same two circuits (*Tr* ii/224). It is destined to acquire still another name in the dynamical theory yet to come. In this sense, Part iv of the *Treatise* becomes the record of the successive revelations of M.

What is this ubiquitous quantity M? In Felici's experiments, as in Neumann's theory, it appears formally as a *potential*. But a potential appears to characterize a state, and Maxwell pauses to question what sort of state it is that is measured by the curious quantity M. Pursuing this speculation, Maxwell introduces what is in effect an essay on one of Faraday's most characteristic concepts, that of the electrotonic state (*Tr* ii/187; compare pages 14 ff., above). This, Maxwell will show, is proportional to the quantity F, and hence, for a given current in the primary, is proportional in turn to M.

Maxwell was always deeply impressed by the firm grip which the concept of the "electrotonic state" had upon Faraday's imagination. The discussion here in the *Treatise* is paralleled by an account in the second part of the paper "On Faraday's Lines" (*SP* i/188 ff.). In both places Maxwell points out that Faraday had immediately grasped the idea of a state, *changes in which* produced the observed inductive effects. Despite disclaimers from time to time throughout the *Experimental Researches*, he had returned to this same notion with a kind of inevitability, as we saw in Chapter 1 of this study. Maxwell was equally preoccupied with the concept. In fact, the chief original contribution of "Faraday's Lines" had been to show how the electrotonic state could be rendered in mathematical form: it was revealed there in terms of the vector-potential **A** (*SP* i/203). The close relation between the mutual potential *M* and the vector potential **A** was sketched on page 117, above. Maxwell's account of this in the *Treatise* demands to be quoted here:

> The whole history of this idea [that of the electrotonic state] in the mind of Faraday, as shewn in his published *Researches*, is well worthy of study. By a course of experiments, guided by intense application of thought, but without the aid of mathematical calculations, he was led to recognise the existence of something which we now know to be a mathematical quantity, and which may even be called the fundamental quantity of electromagnetism. But as he was led up to this conception by a purely experimental path, he ascribed to it a physical existence, and supposed it to be a peculiar condition of matter, though he was ready to abandon this theory as soon as he could explain the phenomena by any more familiar forms of thought. (*Tr* ii/187)

The odyssey of the "electrotonic state" which Maxwell has described in this paragraph is surely one of the prime illustrations for him of the development of scientific knowledge. Experience led "intense" thought to the recognition of a new and unfamiliar *form of thought*. Essentially, this new form of thought was inevitably mathematical, although it did not at first reveal itself in that mode, and mathematicians themselves did not recognize it. Now, however, we are able to identify it, in the guise of the vector **A**, as the fundamental *quantity* of a mathematical science; Maxwell showed this first in "Faraday's Lines," and will show us again in the section to follow in the *Treatise*. For Maxwell, the fact that **A** is "fundamental" implies that it is at the level of the most basic concepts in terms of which electromagnetism can be understood: that is, Faraday's effort to *understand* had inevitably led him to that *form of thought* which alone opens the phenomena to the understanding.

Faraday ascribed to this concept *physical existence*, Maxwell says, and indeed in Maxwell's ultimate view, rightly so. The vector **A** is by no means a mere mathematical convenience; it is a measure of a dynamical quantity operative in nature. Faraday himself fell back from this new discovery to retreat to explanation in terms of more familiar forms of thought. This is understandable, but it leaves a task to be completed. Fortunately, Maxwell has taken up the inquiry—had, in fact, done so in 1856 in Part II of "Faraday's Lines"—armed with a better preparation than Faraday's to deal with a mathematical concept. He continues his account of the history of **A** in the *Treatise*:

> Other investigators were long afterwards led up to the same idea by a purely mathematical path, but, so far as I know, none of them recog-

> nised, in the refined idea of the potential of two circuits, Faraday's bold hypothesis of an electrotonic state. (*loc. cit.*)

The "other investigators" were perhaps chiefly Franz Neumann (in memoirs of 1845 and 1847, and hence some fifteen years after Series I),[102] Wilhelm Weber, whose relation to Faraday's work we have discussed earlier, and William Thomson. We have seen the manner in which the vector **A** made its way into Thomson's early papers—certainly without any allusion to its possible relation to Faraday's "electrotonic state" (page 108, above). Maxwell has tacitly identified his own role in this account, for he alone among investigators pointed out, in "Faraday's Lines," this crucial identity between Faraday's hard-won concept and the mathematicians' vector potential. Here is the way he alludes to that relation at this point in the *Treatise*:

> The scientific value of Faraday's conception of an electrotonic state consists in its *directing the mind to lay hold of a certain quantity, on the changes of which the actual phenomena depend*. Without a much greater degree of development than Faraday gave it, this conception does not easily lend itself to the explanation of the phenomena. We shall return to this subject again [i.e., in the course of the exposition of Maxwell's own dynamical theory]. (*Tr* ii/188; emphasis added)

We recognize, in the italicized passage, both the intuitive expectation which led Faraday in the search for the electrotonic state (page 15, above), and the potential which resulted from the Felici experiments.

Maxwell in the *Treatise* will finally provide the development Faraday was unable to give his idea, and with this added development the concept will lend itself supremely to providing a certain kind of explanation. It is Maxwell's hope to mate the insight of the discoverer with the abstract reasoning of the mathematician, and thus to provide a solid foundation in human understanding for a "dynamical theory of electromagnetism."

Having taken this look forward to the explanation he hopes to provide, Maxwell once again takes a deliberate step backward in returning to the idea of "lines of force," in order to complete this initial account in terms that Faraday himself had been able to use successfully. The law of induction was stated earlier in these inferior terms. The real law of induction does not appear in these first chapters, which are merely preliminary to Maxwell's own dynamical theory.

Maxwell's reason for recognizing the electrotonic state (in the form of the vector potential **A**) as more fundamental than the lines of force was articulated in "Faraday's Lines." It is, very simply, that the lines of force

102. F. E. Neumann, *op. cit.*, and "Ueber ein allgemeines Princip der mathematischen Theorie inducierter elektrischen Ströme" (1847), *Gesammelte Werke*, *3*, 347 ff. Also in *Ostwalds Klassiker, Nr. 36*. On Maxwell's use of the vector **A**, see Alfred Bork, "Maxwell and the Vector Potential," *Isis*, *58* (1967), pp. 210 ff.

passing *through* a closed circuit are still remote from the conductor in which the current is observed, and hence cannot to the mind's satisfaction account for the phenomena in the circuit itself. Laws formulated in terms of the flux through a circuit therefore merely introduce action-at-a-distance all over again, in a new form—the distance between the open center of the loop, through which the flux is passing, and the conductor itself, which is mysteriously acted upon by this remote flux. The vector **A**, on the other hand, is evaluated along the circuit itself; it exists at the place at which it has its effect, and the last vestige of action-at-a-distance is eliminated. The former method is therefore "artificial," the latter "natural":

> We have now obtained in the functions α_0, β_0, γ_0 [the components of the vector potential **A**] the means of avoiding the consideration of the quantity of magnetic induction which *passes through* the circuit. Instead of this artificial method we have the natural one of considering the current with reference to quantities existing in the same space with the current itself. To these I give the name of *Electro-tonic functions*, or *components of the Electro-tonic intensity*. (*SP* i/203, Maxwell's emphasis)

The objection applies equally, of course, to the law of induction in Maxwell's Chapter III, and to the law of mechanical action in Chapter I, which was expressed for Faraday's sake in terms of the surface integral of the magnetic induction through any surface bounded by the circuit (page 131, above). It, too, is artificial, and will be replaced by a more insightful principle in the dynamical theory to come.

The entire account we have been getting in these opening chapters of Part iv is a mathematical transcription of the level of understanding Faraday had himself mastered, and hence is introductory and provisional. It is nonetheless significant that Maxwell guides us by this route: the access to the dynamical theory is through the theory of lines of force. The history of science must be rehearsed, Maxwell apparently believes, in the mind's progress toward an ultimate understanding. For Faraday, the lines of force were "always in his mind's eye when contemplating his magnets or electric currents" (*Tr* ii/188). In terms of the intellectual "eye-knowledge" Maxwell is seeking, it is perhaps crucial to fix this image of Faraday's firmly in the mind before passing on to that other, the electrotonic state, in which Faraday firmly believed, but which he was never able to focus to his satisfaction.

To close this preliminary study, Maxwell now draws together Faraday's own explanation of the phenomena of induction in terms of the lines of force. Maxwell draws the reader's attention to other passages in the *Treatise* in which the lines of force have already proved useful, and then summarizes their use in Faraday's view of induction. He identifies (*Tr* ii/188–89) three modes of reasoning in terms of the lines, in as many serieses of the *Experimental Researches*.

> [(1) First Series:] [T]he direction of the current in a conducting circuit, part of which is moveable, depends on the mode in which the moving part cuts through the lines of magnetic force. [*cf. XR* i/32]
>
> [(2) Second Series:] [T]he phenomena produced by variation of the strength of a current or a magnet may be explained, by supposing the system of lines of force to expand from or contract towards the wire or magnet as its power rises or falls. [*cf. XR* i/63, 68]
>
> [(3) Twenty-eighth Series:] [T]he moving conductor, as it cuts the lines of force, sums up the action due to an area or section of the lines of force. [*cf. XR* iii/332, 335, 345]

The references to the *Experimental Researches* are those which Maxwell himself gives in the text. He thereby quickly reminds us of three stages in Faraday's progressive vision of the lines—first as objects inferred from the phenomena of induction, then gaining independence of existence as entities (in the second stage), and unifying into a system characterized by a constant total power measured by the moving wire (third stage).

What Faraday was not able to do, Maxwell points out, was to give quantitative expression to his principle—for lack, he says, of a definition of electromotive force. If my own remarks in Chapter 1 are valid, it might seem more accurate to say that Faraday had no interest, no impulse toward a law in quantitative form, and that it was not so much the lack of a measure of electromotive force, as a lack of a notion of a quantitative law, which held Faraday back from producing what we now call "Faraday's law of induction" in proper mathematical form.

Finally, Maxwell pauses to point out an important terminological shift that obscures the interpretation of Faraday:

> Instead of speaking of the number of lines of *magnetic force*, we may speak of the *magnetic induction* through the circuit, or the surface-integral of magnetic induction extended over any surface bounded by the circuit. (*Tr* ii/189; emphasis added)

This sounds like a simple equivalence, but behind it lies the basic distinction between *magnetic force* and *magnetic induction*, the vectors **H** and **B**. The distinction is not articulated by Faraday so far as I am aware, but it has been carefully made in the *Treatise*, following Thomson, who apparently introduced it.[103] The difference is enormous, of course, whenever iron is present. Faraday is fully aware of the effects of iron, but has not identified the two concepts separately; Thomson did so, in terms of his "crevass" definitions, when he asked himself what would be the magnetic force (force on a test pole) inside a solid mass of magnetized iron—and recognized that the answer would depend crucially on the shape of the hole that was to

103. Chapter II of Part iii of the *Treatise*; cf. Thomson in the "Mathematical Theory of Magnetism" (1849). (*EM*: 366 ff.).

admit the test body. The shift from a test pole to a moving wire, which was very much a part of Faraday's own motion of thought, makes it possible to trace the lines—as lines of induction, **B**—through the body of the magnet or the bar of iron, and hence to confirm their continuity. Thus magnetic induction had in Faraday's own thought virtually displaced magnetic force, and the principle of continuity of the vector **B** had been established, even though the terminological refinements were left to others. By the end of the *Experimental Researches*, the notion of the isolated pole was foreign to Faraday's thinking, and magnetic force **H** would hardly have a place in translating Faraday's later thoughts.

With this refinement, Maxwell adjourns his exposition of "Faraday's method"; he will resume it in the fourth chapter, on self-induction, but no additional theory will be introduced there. Maxwell will turn immediately to questions of the significance of self-induction: does it, he will ask, point to the existence of true momentum or energy in the electric current? The answer leads him directly into the dynamical theory itself, and hence Maxwell's Chapter IV is better considered in its relation to dynamical theory, at the beginning of Part Two of this study.

Before closing his Chapter III, however, Maxwell interrupts the interpretation of Faraday's thought to, as he says, "enumerate" other theories of induction. Lenz, F. E. Neumann, Helmholtz, Thomson, and Weber thereupon pass in hurried review. Of these theories, perhaps only that of Franz Neumann demands our attention. Maxwell gives great credit to him, in a passage of considerable interest to students of Maxwell:

> On this [Lenz's] law F. E. Neumann ... founded his mathematical theory of induction, in which he established the mathematical laws of the induced currents due to the motion of the primary or secondary conductor. He shewed that the quantity *M*, which we have called the *potential* of the one circuit on the other, is the same as the *electromagnetic potential* of the one circuit on the other, which we have already investigated in connection with Ampère's formula. We may regard F. E. Neumann, therefore, as having completed for the induction of currents the mathematical treatment which Ampère had applied to their mechanical action. (*Tr* ii/190; emphasis added)

It is Neumann, then, who showed what Maxwell has as yet only hinted: the crucial identity of the mechanical potential of two circuits (i.e., their "electromagnetic potential," which Maxwell discussed in his Chapter I and alluded to in Chapter II) and the electromotive potential (which in the quotation above Maxwell calls "the quantity *M*" and which we have discussed, in the present chapter, in relation to the experiments of Felici. Maxwell's debt to Neumann is therefore very great: we have seen already something of the role the potential *M* is to play, as a mathematical key to Faraday's thought.

Why, then, does Maxwell here dismiss Neumann so briefly? I think it is most likely that Maxwell is in the best sense impatient to move on to the dynamical theory—the way has been paved, and a digression into a further study of a formal theory would not advance the real work of the *Treatise.* Maxwell draws heavily from Neumann's result; but it is in truth a purely formal debt, over which the *Treatise* need not delay. Maxwell is resolved to proceed with the unfolding of a single theme—that of Faraday and the electrotonic state.

PART TWO

Maxwell's Dynamical Theory of Electromagnetism

Chapter 4
Maxwell's Exposition of the Lagrangian Equations

LAGRANGIAN THEORY IN THE *TREATISE*

Maxwell builds gradually toward his own theory of the electromagnetic field in the *Treatise*. As we have seen, the first four chapters of Part iv have been confined to a relatively literal interpretation of Faraday, and to an excursion into Ampère's action-at-a-distance theory—the latter as a way of taking the measure of the alternative which Maxwell rejects. These chapters constitute in effect a translation of Faraday at the first level. When this introductory exercise has reached a certain point, at the end of Chapter IV, Maxwell judges that the reader is ready to begin anew, on what is to be Maxwell's own firmest statement, his "dynamical theory" of the field. The transition comes with almost perplexing abruptness. All that has gone before is suddenly erased with this announcement:

> What I propose now to do is to examine the consequences of the assumption that the phenomena of the electric current are those of a moving system, the motion being communicated from one part of the system to another by forces, the nature and laws of which we do not yet even attempt to define, because we can eliminate these forces from the equations of motion by the method given by Lagrange for any connected system.
>
> In the next five chapters of this treatise I propose to deduce the main structure of this theory of electricity from a dynamical hypothesis of this kind.... (*Tr* ii/198)

Thereafter, as the "dynamical theory" unfolds through a sequence of five chapters, we move in a realm of mathematical abstraction in which the old concepts are cautiously introduced again, one by one, re-defined and cleansed of dubious physical connotations.

As a literary device, this is a bold move. It is as if, in a novel, the movement were interrupted at the height of its development, and the whole scene suddenly viewed anew through the eyes of an intruder. The characters whom we had best known we now recognize only with the greatest difficulty, and the conclusion of the story is to be in terms which could hardly have been anticipated earlier.

The "intruder" in the present case is Joseph Louis Lagrange, whose *Mécanique analytique* of 1788 was introduced to the British scientific world only after a long delay which reflected the general lack of contact in England with the analytic methods of the Continent.[1] Maxwell apparently first made use of Lagrangian methods in his "Dynamical Theory of the Electromagnetic Field" in 1864; there he refers only to Lagrange as source. He might indeed have met these methods in Arthur Cayley's "Report on Theoretical Dynamics," presented to the British Association in 1857, which Maxwell cites as a principal reference at the point in the *Treatise* at which the exposition of Lagrange's method begins (*Tr* ii/199). But Maxwell himself gives greatest credit to Thomson and Tait's major joint work, the *Treatise on Natural Philosophy*, for having made this method serviceable in England, and to the student of nature.[2] He makes this clear in a review of the second edition of the *Treatise*, in 1879:

> The credit of breaking up the monopoly of the great masters of the spell, and making all their charms familiar in our ears as household words, belongs in great measure to Thomson and Tait. The two northern wizards [i.e., of Glasgow and Edinburgh, respectively] were the first who, without compunction or dread, uttered in their mother tongue the true and proper names of these dynamical concepts which the magicians of old were wont to invoke only by the aid of muttered symbols and inarticulate equations. And now the feeblest among us can repeat the words of power and take part in dynamical discussions which but a few years ago we should have left for our betters. (*SP* ii/182)

Maxwell's application of Lagrange's equations to electromagnetism in 1864 came at the time when Thomson and Tait's treatise was taking shape, and it seems likely that Maxwell's attention was called to the subject by their interest in it.[3]

Maxwell's enthusiasm for the new method becomes very evident in the *Treatise*, and Lagrange is adopted, along with Faraday and in a curious way Ampère as well, as a guide in the composition of Maxwell's work. As we have

1. Joseph-Louis Lagrange, *Mécanique analytique* (1811) (2 vols.; Paris: Albert Blanchard, 1965). This is a reprint of the fourth edition (Paris: 1888–89), with notes by J. Bertrand and G. Darboux. On the study of mathematics at Cambridge, see Walter W. Rouse Ball, *A History of the Study of Mathematics at Cambridge* (Cambridge: 1889), especially pp. 118 ff. See also J. T. Merz, *History, 1,* pp. 232 ff.

2. William Thomson and Peter Guthrie Tait, *A Treatise on Natural Philosophy: Volume 1* (Oxford: 1867); no further volumes ever appeared, but the first volume was enlarged in a second edition in two parts in 1878. The final, 1912, edition was reprinted under the title *Principles of Mechanics and Dynamics* (2 vols.; New York: Dover Publications, [1962]). Except where otherwise noted, I shall refer to this reprint edition.

3. Maxwell's intimate friendship with Tait, which began at the Edinburgh Academy, and thereafter kept them in close touch on many scientific matters, is recounted both by Tait and by Campbell (Knott, *Life of Tait,* pp. 261 ff.; Campbell, *Life of Maxwell,* pp. 62–63 and p. 191). Little correspondence between Maxwell and Tait has apparently been preserved from the period in question, so my impression that Maxwell would have been hearing of the progress of Thomson and Tait's project remains unproved.

already seen, each new guide means a new strategy—and Maxwell greets the Lagrangian strategy as peculiarly adapted to become the vehicle of his own electromagnetic theory. Another passage from the review quoted above suggests the significance of Lagrange's equations in Maxwell's view:

> [W]hen we have reason to believe that the phenomena which fall under our observation form but a very small part of what is really going on in the system, the question is not—what phenomena will result from the hypothesis that the system is of a certain specified kind? but—what is the most general specification of a material system consistent with the condition that the motions of those parts of the system which we can observe are what we find them to be?
>
> It is to Lagrange, in the first place, that we owe the method which enables us to answer this question *without asserting either more or less than all that can be legitimately deduced from the observed facts*. But though this method has been in the hands of mathematicians since 1788, when the *Mécanique analytique* was published, and though a few great mathematicians, such as Sir W. R. Hamilton, Jacobi, &c., have made important contributions to the general theory of dynamics, it is remarkable how slow natural philosophers at large have been to make use of these methods. (*SP* ii/781; emphasis added)

For Maxwell, many different styles have their places in the rhetoric of mathematical physics, and he adopts distinctly different modes even in successive sections of the *Treatise*. We have pointed out the intuitive mode, leaning heavily on illustration, with which the exposition of electromagnetic theory began (page 185, above). Now in the statement of the "dynamical theory" on which he is to take his stand, the primary goal is precision. In terms of the alternatives he set before the reader earlier, represented by Ampère and Faraday, his choice is now more for the spare style of Ampère than the full and suggestive manner of Faraday. In this second major phase of the exposition of his theory, the phase which corresponds not to the nascent idea but to rigorous demonstration of the kind exemplified in the *Théorie mathématique*, Maxwell becomes almost as much a follower of Ampère as of Faraday; it has perhaps been too little appreciated how much Maxwell on balance owes to Ampère. At the same time, Maxwell keeps the full range of his science in view, and he labors to reconcile each style to the legitimate demands of the other; it is as if he sought a reconciliation of the styles, just as he looked for a point of connection of the concepts, of these two masters (page 162). Here, the abstract formulation will be gradually adjusted and adapted, so that the physical imagery of Faraday will gradually emerge once again from it. This is indeed the great trick which Maxwell has determined to bring off in Part iv of the *Treatise*: *to reconstruct Faraday's concepts* out of a new beginning in abstract, analytic mechanics—and thereby to achieve a "translation" of Faraday's thought on a more secure intellectual foundation.

It must be added immediately that the *Treatise* will not conclude with this Lagrangian theory, and that in Maxwell's hierarchy of methods the dynamical theory is not the ultimate theory: it is rather the most precise and fundamental general statement—in terms of the quotation above, "the most general specification of a material system" consistent with observed phenomena. Beyond the *dynamical* theory will come a more tentative *physical* theory, in Chapter IX of Part iv of the *Treatise*; the hypothesis of the electromagnetic nature of light is likewise beyond the dynamical theory. Both of these hypotheses fall outside the scope of the present study.

In Chapter IV of Part iv of the *Treatise*, Maxwell seeks the significance of the phenomena of self-induction. As is well known, the tendency of a current forced through a large coil to increase slowly when the switch is closed, and to continue to flow after the source of power is removed, bears an apparent analogy to the momentum of a moving body. As Maxwell points out, Faraday had drawn this comparison (*Tr* ii/195). On the other hand, it cannot be literally pursued, as if the current were a fluid in the wire, since it is not the length of the wire but the configuration of the coil which determines the amount of this apparent inertia.[4] Acknowledging this difficulty, Maxwell nonetheless insists that we have here a clue to the operation of some form of mechanical system, set in motion with the phenomenon which we call the "flow of current":

> It is difficult, however, for the mind which has once recognised the analogy between the phenomena of self-induction and those of the motion of material bodies, to abandon altogether the help of this analogy, or to admit that it is entirely superficial and misleading. The fundamental dynamical idea of matter, as capable by its motion of becoming the recipient of momentum and of energy, is so interwoven with our forms of thought that, whenever we catch a glimpse of it in any part of nature, we feel that a path is before us leading, sooner or later, to the complete understanding of the subject. (*Tr* ii/196–7)

This "path" to an intellectually satisfying *dynamical* understanding of electromagnetism is the one which Maxwell will pursue in the sequence of chapters here beginning. The idea of momentum presents difficulties, but there is another, indisputable token that we are in fact dealing with a material system. Maxwell says of the electric current:

> We have already shewn that it has something *very like momentum*, that it resists being suddenly stopped, and that it can exert, for a short time, a great electromotive force.

4. Thomson had discussed this point in an article entitled "Magnetism, Dynamical Relations of" in Nichol's *Cyclopedia of the Physical Sciences* (2nd ed.; London: 1860), which is for the most part drawn from an earlier article "On the Mechanical Values of Distributions of Electricity, Magnetism, and Galvanism" (1853) (*Mathematical and Physical Papers*, *1*, pp. 521 ff.). The passage concerning inertia is in turn reprinted in a footnote at (*EM*: 447).

> But a conducting circuit in which a current has been set up has the power of doing work in virtue of this current, and this power cannot be said to be something very like energy, for it is *really and truly energy*. (*Tr* ii/197; emphasis added)

This is the rock on which Maxwell builds; and the concept of matter in terms of which the field is seen as a material system, is that of the ultimate vehicle for the conveyance of energy.

The transition from the "first level" of translation of Faraday in Chapters I–IV to Maxwell's own "dynamical theory" of Chapters V–IX is made in the following argument, which I have taken the liberty of dividing into numbered steps:

> [1] [A] conducting circuit in which a current has been set up has the power of doing work in virtue of this current, and this power ... is really and truly energy.
>
> [2] ... therefore ... a system containing an electric current is a seat of energy of some kind;
>
> [3] and since we can form no conception of an electric current except as a kinetic phenomenon, its energy must be kinetic energy, that is to say, the energy which a moving body has in virtue of its motion.
>
> [4] [T]he electricity in the wire cannot be considered as the moving body ... since the presence of other bodies near the current alters its energy.
>
> [5] We are therefore led to enquire whether there may not be some motion going on in the space outside the wire.
>
> [6] Therefore I propose ... to examine the consequences of the assumption that the phenomena of the electric current are those of a moving system. (*Tr* ii/197–98)

The first chapter of the new sequence is devoted to Maxwell's own exposition of Lagrange's theory of the motion of a connected mechanical system, and we shall turn to this first, in the remainder of the present chapter of this study. The meaning of the term "dynamical" will be examined in Chapter 5 of this study, and Maxwell's development of his own dynamical theory of electromagnetism will be traced in Chapter 6.

When Maxwell refers to "the equations of motion of a connected system" in the title of his Chapter V, it is Lagrange's equations as introduced in the *Mécanique analytique* that he has in mind, as he immediately makes clear (*Tr* ii/199). There are two respects in which Lagrange's analytic theory exerts a strong influence on Maxwell's *Treatise*.

First, the equations developed by Lagrange fundamentally alter the mode of approach to problems concerning the motion of a system: Lagrange showed how to eliminate the inner connections of a system from explicit consideration, permitting attention to focus on a reduced set of variables just equal in number to the actual degrees of freedom of the system as a whole. These independent variables may be, in terms of an earlier quotation, "those parts of the system which we can observe"—what are often called in our contemporary discussions the "observables." In Maxwell's problem, these are the currents and the positions of the conductors. Only by means of Lagrange's technique would it be possible to carry through Maxwell's program and construct a general account of a connected mechanical system on the basis of this slender evidence.

Second, concomitant with this generality of the Lagrangian account—a generality which makes it possible to describe the mechanical system "without asserting either more or less than all that can be legitimately deduced from the observed facts," as Maxwell claimed for the method in the quotation earlier—there arises a level of abstraction which was a source of pride for Lagrange, but which becomes a pressing problem for Maxwell.

It was Lagrange's special boast to have reduced the science of motion to the precise reasoning of analytic mathematics—to have rescued it from the uncertain realm of geometrical arguments in which diagrams and the visual imagination play a role often difficult to assess. His Preface makes this famous claim:

> On ne trouvera point de Figures dans cet Ouvrage. Les méthodes que j'y expose ne demandent ni constructions, ni raisonnements géométriques ou méchaniques, mais seulement des opérations algébriques, assujeties à une marche réguliere et uniforme. Ceux qui aiment l'Analyse, verront avec plaisir la Méchanique en devenir une nouvelle branche, et me sauront gré d'en avoir étendu ainsi le domaine.[5]

"The aim of Lagrange was to bring dynamics under the power of the calculus," Maxwell says (*Tr* ii/199). Anyone who has wrestled with the imperfect geometrical arguments of Newton's *Principia* will realize the value of this revision—and Maxwell was not only a student but a teacher of the *Principia*. Yet with this transformation of mechanics into an analytic science, there began a reign of symbols and equations about which Maxwell was deeply troubled. For him, Lagrange epitomizes both the impressive power of analytic mathematics, and the obscurity which accompanies its symbols. As we have seen, Maxwell's own impulse is to the use of just those diagrams which Lagrange is glad to have exorcised. Lagrange thus presents Maxwell with a challenge: to make use of the full effectiveness of the analytic techniques, recognizing their power and precision, while at the same time finding means to illuminate the mathematical symbols so that they may once again convey meaningful ideas.

5. Lagrange, *Mécanique analytique, 1*, p. i.

This is one of the governing problems in terms of which the *Treatise* was conceived. With respect to other representatives of the analytic school, such as Ampère and Poisson, Maxwell was able to preserve a certain distance. Now he is committed to *employing* for his own theory a technique even more abstract than that of Ampère. Maxwell's resolution of this problem will bring into relief his convictions concerning the right relation of abstract mathematics and the physical imagination.

Given the respect Maxwell accords to the role of Thomson and Tait's *Treatise* in the formulation of Lagrangian theory—and especially if I am right in my surmise that Maxwell was guided by their work in the development of his own thought about the method—it is indeed interesting that his own exposition follows quite a different plan from theirs. Since I think that this divergence is deliberate and bears upon Maxwell's concept of the foundations of dynamical explanation, I should like first to sketch the development of Lagrange's equations as given by Thomson and Tait, and then that given by Maxwell, in order to be able to compare the two.

THOMSON AND TAIT'S DERIVATION OF LAGRANGE'S EQUATIONS

There are really two phases to Lagrange's theory. Lagrange first shows how to solve the problem of the motion of a system whose parts are related by given equations of connection, or "constraint." If the system consists of n bodies, Newton's second law, applied to each of three coordinate directions for each of the bodies, would yield $3n$ equations of motion for the system. These are not, however, independent equations, since on account of the constraints there are not $3n$ independent variables. Lagrange produced a very simple systematic procedure for eliminating the constraints from the equations of motion, yielding a set of equations which can be solved. If there are i constraints among the $3n$ variables, the result is a set of $(3n-i)$ independent equations of motion.[6]

Although the possibility of accomplishing this is crucial, the method for doing so does not concern Maxwell in the *Treatise*, and we may pass on to the second phase of Lagrange's work, which is to write a new set of equations of motion of a form quite different from Newton's Second Law, in which the independent variables are any set of $(3n-i)$ independent coordinates—the so-called "generalized coordinates." It will be necessary to establish these equations carefully, as they are the basis of Maxwell's electromagnetic theory. We first follow Thomson and Tait, whose exposition is based on that of Lagrange.[7]

6. Thomson and Tait, *Treatise*, pp. 268 ff.

7. Thomson and Tait, *op. cit.*, *1*, pp. 318 ff. Concerning a difference between this derivation and that of the first edition, see p. 209, below.

We begin by transforming the equations of motion of a set of free particles from their expression in Cartesian coordinates to a new set of expressions in terms of a new set of variables, the relation being given between the new variables and the old. If the new variables are the generalized coordinates of Lagrange, the new equations are the Lagrangian equations. If there are constraints in the system, these will simply reduce the number of the generalized coordinates.

Suppose that we have a system of masses m_1, m_2, ... m_k, with Cartesian coordinates x_1, y_1, z_1, x_2, ..., and let a new set of coordinates be defined in relation to the Cartesian coordinates by a set of equations. Then we will have a set of relations giving the Cartesian coordinates in terms of the new coordinates φ, χ, ψ, ...:

$$x_1 = x_1(\varphi, \chi, \psi, \ldots)$$
$$x_2 = x_2(\varphi, \chi, \psi, \ldots)$$
$$y_1 = y_1(\varphi, \chi, \psi, \ldots) \ \ldots, \text{etc.}$$

When the description of the motion of the set of masses is thus shifted from Cartesian coordinates to new "generalized" coordinates, corresponding "generalized" measures arise to replace the familiar quantities, velocity, momentum, and force. Thus if "generalized velocity" is defined as $\dot{\varphi}$, the time-rate of variation of a generalized coordinate φ, the new generalized "velocity" will not necessarily be a true velocity in the Newtonian sense. A generalized "velocity" is only *metaphorically* a velocity (a generalized coordinate may, for example, be an angle; the corresponding generalized velocity would be an angular velocity—but an angular "velocity" is not literally a velocity in the sense of the laws of motion).

What is to be meant, then, by the term "generalized force"? The answer is determined by observing that the quantity of work is independent of the coordinate system in which measurements are made, and at the same time by deciding that the formal definition of "work" shall be unchanged: the work will be the product of "force" and "displacement," whether these are Newtonian or generalized quantities. Then "generalized force" will be that quantity which in any given case must be multiplied by a generalized displacement to obtain an actual quantity of work. Let us work this out analytically. In Cartesian coordinates,

$$W = \sum_i \mathbf{f}_i \cdot \mathbf{ds}_i = \sum_i \left(f_{i_x} dx_i + f_{i_y} dy_i + f_{i_z} dz_i \right).$$

Using the set of transformation equations, we can shift this expression into terms of the generalized displacements $d\varphi$, $d\chi$, $d\psi$, ...:

$$dx_i = \frac{\partial x_i}{\partial \varphi} d\varphi + \frac{\partial x_i}{\partial \chi} d\chi + \frac{\partial x_i}{\partial \psi} d\psi \cdots .$$

Then

$$W = \sum_i \left(f_{i_x} \frac{\partial x_i}{\partial \varphi} d\varphi + f_{i_x} \frac{\partial x_i}{\partial \chi} d\chi + f_{i_x} \frac{\partial x_i}{\partial \psi} d\psi + \cdots \right)$$
$$+ \sum_i \left(f_{i_y} \frac{\partial y_i}{\partial \varphi} d\varphi + f_{i_y} \frac{\partial y_i}{\partial \chi} d\chi + \cdots \right)$$
$$+ \sum_i \left(f_{i_z} \frac{\partial z_i}{\partial \varphi} d\varphi + \cdots \right)$$

By our agreement concerning the definition of "generalized force," the quantities multiplying the generalized displacements in this equation must be identified as the "generalized forces," corresponding, as components of the total force, to the generalized coordinates by which they are multiplied. For example, the generalized force corresponding to the φ-component of motion of the first body will be:

$$\Phi_1 = f_{1_x} \frac{\partial x_1}{\partial \varphi} + f_{1_y} \frac{\partial y_1}{\partial \varphi} + f_{1_z} \frac{\partial z_1}{\partial \varphi} + \cdots .$$

Note that Φ_1 *is not necessarily a Newtonian motive force*—it might be a torque, for example, if φ is an angular displacement; but it might equally well be an otherwise anonymous quantity. Nonetheless, as we shall now see, a *law of motion* can be written in which this new quantity, generalized beyond recognition, takes the place of Newtonian force. The law will, then, be a "force" law, though only in a metaphorical sense.

Thomson and Tait give different derivations of this Lagrangian law of motion in the first and second editions of their *Treatise on Natural Philosophy* (1867 and 1879, respectively). The second, as they point out, is simpler in that it proceeds directly from Newton's Second Law, but as the first is essentially that given by Lagrange, and is thus the one which Maxwell had before him while writing the *Treatise* (whether he was looking at Thomson and Tait, or the *Mécanique analytique*), it will be best to outline that proof here.[8] As I have pointed out, the proof here is altogether distinct from the exposition given by Maxwell in Chapter V of the *Treatise*, and it should be mentioned also that there is still another way of deriving this Lagrangian force law, namely as a consequence of Hamilton's Principle, from which it follows very elegantly by means of the calculus of variations.[9]

8. The difference is minor; in the revised proof Newton's Second Law is used directly, without the introduction of the variations which occur in the account given here. *Cf.* Thomson and Tait, *op. cit.*, *1*, p. 304, and §329 of the first edition.

9. Thomson and Tait, *op. cit.*, *1*, pp. 328 ff. Two excellent texts on this general subject are Robert Lindsay and Henry Margenau, *Foundations of Physics* (2nd ed., New York: Dover Publications, 1957), Chapter III; and C. Lanczos, *The Variational Principles of Mechanics* (Toronto: University of Toronto Press, 1949). Both place the subject in the context of its historical development. Note also Larmor, writing as editor, in Maxwell's *Matter and Motion* (London: Society for Promoting Christian Knowledge, 1876), pp. 145 ff.

Newton's Second Law applied to a set of k bodies will yield $3k$ equations:

$$f_{1_x} = m_1\ddot{x}_1 \qquad f_{1_y} = m_1\ddot{y}_1 \qquad \cdots$$
$$f_{2_x} = m_2\ddot{x}_2 \qquad f_{2_y} = m_2\ddot{y}_2 \qquad \cdots$$
$$f_{k_x} = m_k\ddot{x}_k \qquad \cdots$$

Suppose we give the system a motion corresponding, after an interval of time dt, to a set of displacements δx_1, δy_1, δz_1, δx_2, δy_2, … . The total work done will be given by the summation:

$$W = \sum_i \left(f_{i_x}\delta x_i + f_{i_y}\delta y_i + f_{i_z}\delta z_i \right)$$
$$= \sum_i m_i \left(\ddot{x}_i \delta x_i + \ddot{y}_i \delta y_i + \ddot{z}_i \delta z_i \right).$$

By using the equations for the transformation from Cartesian to generalized coordinates (page 208, above), we know that we can represent the same work in terms of the generalized displacements and certain functions which we may denote Φ, Χ, and Ψ:

$$\sum_i m_i \left(\ddot{x}_i \delta x_i + \ddot{y}_i \delta y_i + \ddot{z}_i \delta z_i \right) = \Phi\delta\varphi + \mathrm{X}\delta\chi + \Psi\delta\psi.$$

By definition, these quantities Φ, Χ, and Ψ will be the "generalized forces." To find out what they are, Thomson and Tait simply carry out the substitution in terms of the new coordinates in the left-hand side of this equation, and then identify terms. The result for the generalized force associated with the coordinate φ is, for example:

$$\Phi = \frac{d}{dt}\sum_i m_i \left(\dot{x}\frac{\partial x}{\partial \varphi} + \dot{y}\frac{\partial y}{\partial \varphi} + \dot{z}\frac{\partial z}{\partial \varphi} \right)$$
$$= \sum_i m_i \left(\dot{x}\frac{\partial \dot{x}}{\partial \varphi} + \dot{y}\frac{\partial \dot{y}}{\partial \varphi} + \dot{z}\frac{\partial \dot{z}}{\partial \varphi} \right).$$

Now we know that the expression for kinetic energy in Cartesian coordinates is:

$$T = \tfrac{1}{2}\sum_i m_i \left(\dot{x}_i^2 + \dot{y}_i^2 + \dot{z}_i^2 \right),$$

which would be transformed into generalized coordinates in this way:

$$T = \tfrac{1}{2}\sum_i m_i \left[\left(\frac{\partial x}{\partial \varphi}\dot{\varphi} + \frac{\partial x}{\partial \chi}\dot{\chi} + \frac{\partial x}{\partial \psi}\dot{\psi} \right)^2 + \left(\frac{\partial y}{\partial \varphi}\dot{\varphi} + \cdots \right)^2 + \left(\frac{\partial z}{\partial \varphi}\dot{\varphi} + \cdots \right)^2 \right].$$

Differentiating this expression for T with respect to φ and $\dot{\varphi}$ and comparing the result with the equation for Φ reveals that the generalized force Φ can be very simply represented in terms of derivatives of the kinetic energy:

$$\Phi = \frac{d}{dt}\left(\frac{\partial T}{\partial \dot{\varphi}}\right) + \frac{\partial T}{\partial \varphi}.$$

This is Lagrange's equation of motion for the generalized coordinate φ. Similar equations apply of course to motion with respect to the other generalized coordinates; and the set of equations, equal in number to the degrees of freedom of the system, are Lagrange's equations of motion for the system.

If the generalized force Φ is obtainable from a potential function $V = V(\varphi, \chi, \psi, \ldots)$, we may write

$$\Phi = -\frac{\partial V}{\partial \varphi}.$$

Then if we define $L = (T - V)$, the equation of motion becomes:

$$\frac{d}{dt}\left(\frac{\partial L}{\partial \dot{\varphi}}\right) - \frac{\partial L}{\partial \varphi} = 0.$$

Today L is referred to as the "Lagrangian" of the system. Maxwell, however, is working throughout with only kinetic energies; so he does not use this last form in the dynamical theory of the electromagnetic field.

By analogy with simple cases such as that of the linear motion of a single particle, for which T has no dependence on position, we see that the term

$$\frac{d}{dt}\left(\frac{\partial T}{\partial \dot{\varphi}}\right)$$

is *analogous to* the "time rate of change of momentum" in Newtonian theory. This leads us, again by analogy, to define

$$p_\varphi = \frac{\partial T}{\partial \dot{\varphi}}$$

as the "generalized momentum" corresponding to the φ-coordinate.

We have seen, then, that in the Lagrangian law of motion, the coordinates are not in general linear, their increments do not in general represent linear displacements, their time-rates are not Newtonian velocities, and Φ is not in general a Newtonian force.

What about the most fundamental quantity in the Newtonian world-view, the "quantity of matter"? It of course appears as the quantities m_i in the expression for the kinetic energy. But if we complete the transformation of that expression by expanding the square, and then regroup the terms, we see that we will have a homogeneous quadratic function of the variables $\dot{\varphi}$, $\dot{\chi}$, $\dot{\psi}$, … , *i.e.*, an expression of the form

$$T = \tfrac{1}{2}\left[(\varphi,\varphi)\dot{\varphi}^2 + (\chi,\chi)\dot{\chi}^2 + \cdots + 2(\varphi,\chi)\dot{\varphi}\dot{\chi} + 2(\varphi,\psi)\dot{\varphi}\dot{\psi} + \cdots\right].$$

Here the symbols (φ, χ), etc., simply denote coefficients of the corresponding terms, each coefficient being an expression compounded of the partial derivatives of the old coordinates with respect to the new. They are analogous to masses, inasmuch as they are coefficients of the squared "velocities"; but the literal Newtonian masses have been swallowed up in this transformation. *The Lagrangian theory opens the way to a reconsideration of the concept of the "mass" of a physical system.* As we shall see, in the little book *Matter and Motion* Maxwell undertakes to formulate this new understanding of mass or matter, and it is in this enlarged sense only that Maxwell's "dynamical theory of electromagnetism" can be said to be a theory of the motion of a "material" system.[10] We shall return to the criticism of the basic concepts of dynamics, for which the way is prepared by the Lagrangian law of motion, in Chapter 5, below.

It would be difficult to convey the power of Lagrange's equations more effectively than did Maxwell in his review of Thomson and Tait, to which we have referred. The image introduced there of a group of bell-ringers, pulling upon ropes connected to an inaccessible set of bells, is this:

> In an ordinary belfry, each bell has one rope which comes down through a hole in the floor to the bellringers' room. But suppose that each rope, instead of acting on one bell, contributes to the motion of many pieces of machinery, and that the motion of each piece is determined not by the motion of one rope alone, but by that of several, and suppose, further, that all this machinery is silent and utterly unknown to the men at the ropes, who can only see as far as the holes in the floor above them.
>
> Supposing all this, what is the scientific duty of the men below? They have full command of the ropes, but of nothing else. They can give each rope any position and any velocity, and they can estimate its momentum by stopping all the ropes at once, and feeling what sort of tug each rope gives. If they take the trouble to ascertain how much work they have to do in order to drag the ropes down to a given set of positions, and to express this in terms of these positions, they have found the potential energy of the system in terms of the known coordinates. If they then find the tug on any one rope arising from a velocity equal to unity communicated to itself or to any other rope, they can express the kinetic energy in terms of the co-ordinates and velocities.
>
> These data are sufficient to determine the motion of every one of the ropes when it and all the others are acted on by any given forces. This is all that the men at the ropes can ever know. If the machinery above has more degrees of freedom than there are ropes, the coordinates which express these degrees of freedom must be ignored. There is no help for it. (*SP* ii/783–84)

Here the linear displacements of the ropes are the generalized coordinates. In this, the example does not quite go to the extreme case, and they will feel a Newtonian tug. But if the ropes were wound on winches, the

10. Maxwell, *Matter and Motion* (London: Society for Promoting Christian Knowledge, 1876). Reprinted New York: Dover Publications (1952).

angular displacements of the handles would serve as well for coordinates, while the tugs would be supplanted by twists. The theory applies in any case, though the equations may become complicated—as they would, for example, if the drums of the winches were mounted eccentrically to their axes.

The Lagrangian method in dynamics is an illustration of the ultimate rhetorical instrument, where precision of expression is the goal, for it expresses precisely "... all that the men at the ropes can ever know ... ," and no more. By employing just those variables which can be measured and controlled, Lagrange is able to say just what is physically admissable. Yet the theory leaves the way open to pursue the analysis further, in terms of additional independent variables, if the empirical situation at any time permits—if in effect a trap-door opens in the ceiling, to permit some glimpse of the workings of the bells. Hence it comes close to the controlling rhetorical principle of the *Treatise*:

> We must therefore seek for a mode of expression which shall not be capable of expressing too much, and which shall leave room for the introduction of new ideas as these are developed from new facts. (*Tr* ii/7)

MAXWELL'S EXPOSITION OF LAGRANGIAN THEORY

Despite Maxwell's expressed satisfaction with the work of Thomson and Tait, his own presentation of the Lagrangian theory in Chapter V of the *Treatise* is distinctly different. There is an air of simplicity about Maxwell's argument—fewer symbols are in play than in the laborious process we have displayed above—but there is at the same time an aspect of indirectness and obscurity, almost of sleight-of-hand, which would hardly lead us to regard Maxwell's argument as an improvement over that of Thomson and Tait if direct demonstration were the whole objective. Maxwell explains at length in the first and last sections of the chapter, that simplicity is not his aim. He has no fault to find with the demonstrations as they have been carried out by others, and he does not propose to improve upon them. But they are in the language of the mathematician, and they therefore call for translation into the language of dynamics if they are to be "in a state fit for direct application to physical questions" (*Tr* ii/199). His purpose, then, is not simply to repeat what has been done adequately by others, or to produce an elegant derivation of a mathematical result, but to reshape the theory from what he calls "a physical point of view" (*Tr* ii/199).[11]

11. Maxwell wrote to Tait (June 30, 1872): "... I have been overhauling the Equations of motion and have got a way of deducing them ... from the variables their velocities and the forces action *on them* alone. This is done by beginning with an impulsive force. It constitutes an improvement in my book, and a preparation for Electrokinetics and Magnetic action on light" (H.414). See "On the Proof of the Equations of Motion of a Connected System" (1873) (*SP* ii/308 ff.).

Curiously, this does not mean that the chapter will include physical analogies or illustrations of the sort we have met in the preceding chapters; one physical example of the 1864 paper was pruned when the corresponding section of the *Treatise* was written (*SP* i/537–38). The mechanical model of a coupled system which appears at (*Tr* ii/228) was constructed for Maxwell's lectures and was inserted into the third edition of the *Treatise* by J. J. Thomson as editor; there is no mention of it or anything similar by Maxwell here. The phrase "physical point of view" must be understood in some other sense.

I believe that a clue is to be found by considering the nature of Maxwell's divergence from Thomson and Tait, and Lagrange, in deriving the new equations of motion. He contrives to do this in such a way that *the term "mass" does not occur*. This might seem a perverse way to introduce dynamical thinking, but note that at the same time the terms "kinetic energy," "coordinates," "velocities," "momenta" and "force" are used freely. These terms, in this chapter, regularly refer to the new generalized quantities—*nowhere do Cartesian coordinates or the strictly Newtonian quantities appear*. The new quantities, then, instead of being treated as purely mathematical transformations of quantities which were physically understood in the Newtonian usage, are treated from the beginning as in themselves meaningful dynamical terms. Clearly Maxwell is undertaking to give immediate dynamical meaning to the *new* terms, which appear from the beginning in their own right, and not as transformations of anything else.

Certain of the new relations, then, which would otherwise be derived, must be asserted as principles. I think this observation goes a long way toward clearing up the logic of the argument. It is characteristic of Maxwell that he does not point this out explicitly to the reader, but leaves the terms in which he has framed the argument to speak for themselves. If he is right, and they do have intuitive meaning, any framework of explanation will only be a distraction. The book *Matter and Motion*, which in this sense is of a piece with Chapter V of part iv of the *Treatise*, consistently speaks only from within the new point of view. What this means for Maxwell metaphysically will become apparent in Chapter 5, below.

Maxwell begins by asserting as a principle an equation which looks like Newton's Second Law of Motion, but which is in truth referring to generalized coordinates:

> When a body moves in such a way that its configuration, with respect to the force which acts on it, remains always the same … the moving force is measured by the rate of increase of the momentum. If F is the moving force, and p the momentum,
>
> $$F = \frac{dp}{dt}$$
>
> whence

$$p = \int F dt .$$

(*Tr* ii/201)

Except for this relation between them, F and p remain undefined terms—we are, in effect, invited to form ideas of the generalized terms directly. In general, they are not Newtonian force and momentum, and thus p is not mv, though Maxwell does not trouble our passage into the new dynamics by pointing this out. He is proposing new ground as *terra firma*.

Of the two quantities, "force" and "momentum," it is the former which is treated as intuitively prior, and the statement which follows that quoted above is to be regarded as Maxwell's definition of momentum:

> The time-integral of a force is called the Impulse of the force; so that we may assert that the momentum is the impulse of the force which would bring the body from a state of rest into the given state of motion. (*Tr* ii/202)

Compare the suggestion concerning the bell-ringers (page 212, above): "... they can estimate its momentum by stopping all the ropes at once, and feeling what sort of tug each rope gives." The *tug* is an impulse, the negative of the impulse which would bring the system from rest into motion. By taking *force* as fundamental, Maxwell makes it unnecessary to state a definition of momentum as the product of "mass" and velocity, and thereby makes it possible to omit the term "mass" altogether in establishing the foundations of dynamics.

The qualification which limits the definition of momentum in the integral expression above warrants some discussion. Maxwell asserts that "... its configuration, with respect to the force which acts on it, remains always the same." Taking the term "configuration" to mean the relative positions of the parts of the system, or its shape, the condition is that the shape remain constant, the force being related to it in a constant way. This condition is not in general fulfilled in a complex body, for the configuration is constantly changing in a way that depends upon the relative velocities of the parts. Nevertheless, the required restriction can be met by means of a stratagem.

Let us, with Maxwell, denote the generalized coordinates by q_i and the velocities $\dot{q}_i$. If there is an increment in some coordinate dq_i, then for continuous motion (which the discussion assumes) it must be expressible in the form

$$dq_i = \dot{q}_i dt$$

and $dq_i \to 0$ as $dt \to 0$. On the other hand we may (with Newton—compare Book I, Prop. I of the *Principia*) envision impulsive forces, by which a finite increment of momentum is imparted in an interval of time as small as we please, F being imagined to increase as necessary so that the integral

$$\Delta p = \int f dt$$

remains constant as $dt \to 0$. Maxwell extends his definition of momentum to include the case of changing configurations by means of this *instantaneous impulse*; his definition, which I shall divide into two stages, is this:

> [I] The limiting value of the impulse, when the time is diminished and ultimately vanishes, is defined as the *instantaneous* impulse ...
>
> [II] and the momentum p, corresponding to any variable q, is defined as the impulse corresponding to that variable, when the system is brought instantaneously from a state of rest into the given state of motion. (*Tr* ii/202)

The point of this definition is that the integration $\int F dt$ is performed while all the q_i are held constant—before the configuration "has a chance" to change. It is, then, a *partial* integration. We are given a system in a certain state (q_i', $\dot{q}_i'$), and we ask: what is to be meant by the momentum corresponding to a given variable q_i in this state? Maxwell's answer is that the momentum p_i corresponding to the variable q_i is the instantaneous impulse required of the force F_i when the system is brought from rest into the given state of motion—with all variables held constant at the values they have in that state (q_i'). Though it may be hard to visualize bringing the system into motion in the given configuration by means of the requisite assortment of simultaneous hammer-blows, we can always think of the inverse process, "stopping all the ropes at once," as Maxwell has suggested. The force will in general be infinite, but the physically significant quantity is the impulse; the impulses will be finite, and each will measure the momentum of the corresponding coordinate.[12]

Maxwell next introduces the concepts of work and energy, assuming initially that "the work done by the force F_1 during the impulse is the space-integral of the force" (*Tr* ii/203). In expressing the definition in this way, he is evidently borrowing a concept from the physics of Cartesian coordinates which he will immediately extend, for the "space" is now represented by generalized coordinates; he writes:

$$W = \int F_i \, dq_i \, .$$

In effect, he begins with a generalized concept of work. He now derives an alternative expression for the work done by a "very small" (read: "infinitesimal") impulse, using a mean-value argument. The generalized velocity changes during the impulse, but if we write its least and greatest values as $\dot{q}_i''$ and $\dot{q}_i'$, respectively, the work done during the impulse will lie between

12. Note a resemblance to the "Dirac delta-function": a function which in the limit is infinite at a single value of the independent variable and zero elsewhere, such that the integral of the function has a finite value.

the limits

$$\dot{q}_i{}'' \int F_i\, dt \;\leq\; W \;\leq\; \dot{q}_i{}' \int F_i\, dt$$

or

$$\dot{q}_i{}'' dp_i \;\leq\; W \;\leq\; \dot{q}_i{}' dp_i\,.$$

As the impulse is diminished, $\dot{q}_i{}''$ and $\dot{q}_i{}'$ approach the value $\dot{q}_i$, and we have:

$$dW_i = \dot{q}_i\, dp_i\,.$$

This is in effect a reconstruction of our concept of work, now understood in terms of velocity and impulse; it leaves out of view quasi-static processes and potential energies, even if strictly speaking it does not exclude them: "*the work done by a very small impulse is ultimately the product of the impulse and the velocity*" (*Tr* ii/203).

The apparently kinetic character of the electromagnetic processes invites a return to the Cartesian project: to understand nature in terms of "matter and motion," without recourse to latent "potential" energies. Maxwell emphasized the kinetic aspect of the current in stage 3 of the reasoning which led us into the dynamical theory (page 205, above), asserting:

> [I]ts energy must be kinetic energy, that is to say, the energy which a moving body has in virtue of its motion. (*Tr* ii/197)

With this goal in view, Maxwell assumes that the "work done," which he has just expressed in terms of impulse and velocity, will be kinetic energy:

> When work is done in setting a conservative system in motion, energy is communicated to it, and the system becomes capable of doing an equal amount of work against resistances before it is reduced to rest.
>
> The energy which a system possesses in virtue of its motion is called its Kinetic Energy, and is communicated to it in the form of the work done by the forces which set it in motion. (*Tr* ii/203)

Convinced that an account of a connected system need only consider such kinetic energy, Maxwell writes the above work as an increment in kinetic energy dT; then for such infinitesimal processes throughout the system we will have:

$$dT = \sum_i \dot{q}_i\, dp_i\,.$$

Maxwell now deduces, quite simply, a result which Thomson and Tait obtained only by a cumbersome manipulation of the equations of transformation from Cartesian to generalized representation.[13] Having *begun* with generalized coordinates, Maxwell has relatively little to do. Since in fact $dq_i = \dot{q}_i\, dt$, we may equally well represent T as a function of p and q instead of p and $\dot{q}$. Then, taking $T = T(p_i, q_i)$, we have the expression for the total derivative:

13. Thomson and Tait, *Treatise, 1*, p. 290, equation (10).

$$dT = \sum_i \left(\frac{\partial T}{\partial p_i} dp_i + \frac{\partial T}{\partial q_i} dq_i \right).$$

Since an expression such as $\partial T / \partial p_i$ will be completely different in general when T is represented as a function of (p, q) instead of $(p, \dot{q})$, Maxwell introduces the device of distinguishing the representations of T by subscripts. Anticipating a third form to be used later, we may here list the notations Maxwell uses:

$$T_{p\dot{q}} = T(p, \dot{q})$$
$$T_p = T(p, q)$$
$$T_{\dot{q}} = T(\dot{q}, q).$$

Of course T is the same quantity in every case, the total kinetic energy of the system; the subscripts merely identify the choice of representation, and hence distinguish the functions by which T may be calculated.[14] With this refinement in notation, the equation above becomes

$$dT_p = \sum_i \left(\frac{\partial T}{\partial p_i} dp_i + \frac{\partial T}{\partial q_i} dq_i \right).$$

Let us apply this equation to the special case—which happens to be that case by which "momentum" is defined—in which the process is one of instantaneous impulse. As we noted earlier, the continuity of the motion requires that $dq = \dot{q}\, dt$ go to zero as dt goes to zero, while we have contrived that dp shall *not* go to zero in the same interval. Hence the dependence of dT upon dq as expressed in the second term makes no contribution in the instantaneous process, and dT here becomes simply

$$dT_p = \sum_i \frac{\partial T}{\partial p_i} dp_i \, .$$

But under the same restriction to an instantaneous impulse, we had (*Tr* ii/203)

$$dT_p = \sum_i \dot{q}_i \, dp_i \, .$$

Hence

$$\dot{q}_i = \frac{\partial T_p}{\partial p_i} \, .$$

Maxwell is now ready to discard the stratagem of the instantaneous impulse; he says of this equation for $\dot{q}_i$:

> [T]he variables, and the corresponding velocities and momenta, depend on the actual state of motion of the system at the given instant, and not on its previous history.

14. Although their notations were different, the distinction of notations for different functional representations of T was made by Thomson and Tait (*op. cit.*, *1*, p. 306).

> Hence, the equation ... is equally valid, whether the state of motion of the system is supposed due to impulsive forces, or to forces acting in any manner whatever.
>
> We may now therefore dismiss the consideration of impulsive forces. (*Tr* ii/204)

Maxwell transcribes the equation for $\dot{q}$ into words, and I think this is significant; for him, it is not simply a mathematical equivalence, but a refinement of our *concept* of generalized velocity: "*the velocity corresponding to the variable q is the differential coefficient of* T_p *with respect to the corresponding momentum p*" (*Tr* ii/204). It would clarify things, perhaps, to add to this verbal statement a reminder of what our partial-derivative notation makes clear, that the derivative is taken at constant configuration. Even though we discard the device of the instantaneous impulse, its role is still implicit here, for the "constant configuration" is visualized physically through the device of an impulse too sudden to permit a shift of configuration.

Maxwell began with a force law that looked so familiar from Newtonian physics that we were willing to accept it without question: "the moving force is measured by the rate of increase of momentum" (page 214, above). But he was careful to guard his statement with a limitation to motions with constant configuration. The situation is not so simple in other cases and, indeed, the law which we were willing to accept as the translation of Newton's Second Law of Motion is not in general valid. It is time now for Maxwell to remove the mask of innocence, and reveal the complexity latent in the law which he earlier would have us regard as a first principle.

In § 561 Maxwell considers the effect of letting the system "move in any arbitrary way, subject to the conditions imposed by its connexions" (*Tr* ii/205). He first corrects the expression for dT_p. The second term of the total differential cannot now be ignored, and we have instead (schematizing the argument of § 561):

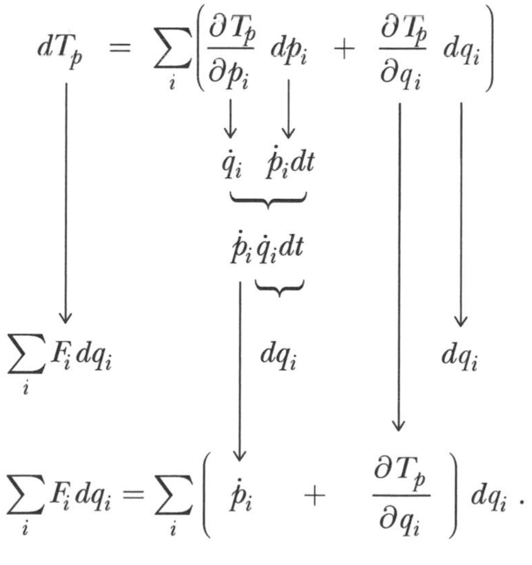

$$dT_p = \sum_i \left(\frac{\partial T_p}{\partial p_i}\, dp_i + \frac{\partial T_p}{\partial q_i}\, dq_i \right)$$

$$\frac{\partial T_p}{\partial p_i} \downarrow \dot{q}_i \qquad dp_i \downarrow \dot{p}_i\, dt \qquad \underbrace{\dot{q}_i \;\; \dot{p}_i\, dt}_{\dot{p}_i \dot{q}_i\, dt} \qquad \underbrace{\dot{q}_i\, dt}_{dq_i}$$

$$dT_p \downarrow \sum_i F_i\, dq_i \qquad\qquad dq_i \downarrow dq_i$$

$$\sum_i F_i\, dq_i = \sum_i \left(\dot{p}_i + \frac{\partial T_p}{\partial q_i} \right) dq_i \,.$$

And because the dq_i are independent,

$$F_i = \dot{p}_i + \frac{\partial T_p}{\partial q_i}.$$

This is the corrected law of force. To the original law, we had to add a second term: "The second part is the rate of increase of the kinetic energy per unit of increment of the variable, the other variables and all the momenta being constant" (*Tr* ii/205).

It is interesting that the preceding argument was rejected both by J. J. Thomson, as editor of the third edition of the *Treatise*, and by Sir Joseph Larmor, in a note to *Matter and Motion*, though for different reasons.[15] Thomson says:

> This proof does not seem conclusive as dq is assumed to be equal to $\dot{q}\,dt$, that is to $(\partial T_p/\partial p)\delta t$, so that all we can legitimately deduce … is
>
> $$\sum\left(\frac{\partial p_i}{\partial t} + \frac{\partial T_p}{\partial q_i} - F_i\right)\frac{\partial T_p}{\partial p_i} = 0. \qquad \begin{matrix}[i \leftrightarrow r]\\ [F \leftrightarrow p]\\ \left[\frac{\partial}{\partial} \leftrightarrow \frac{d}{d}\right]\end{matrix}$$

Thomson's objection does not seem satisfactory. He denies that the coefficients of the dq_i are independent of the dq_i themselves, because of the apparent connection through the equations $q_i = \dot{q}_i\,dt$. But the choice of the k functions $\dot{q}_i(t)$ is completely arbitrary, leaving any combination of the q_i and $\dot{q}_i$ possible, and this, it seems to me, meets the mathematical requirement of independence, permitting equation of the coefficients.

Larmor pointedly rejects Thomson's objection: "The equation is then derived correctly, as the variations dq_i are fully arbitrary." But he introduces what would seem to be a more fundamental objection. He protests that Maxwell appears to have derived the full set of equations of motion (*i.e.*, one equation for each value i from $i = 1$ to $i = k$) from the principle of the Conservation of Mechanical Energy alone, something Lagrange had already declared to be impossible—the Conservation of Energy alone does not determine a motion.

The question raised by Larmor is, then, on what principle or principles does Maxwell's derivation rest? To me it seems clear that the argument depends upon the equation

$$p = \int F\,dt\,;$$

and while this was introduced as a definition of momentum, it is equivalent to the assumption of Newton's Second Law (properly generalized) as a first

15. Larmor, in Maxwell, *Matter and Motion*, p. 158. Thomson's note appears in the third edition of the *Treatise* at (*Tr* ii/205).

principle. Nonetheless, Larmor's objection serves well to remind us of the difficulty and importance of keeping track of first principles. Maxwell has deliberately led us over shifting ground: he is guiding us to new habits of thought, or rather to the recognition of new *principles* which, once discovered, have a status more significant than mere habits. In the process of transition—which is essentially a process of learning—it is necessary that old modes of thought be dissolved, and concepts held plastic for a time. No wonder that in this far-reaching revision of the fundamental science, first principles become difficult to identify.

It is in this spirit that Maxwell has refused to repeat for us the customary formulation for kinetic energy, namely,

$$T - \frac{1}{2}\sum_i m_i v_i^2 .$$

He later points out the indirection he has employed: "In the preceding investigations we have avoided the consideration of the form of the function which expresses the kinetic energy in terms either of the velocities or of the momenta" (*Tr* ii/208).

Finally, however, he must produce an expression for T, though of course in generalized terms. To do so Maxwell employs a device which he has used earlier to calculate the energy of an electrostatic system; this is the device of a proportionally growing system, such as a condenser being charged. The factor 1/2 then appears, as it did in the expression for electrical energy, as the result of integrating $\int_0^1 n\,dn = \frac{1}{2}$, where n is the scale factor measuring the growth of the system. Of course we have perfect freedom to prescribe whatever history we wish for the system since the variables are independent of one another, and the result will be a function of the state of the system, quite independently of the mode by which it was achieved. The argument is presented in § 562; we will reproduce it in schematic form, as we did previously for the argument in § 561.

We seek an expression for $T_{p\dot{q}}$, as a function of the variables p_i, $\dot{q}_i$. Let p_i and $\dot{q}_i$ denote the values of these variables in the state in question, and let p_i' and $\dot{q}_i'$ denote the values in another state proportional to the first, in the sense that

$$p_i' = np_i ,$$

$$q_i' = nq_i .$$

Now suppose that the state $(p_i', \dot{q}_i')$ arises by the growth of n from $n = 0$ to $n = 1$. Beginning then with the relation

$$dW_1 \;=\; F_1\,dq_1 \;=\; dT_1 ,$$

Maxwell will derive the desired expression. In the course of the argument, it is to be kept in mind that n is the varying quantity. We have:

$$dW_1 = F_1 dq_1 = \underbrace{dT_1}_{\dot{q}_1 \cdot dp_1'}$$

$$n\dot{q}_1' \cdot p_1 dn$$

$$dT_1 = p_1 \dot{q}_1 \cdot n\,dn$$

$$T_1 = \int_0^1 p_1 \dot{q}_1 \cdot n\,dn$$

$$T_1 = p_1 \dot{q}_1 \underbrace{\int_0^1 n\,dn}_{\frac{1}{2}}$$

$$T_{p\dot{q}} = \frac{1}{2}\sum_i p_i\,\dot{q}_i\ .$$

At this point, Maxwell is ready to derive Lagrange's equations of motion (*Tr* ii/207, page 211, above). He uses the simple device of equating the three representations of the kinetic energy (page 218, above), as

$$T_p + T_{\dot{q}} - 2T_{p\dot{q}} = 0\,.$$

The rest of the proof is entirely formal, employing only relations which we have already established. The result is:

$$F_i = \frac{d}{dt}\frac{\partial T_{\dot{q}}}{\partial \dot{q}_i} - \frac{\partial T_{\dot{q}}}{\partial \dot{q}_i}\,.$$

This is the equation of motion upon which Maxwell will build his electromagnetic theory. He will work, therefore, with the kinetic energy in the representation $T_{\dot{q}} = T(q, \dot{q})$. Up to this point, he has derived an explicit expression only for the form $T_{p\dot{q}}$. To transform the representation, Maxwell argues in § 565 in the following way:

We have just shown that $T_{p\dot{q}}$ is given by

$$T_{p\dot{q}} = \frac{1}{2}\sum_i p_i \dot{q}_i\,,$$

that is,

$$T_{p\dot{q}} = \frac{1}{2}\sum_i \frac{\partial T_{\dot{q}}}{\partial \dot{q}_i}\dot{q}_i\,.$$

It follows, Maxwell then asserts, that $T_{\dot{q}}$ is a homogeneous second-degree function of the $\dot{q}_i$. To show this, we may begin with such a function and show that it satisfies the equation above. We may tentatively write $T_{\dot{q}}$ as a homogeneous second-degree function of the $\dot{q}_i$ as follows:

$$T_{\dot{q}} = P_{11}\dot{q}_1^{\,2} + P_{22}\dot{q}_2^{\,2} + \cdots + P_{12}\dot{q}_1\dot{q}_2 + \cdots = \frac{1}{2}\sum_{ij} P_{ij}\dot{q}_i\dot{q}_j,$$

in which each of the P_{ij} is in general a function of all the q_i, but does not contain the $\dot{q}_i$ (otherwise, the degree of the function would rise). Evidently $P_{ij} = P_{ji}$ since the order of the factors $\dot{q}_i\dot{q}_j$ is indifferent. The factor 1/2 corrects the summation notation for the fact that each term occurs twice in the double sum. Differentiating, we get partial derivatives of the form

$$\frac{\partial T}{\partial \dot{q}_i} = 2P_{ii}\dot{q}_i + P_{i1}\dot{q}_1 + P_{i2}\dot{q}_2 + \cdots ;$$

thus

$$\frac{1}{2}\dot{q}_i\frac{\partial T}{\partial \dot{q}_i} = P_{ii}\dot{q}_i^{\,2} + \frac{1}{2}P_{i1}\dot{q}_i\dot{q}_1 + \frac{1}{2}P_{i2}\dot{q}_i\dot{q}_2 + \cdots.$$

And since $P_{i1} = P_{1i}$,

$$\sum_i \frac{1}{2}\dot{q}_i\frac{\partial T}{\partial \dot{q}_i} = \frac{1}{2}\sum_i P_{ij}\dot{q}_i\dot{q}_j .$$

T_q is therefore of the form assumed, for when transformed this expression satisfies the equation for $T_{p\dot{q}}$; and it is clear, from the process by which we have established this fact, that this is the most general form which will do so.

Furthermore, differentiating a second time,

$$\frac{\partial^2 T}{\partial \dot{q}_i^{\,2}} = 2P_{ii} \quad \text{and} \quad \frac{\partial^2 T}{\partial \dot{q}_i \partial \dot{q}_j} = 2P_{ij} .$$

Therefore:

$$T_{\dot{q}} = \frac{1}{2}\left(\frac{\partial^2 T}{\partial \dot{q}_1^{\,2}}\dot{q}_1^{\,2} + \frac{\partial^2 T}{\partial \dot{q}_2^{\,2}}\dot{q}_2^{\,2} + \cdots + \frac{\partial^2 T}{\partial \dot{q}_1 \partial \dot{q}_2}\dot{q}_1\dot{q}_2 + \cdots\right).$$

Following "treatises on the dynamics of a rigid body," by which I think we are to understand primarily that of Routh,[16] Maxwell terms the coefficients P_{ij} "products of inertia" when $i \neq j$. In the special case $i = j$, he names the coefficients P_{ii} "moments of inertia." Once again, however, Maxwell is concerned to extend the familiar concept: "We may extend these names to the more general problem which is now before us, in which these quantities are not, as in the case of a rigid body, absolute constants, but are functions of the variables q_1, q_2, etc." (*Tr* ii/209).

16. Edward J. Routh, *Elementary Treatise on the Dynamics of a System of Rigid Bodies* (2nd ed., Cambridge: 1868). The definitions that follow are given at the outset of the second edition; I have not seen the first edition. Routh and Maxwell took their degrees at Cambridge in the same year, sharing the Smith's Prize (Campbell, *Life*, p. 124).

With this, we complete a survey of Maxwell's exposition of the methods which he calls here "pure dynamics." Maxwell's retrospective statement of the purpose of the chapter (*Tr* ii/210) brings together his convictions about both the nature of science and its history. He has been concerned with the relation of analytic methods to thought clothed in intelligible words. At the same time, his conception of the role of the *Treatise* in the general opening of a new science is very clear.

For Maxwell, it is not by chance, but is in the nature of things, that Lagrange's work contained "no figures," and that Lagrange and most of his followers worked as mathematicians:

> [T]hey have endeavoured to banish all ideas except those of pure quantity, so as not only to dispense with diagrams, but even to get rid of the ideas of velocity, momentum, and energy, after they have been once for all supplanted by symbols in the original equations. (*Tr* ii/210)

Science, for Maxwell, is not fundamentally empirical—at least, not in any direct Baconian sense. As we shall see in the next chapter, the fundamental ideas of science are, in Maxwell's view, *a priori* constructions of the human mind and reflect the necessary laws of the mind's own functioning. Mathematics therefore necessarily leads, and empirical science follows:

> [T]he development of the ideas and methods of pure mathematics has rendered it possible, by forming a mathematical theory of dynamics, to bring to light many truths which could not have been discovered without mathematical training. (*Tr* ii/210)

We recognize in this Maxwell's characterization of his own relation to Faraday, as he has discussed it in respect to the "electrotonic state" (page 194, above). There are truths in Faraday's work which wait to be brought to light; Lagrange, on the other hand, has developed "ideas and methods of pure mathematics" which wait to illuminate them. It is Maxwell's rôle to act as intermediary, bringing Lagrange's theory to bear upon the forms of thought which Faraday has evolved from his intense study of nature. In so doing, Maxwell will accomplish the translation of Faraday at a deeper level than that which he achieved in the first chapters of the *Treatise*. His reformulation of dynamical theory in the present chapter has been a necessary preliminary step. He understands the reformulation of Lagrange in this way:

> [I]f we are to form dynamical theories of other sciences, we must have our minds imbued with these dynamical truths as well as with mathematical methods.
>
> In forming the ideas and words relating to any science, which, like electricity, deals with forces and with their effects, we must keep constantly in mind the ideas appropriate to the fundamental science of

> dynamics, so that we may, during the first development of the science, avoid inconsistency with what is already established. (*Tr* ii/210)

We see that this is what Maxwell has attempted in Chapter V: to transform Lagrange's "mathematical methods" into *ideas* "appropriate to the fundamental science of dynamics":

> In order to be able to refer to the results of this Lagrangian analysis in ordinary dynamical language, we have endeavoured to retranslate the principal equations of the method into language which may be intelligible without the use of symbols. (*Tr* ii/210)

Once again, we see the task of the *Treatise* expressed as a problem of language, and Faraday's plea for a science freed from its "hieroglyphs" endorsed as a legitimate demand (page 49, above). Lagrange must first be translated—out of pure mathematics, into dynamics—in order that through him Faraday in turn can be translated.

In this chapter, Maxwell has tried to interpret Lagrange's equations, not as others had done, through an equivalence to Newton's concepts and laws, but by means of a reformation of the foundations and the vocabulary of mechanics. He has tried to find new ideas in the new equations, and to express this new body of thought in working form. The new first principles will be appropriate not to a mechanics of the hard, massy bodies which Newton and Locke take as primary, but to the electromagnetic field as a continuous mechanical system, in which "mass" cannot be identified except as the bearer of momentum or energy. But though Maxwell has been pressed to this work of interpreting Lagrange by the facts of electromagnetism and by Faraday's thoughts arising from his own experimentation, it has not been a work of empirical science. It has been a work of mind, reflecting on its own productions.

With this preparation, we are ready to begin "forming the ideas and words" relating to the new science of electricity. This is the place of the *Treatise* in the history of science, as Maxwell intends it. By no means a definitive work, it is to give initial form to the thoughts Faraday has evolved, by bringing the reconstituted, generalized dynamics of Lagrange to bear upon them. The result will be more like a chemical reaction than a decisive argument. Curiously, it will be a step from the *a priori* into the emprirical, rather than the reverse. Although the work has been invited by Faraday's *Experimental Researches,* it actually consists in moving from pure mathematics toward the phenomena, by a process of deduction from the dynamical equations derived in this chapter.

These statements will become clearer—and at the same time properly qualified—in the following chapter, as we examine the metaphysical context in which Maxwell understands the term "dynamics."

Chapter 5
The Meaning of "Dynamical Theory" for Maxwell

PHYSICS

"Dynamical" is a very important term for Maxwell, and a "dynamical theory" is for him a theory of a definite and special kind, regularly distinguished in his work from theories in other modes. A fundamental distinction in Maxwell's writing is that between a *dynamical* and a *physical* theory. This is a dichotomy which does not necessarily arise in physics itself, but its interpretation is one of the most important problems for a student of Maxwell. Its roots, as I hope to show shortly, are metaphysical, but its manifestations in Maxwell's "strategy" are everywhere in his work; and it will, I think, be best to begin by examining some instances of the distinction in practical alternatives as he sees them.

The *Treatise on Electricity and Magnetism* was incubated in three major papers (page 40, above). They are "On Faraday's Lines of Force" (read in 1855–56); "On Physical Lines of Force" (1861–62); and "A Dynamical Theory of the Electromagnetic Field" (1864–65). Each of these papers takes a different approach to the understanding of electrical phenomena. The first proposes what Maxwell calls a "physical analogy"—not as a theory of electromagnetism, nor in order to explain the phenomena, but to give a certain intelligibility to Faraday's conception of "lines of force" by bringing that imagery into formal correspondence with the kinematic concept of an ideal fluid. This is a mathematical, not a physical fluid, to which certain arbitrary but mutually consistent properties have been assigned (it is, for example, both frictionless and devoid of inertia).[1]

The second paper has a quite distinct purpose, though the intentions of all three papers are linked, with the *Treatise*, as related stages in the

1. The imaginary fluid is derived from "caloric," which Maxwell abandoned as a physical analogue because of developments in the theory of heat: "[W]hile the mathematical laws of the conduction of heat derived from the idea of heat as a substance are admitted to be true, the theory of heat has been so modified that we can no longer apply to it the idea of substance. It is for this reason that in choosing a concrete form in which to develop the analogy of attraction I have not taken that of heat, as pointed out by Professor Thomson, but at once assume a purely imaginary fluid as the vehicle of mathematical reasoning." (Manuscript version of "Faraday's Lines," H.491. Compare pages 101 ff., above.) Thomson's reluctance to abandon the notion of heat flow is chronicled by his biographer; see S. P. Thomson, *Life of Lord Kelvin*, Chapter VI, especially pp. 266 ff.

approach to the resolution of a scientific enigma. In "Physical Lines," Maxwell offers a *physical* or *mechanical* hypothesis. He shows that phenomena corresponding systematically to those of electromagnetism could be produced by a mechanical system of a certain complex form, which he describes in detail. The system combines the properties of two kinds of medium: (1) a system of vortices provided with idling wheels at their perimeters of contact, and (2) a solid medium in a state of stress. A paper of Thomson's, "On a Mechanical Representation," was discussed in Chapter 2 of the present study; it will be recalled that this proposed a mathematical analogy between the configuration of a strained elastic solid, and both the electric and magnetic fields (pages 105 ff., above). It is not surprising that Maxwell draws upon this for the elastic-solid aspect of his own hypothesis, taking the displacement of the solid medium as the mechanical explanation of the electric field (in Part III of "Physical Lines," *SP* i/489 ff.). We saw that Thomson had already seen his analogy as a "hint" of a physical theory, while Maxwell in turn was accustomed to think in terms of stress and strain patterns in solids (page 109, above). It is impressive, however, to find that Thomson's paper is not only the source of inspiration for the deformability of Maxwell's proposed medium in its aspect as an elastic solid, but that Thomson's solid analogy is regarded by Maxwell as the prototype of his entire vortex hypothesis (*SP* i/453). This reveals, I believe, the scope of the suggestive power of formal analogies as Thomson and Maxwell were using them. Maxwell has employed this power to produce a tentative hypothesis, describing bodies which might conceivably exist in nature and which, if they did, would account for the phenomena. He by no means asserts that they *do* exist, nor does he even argue for their plausibility. "Faraday's Lines" entertained a purely mathematical concept; "Physical Lines" has rather the status of a physical guess. The pair of papers, it may be added, corresponds to two modes of Faraday's own thought, incorporated respectively in Series XXVIII of the *Experimental Researches* ("On Lines of Magnetic Force"), and in the speculative paper "On the Physical Character of the Lines of Magnetic Force." The first, regarding the lines "as representations" (*XR* iii/328), was read as usual to the Royal Society; the second was sent instead to the *Philosophical Magazine*:

> The following paper contains so much of a speculative and hypothetical nature, that I have thought it more fitted for the pages of the Philosophical Magazine than those of the Philosophical Transactions. (*XR* iii/407)

Maxwell similarly published his own "Physical Lines" in the *Philosophical Magazine*.

The usual course of scientific procedure would be to follow a proposed hypothesis with experiments designed to test it at crucial points, and thereafter to contrive new hypotheses, modified as necessary to meet the empirical demands. Maxwell did try such experiments, as we shall see (pages 279–284, below), but when they yielded no positive results he did not continue to devise hypotheses. The method of hypothesis does not lie in the main line of his program.

For Maxwell, the next stage is the "Dynamical Theory," and there (in his own view) he advances *without hypothesis*. Let us turn to his own words in the "Dynamical Theory" for an explanation of the method of this third paper:

> I have on a former occasion [Maxwell refers to "Physical Lines"] ... attempted to describe a particular kind of motion and a particular kind of strain, so arranged as to account for the phenomena. In the present paper *I avoid any hypothesis of this kind*; and in using such words as electric momentum and electric elasticity in reference to the known phenomenon, I wish merely to direct the mind of the reader to mechanical phenomena which will assist him in understanding the electrical ones. All such phrases in the present paper are to be considered as illustrative, not as explanatory.
>
> In speaking of the Energy of the field, however, I wish to be understood literally. ... On our theory it resides in the electromagnetic field, in the space surrounding the electrified and magnetic bodies, and is in two different forms, which may be described *without hypothesis* as magnetic polarization and electric polarization, or, according to a very probable hypothesis, as the motion and the strain of one and the same medium. (*SP* i/563–64; emphasis added)

In the same paper he crisply defines the meaning of two crucial terms:

> The theory I propose may therefore be called a theory of the *Electromagnetic Field*, because it has to do with the space in the neighbourhood of the electric or magnetic bodies, and it may be called a *Dynamical Theory*, because it assumes that in space there is matter in motion, by which the observed electromagnetic phenomena are produced. (*SP* i/527; Maxwell's emphasis)

In Maxwell's view, then, a dynamical theory makes no hypothesis concerning the existence or properties of specific bodies; it makes only the one overall assertion that the phenomena can be understood in terms of *matter and motion*. Other terms enter the dynamical reasoning, of course, but these two Maxwell regards as fundamental and sufficient. The status of this one universal assertion we shall examine with care; it is clear, however, that Maxwell does not regard it as an "hypothesis" in any ordinary sense. The laws which govern matter and motion are, in simple cases, Newton's Laws of Motion, but more generally they are the laws of pure dynamics which we have just discussed. The fact that neither the term "matter" nor "mass" appeared in that discussion in no way undermines the assertion that it was the subject of the chapter: all that we understand properly by the term *matter* has been said in Chapter V of Part iv of the *Treatise*. Maxwell has swept dynamics clean of any notion of little hard bodies; he has asserted precisely what he intends by the term "mass" all the better for not having uttered the word. We shall examine this point further.

It is a curious fact that Tait did not initially catch the distinction Maxwell intended between the "Physical" and the "Dynamical" papers; two years after the publication of the second, Maxwell had to correct a misconception conveyed in an early version of Tait's *Sketch of Thermodynamics*. In a letter to Tait Maxwell writes:

> There is a difference between a vortex theory ascribed to Maxwell at p 57, and a dynamical theory of magnetism by the same author in Phil. Trans. 1865. The former is built up to show that the phenomena are such as can be explained by mechanism. The nature of the mechanism is to the true mechanism what an orrery is to the Solar System. The latter is built on Lagrange's Dynamical Equations and is not wise about vortices.[2]

Both the "Physical Lines" and the "Dynamical Theory" reappear in the *Treatise* but, as others have pointed out, there is a great shift of emphasis toward the dynamical theory.[3] The motion of thought through the sequence of three papers, beginning with "Faraday's Lines," continues into the *Treatise*. The "Dynamical Theory" of 1865 becomes in general the five chapters of the *Treatise* to which the present study is devoted. In the *Treatise*, the ground is laid much more carefully—surely the result of further reflection on Maxwell's part—and certain topics included in the original paper are distributed in other sections of the book. Neither Chapter V of Part iv of the *Treatise*, which we have just discussed, nor Chapter VI, in which the terms of Lagrange's equations are examined in detail for their possible relation to electromagnetic phenomena, appeared in the 1865 paper. On the other hand, the boundary demarcating "dynamics" seems to have been redrawn in the *Treatise*; some topics that were developed in the 1865 paper, particularly the calculation of energy distribution in the field and the electromagnetic theory of light, are in the *Treatise* definitely detached from the dynamical theory and treated separately.

It is not so easy to trace the fate of the "Physical Lines" as it is absorbed into the *Treatise*, but in general the theory of stress in a solid medium is treated in Chapter XI of Part iv, almost directly after the chapters on the dynamical theory (a chapter on units intervenes), while the theory of molecular vortices is included in Chapter XXI on the "Magnetic Action of

2. Knott, *Life of Tait*, p. 215. In Tait's *Sketch of Thermodynamics* (Edinburgh: 1868), pp. 74 ff., the distinction seems to be satisfactorily made. Maxwell in his letter was referring to an earlier pamphlet, as explained by Knott in the passage cited.

3. J. G. Crowther, *British Scientists of the Nineteenth Century* (London: Kegan Paul and Co., 1935), p. 310. Crowther regards it as a misfortune that Maxwell discarded the earlier emphasis on particular models.

Light" (the Faraday effect). This places the hypothesis of vortices in connection with the phenomenon which seemed to Maxwell the best empirical clue to their existence. To borrow Aristotle's terms, with which Maxwell would have been quite familiar, the order of *inquiry* represented by the papers leading up to the *Treatise* is reversed in the order of *knowing* or teaching, represented by the book itself. The Dynamical Theory was the last in the sequence of inquiry, but it stands at the beginning of Maxwell's exposition of his own theory, and in the *Treatise* is followed, rather than preceded, by suggestions of physical hypotheses. It would be appropriate to Aristotle and to Maxwell to identify these two phases as those of *analysis* and subsequent *synthesis*, or *induction* and subsequent *deduction*.

I do not suppose that the distinction between "dynamical" and "physical" methods, as stated thus far with the help of quotations from Maxwell, is entirely clear. Nevertheless we can see that if the distinction is to have the fundamental significance which Maxwell evidently attributes to it—if it is to be more than a mere distinction of degree between broad and specific hypotheses, or between relatively safe and relatively risky hypotheses—then the principle of the distinction must be found in a philosophy of science. For Maxwell, that is indeed where the principle is found, as we shall see in a moment. But perhaps further bearings should be taken from other statements of Maxwell's concerning scientific practice before we follow him into the metaphysical foundation of this methodology.

Curiously, the distinction between "dynamical" and "physical" theory relates to the contrast Maxwell drew between Ampère and Faraday (discussed in Chapter 3 of the present study, pages 176 ff., above). It is Ampère who wrote in the synthetic mode of the completed investigation, while Faraday was, above all men, the inquirer and the physicist. Ampère assumes only the laws of motion and, as we have seen, a strictly minimal selection from the phenomena of electrodynamics. He explicitly rejects the method of hypothesis. In this sense, Maxwell follows Ampère's style rather than Faraday's in presenting his own dynamical theory. We might interpret Maxwell's effort in these chapters of the *Treatise* as an extension of the method of Ampère from the context of the force laws of Newton, to the new dynamics of Lagrange, and from problems of pairs of interacting bodies to those of a continuously connected system. Precisely in the manner of Ampère, Maxwell begins his dynamical theory with the assumption of the universal laws of motion—merely substituting Lagrange for Newton. Like Ampère, he then employs a strict minimum of decisive empirical evidence to determine rigorously the application of those laws of motion to the science in question.

Faraday, by contrast, worked and spoke in the realm of phenomena. His scientific language, as we have seen, was woven of the imagery of his experiments, and he found his way, not by means of vast universal laws, but by visions born of the phenomena before him. Maxwell evidently feels that each of these models is important in the composition of the *Treatise*. There is no question, of course, concerning Maxwell's overwhelming love for Faraday: the beginning of the *Treatise* and the end are his. We have seen that the first chapters of Part iv are an initial exploration of Faraday's thought, with only a digression into Ampère's theory, and that almost as a joke rather than a serious study. Yet the dynamical theory, which is the bedrock upon which Maxwell himself builds, is clearly in the style of Ampère. Maxwell's reconciliation of the virtues of these two opposite scientific styles lies in the relation he understands between "dynamical" and "physical" theories. The strategy of the *Treatise* in this respect is to begin inductively with physical reasoning, following rather closely Faraday's own formulations; then to shift radically to the mode exemplified by Ampère, bringing an abstract mathematical theory to bear on the electromagnetic phenomena in the deductive mode; and, finally, to turn again to physical reasoning, risking hypotheses, and thereby achieving a second translation of Faraday's thought in a much clearer and more fundamental form.

That the dynamical theory of the *Treatise* will be no more than a beginning in this process is clear from the prefatory paragraphs of Chapter V:

> I have applied this method [i.e., the dynamical method] so as to avoid the explicit consideration of the motion of any part of the system except the coordinates or variables, on which the motion of the whole depends [i.e., to avoid hypothesis]. It is doubtless important that the student [for "student" we may safely read "Maxwell"] should be able to trace the connexion of the motion of each part of the system with that of the variables [i.e., to produce a physical theory] but it is by no means necessary to do this in the process of obtaining the final equations, which are independent of the particular form of these connexions [in other words, a physical theory is not necessary to a dynamical theory]. (*Tr* ii/200)

The distinction arises in relation to the question of the electric current:

> [T]here is, as yet, no experimental evidence to show whether the electric current is really a current of a material substance. ...
>
> A knowledge of these things would amount to at least the beginnings of a complete dynamical theory of electricity, in which we should regard electrical action, not, as in this Treatise, as a phenomenon due to an unknown cause, subject only to the general laws of dynamics, but as the result of known motions of known portions of matter, in which not only the total effects and final results but the whole intermediate mechanism and details of the motion, are taken as the objects of study. (*Tr* ii/218)

An important issue, which goes beyond a question of terminology, is raised by the wording of this passage. Is a "physical" theory identical with a "complete dynamical theory"? If so, the distinction of kind has reduced in a different sense to a question of degree: not of degree of certainty of the assumptions made, but of the degree of *completeness* in the sense of filling all gaps of intelligible connection—the completeness sought by Descartes in demanding a continuous medium. This is a problem, however, which can only be resolved in the context of a discussion of Maxwell's philosophy.[4]

Maxwell's objective differs fundamentally from Ampère's in one respect: while Ampère wishes to produce a "finished" treatise, Maxwell wishes only to make a beginning. His aim is to be truly *scientific* in the special, literal sense which he gives to that word, which for him means "science-making" or "science-generating."[5] So he attests at the end of the *Treatise* that his aim has been throughout to go beyond the dynamical theory toward the physical—i.e., to construct a theory which goes beyond itself. A mathematical theory can do this if it *suggests* what Maxwell calls "mental representations" which it does not *per se* assert. Thus the last sentence of the *Treatise* reads:

> Hence all these theories lead to the conception of a medium in which the propagation takes place, and if we admit this medium as an hypothesis, I think it ought to occupy a prominent place in our investigations, and that we ought to endeavour to construct a mental representation of all the details of its action, and this has been my constant aim in this treatise. (*Tr* ii/493)

We shall see how he shapes the Lagrangian mathematical theory to this end in Chapter VI, below.

Maxwell learned from many teachers. Perhaps the most influential of all, particularly with respect to the concept of "dynamical theory," was Newton. As I have pointed out earlier, Maxwell was in a position to know Newton's *Principia* well. He was also a close student of the *Optics*, which guided a large part of Maxwell's own scientific work.[6] Maxwell, in reviewing Thomson and Tait's *Elements of Natural Philosophy* in 1873, wrote:

4. The concept of a complete dynamical theory is delineated in Maxwell's lecture "On the Dynamical Evidence of the Molecular Constitution of Bodies" (1875): "On the other hand, when a physical phenomenon can be completely described as a change in the configuration and motion of a material system, the dynamical explanation of that phenomenon is said to be complete. We cannot conceive any further explanation to be either necessary, desirable, or possible..." (*SP* ii/418)

5. "As a scientific or science-producing doctrine..." *Matter and Motion*, p. 54.

6. Maxwell would very likely have studied the *Principia* under James David Forbes at Edinburgh. That it was part of the regular course of study is attested by Knott (*Life of Tait*, p. 7); sections 1–3 were studied there. Maxwell in turn taught these same sections (through the force law for the conic orbits), plus "a rough view of the Lunar Theory," to his own class of advanced students at Aberdeen in 1858 (Campbell, *Life*, p. 216). Maxwell claimed to be the first to carry out an experiment on color designed by Newton and described in his *Lectiones Opticae*; this was Maxwell's famous "color box" (*SP* ii/230, 269). Maxwell judged that his results with Newton's instrument confirmed Newton's theory (*SP* i/135).

> It is when a writer proceeds to set forth the first principles of dynamics that his true character as a sound thinker or otherwise becomes conspicuous. And here we are glad to see that the authors follow Newton, whose Leges Motus, more perhaps than any other part of his great work, exhibit the unimproveable completeness of that mind without a flaw. (*SP* ii/327)

Maxwell was also a reader of the "Queries" which conclude the *Optics*, and in which Newton's point of view is often revealed; see for example Maxwell's article on "Capillarity" for the *Encyclopædia Britannica* (*SP* ii/542). The excellence of Maxwell's reading of the *Principia* is evidenced by a remark in which he corrects Thomson and Tait's statement of the definition of "quantity of matter." Maxwell rightly grasps, as few modern readers do, that by "density" Newton does not mean mass per unit volume, but percentage of volume filled by matter, reckoned volumetrically. There is only one matter, of invariable inertial effect.[7]

I believe that Maxwell understands Newton's *Principia* thoroughly and takes it as one of the most important models for his own work. Since, if I am right, this will directly affect our understanding of the *Treatise*, and since the *Principia* is so imperfectly known today, I should like to discuss very briefly the significance Newton's plan had for Maxwell.

The work in question is the *Mathematical Principles of Natural Philosophy*. The *principia* then are not physical, but mathematical; physics (natural philosophy) is for Newton a different kind of activity, for which the *Principia* is a necessary prolegomenon. Maxwell, likewise, regularly asserts that abstract mechanics or dynamics is a species of mathematics rather than of physics; for example:

> Mechanics differs from mathematics only by involving the ideas of matter, time, and force in addition to those of quantity and space. The methods employed are the same as in mathematics, and the axioms, or laws of motion, upon which the science is founded are of the same kind as those of geometry.[8]

Newton's *Principia* is, then, concerned with quantities that are measures of physical processes; thus, for example, the *Principia* does not deal with forces, but with quantities which are three different *measures* of force (the absolute quantity, the accelerative quantity, the motive quantity). From the point of view of physics, the "laws of motion" might be regarded as hypothetical, though very well founded; but within the *Principia* itself they are first principles of a mathematical science which proceeds by demonstration in the manner of Euclid. The *Principia* is a work of science in the sense expressed by Aristotle in the *Posterior Analytics*; its mode is demonstration, or

7. Knott, *op. cit.*, p. 195; Thomson and Tait, *Treatise*, p. 220.

8. Maxwell, "Introductory Lecture in King's College, London, 1860" (H.183).

deduction. All the propositions about motion in the first two books follow from the laws. In Newton's view they have nothing to do with the actual world: rather they review the mathematical alternatives for possible worlds. (Newton points out that *force laws*, like that of gravity in our world, might be quite different elsewhere, in other existing worlds; but he does not, I believe, entertain the possibility that the *laws of motion* might be different elsewhere.)[9] Apart from the scholia, the text proper of the first two books of the *Principia* is not concerned with phenomena, or with our world in particular.

It is only in the "System of the World," which is appended to the *Principia* as Book III, that Newton dramatically turns from mathematics to look at a few well-chosen phenomena from observations of our solar system and our planet. There the "Rules of Reasoning" operate to permit induction from the phenomena to empirical laws such as those of Kepler. Given these facts, it is only necessary to insert them into the mathematical propositions of the first two books in order to *deduce* the inverse-square law of universal gravitation. Specifically we have in Book I the proposition that *if* the center of force is at the focus of an elliptical orbit whose apsides abide, we have an inverse-square law of force. By induction from the phenomena, we determine that the apsides of the planetary orbits are nearly stationary. Therefore we *deduce from the phenomena*, with mathematical necessity, that an inverse-square force law obtains. And the carefully developed abstract perturbation theory of Sections IX and XI accounts for the slight motion of the apsides as observed.

All of this is in the mode of analysis, which in my opinion Newton probably learned from his reading of Descartes. The analysis serves to find out the force law—that is, the first principle of the science. With this known, the reasoning then reverses, and the system of the world can be constructed synthetically:

> I offer this work as the mathematical principles of philosophy, for the whole burden of philosophy seems to consist in this—from the phenomena of motions to investigate the forces of nature [analysis], and then from the forces to demonstrate the other phenomena [synthesis].[10]

Compare the following statement of method from Query 31, where Newton points out that the *Principia* and the *Optics* have the same structure:

> By this way of analysis we may proceed from Compounds to Ingredients, and from Motions to the Forces producing them; and in

9. Newton, *Opticks*, Query 31, pp. 403–04.

10. Newton, *Mathematical Principles of Natural Philosophy*, ed. F. Cajori (Berkeley: University of California Press, 1946), pp. xvii–xviii. Compare this passage from the "System of the World": "We have discoursed above on these motions from the Phenomena. Now that we know the principles on which they depend, from those principles we deduce the motions of the heavens *a priori*" (*op. cit.*, p. 420).

> general, from Effects to their Causes, and from particular Causes to more general ones, till the Argument end in the most general. This is the Method of Analysis: and the Synthesis consists in assuming the Causes discover'd, and establish'd as Principles, and by them explaining the Phaenomena proceeding from them, and proving the Explanations.[11]

In the *Principia*, the watershed between analysis and synthesis occurs at Book III, Prop. XI. It is marked by Newton's one, splendid hypothesis: "That the centre of the system of the world is immovable."

The body of the *Principia* is the first two books, which, as I have said, have nothing in particular to do with this world, and have *a fortiori* even less to do with gravity. Newton makes clear in the Preface that he offers the gravitational theory of the "System of the World" merely as an example of the new method. It was in a similar spirit that Descartes offered the *Geometry* as an example appended to the *Discourse on Method.* I think it is true of Maxwell, as it is of Newton and Descartes, that a concern for the fundamental principle and general method outweighs interest in any particular science: all three thinkers are concerned most with the largest questions.

Essentially, I think Maxwell takes Newton's program as the paradigm by which the relation of the mathematical theory of dynamics to its application in physics is to be understood. Maxwell went beyond Newton in the application of this program to at least two problems which are not resolved in the *Principia*: one is the application of the laws of motion to a quantity of bodies so numerous that they must be treated statistically, a problem which Maxwell met in his Adams Prize study of Saturn's rings and extended to the dynamical theory of gases (the foundation of statistical mechanics); the other is the motion of the connected system which constitutes the electromagnetic field, with which we are presently concerned.

For Maxwell, as for Thomson and Tait, Newton is much more than the historic founder of the science of dynamics which is at the foundation of all their work: he remains their master, first among their teachers. Tait is said to have remarked that he and Thomson "rediscovered Newton for the world," and Tait's biographer reports concerning the principle of Conservation of Energy:

> I have heard Tait tell the story of the search after this interpretation [Newton's second interpretation of the Third Law of Motion]. "The Conservation of Energy," he said to Thomson one day, "must be in Newton somewhere if we can only find it." They set themselves to reread carefully the *Principia* in the original Latin, and ere long discovered the treasure in the finishing sentences of the Scholium to Lex III.[12]

11. Newton, *Opticks*, pp. 404–05

12. Knott, *op. cit.*, pp. 190–91. Cf. Thomson and Tait, *Treatise*, p. 246.

Maxwell was equally impressed by this rediscovery of Newton: he writes:

> Newton ... has stated the relation between work and kinetic energy in a manner so perfect that it cannot be improved, but at the same time with so little apparent effort or desire to attract attention that no one seems to have been struck with the great importance of the passage...[13]

The distinction between "dynamics" and "physics" is closely related to the problem of the cause of gravity, which concerned Newton and Maxwell alike. Maxwell's statement in his Royal Institution lecture on "Action at a Distance" (1873) is well known and has helped to call attention to Newton's view, which he cites:

> Newton himself, with that wise moderation which is characteristic of all his speculations, answered that he made no pretence of explaining the mechanism by which the heavenly bodies act on each other. To determine the mode in which their mutual action depends on their relative position was a great step in science, and this step Newton asserted that he had made. To explain the process by which this action is effected was quite distinct a step, and this step Newton, in his *Principia*, does not attempt to make.
>
> But so far was Newton from asserting that bodies really do act on one another at a distance, independently of anything between them, that in a letter to Bentley, which has been quoted by Faraday in this place, he says:
>
> "It is inconceivable that inanimate brute matter should, without the mediation of something else, which is not material, operate upon and affect other matter without mutual contact..." (*SP* ii/315)

Maxwell contrasts Newton's view of these two phases of the problem with the doctrine of Cotes, which widely displaced Newton's, and which left no place for the phase of explanation. Maxwell fully agrees that Cotes' doctrine has served the progress of science up to a certain point:

> But if we leave out of account for the present the development of the ideas of science, and confine our attention to the extension of its boundaries, we shall see that it was most essential that Newton's method [i.e., that of the *Principia*] should be extended to every branch of the science to which it was applicable—that we should investigate the forces with which bodies act on each other in the first place, before attempting to explain how that force is transmitted. No men could be better fitted to apply themselves exclusively to the first part of the problem, than those who considered the second part quite unnecessary. (*SP* ii/317)

The dynamical and physical accounts appear here as two sequential phases in the development of physics, the dynamical taking precedence in a methodological order—compare the organization of Maxwell's *Treatise*,

13. Maxwell, *Theory of Heat* (10th ed., London: 1891), p. 91.

in which the dynamical theory is set out first, and is then followed by chapters advancing hypotheses to explain the dynamical relations. On the two stages, compare Newton:

> How these Attractions may be perform'd, I do not here consider. What I call Attraction may be perform'd by impulse, or by some other means unknown to me. I use that Word here to signify only in general any Force by which Bodies tend towards one another, whatsoever be the Cause. For we must learn from the Phaenomena of Nature what Bodies attract one another, and what are the Laws and Properties of the Attraction, before we enquire the Cause by which the Attraction is perform'd.[14]

For Newton, it is a question of two steps to be taken in the correct sequence. That Newton seriously intends that the advance be made to the second phase is, I take it, quite clear—Maxwell is right in calling Newton as witness. Newton complained of *occult* qualities precisely because they blocked this advance,[15] and in the passage which Maxwell paraphrased in the Royal Institution lecture, Newton had said:

> [T]o derive two or three general Principles of Motion from Phenomena, and afterwards to tell us how the Properties and Actions of all corporeal Things follow from these manifest Principles, would be a very great step in Philosophy, though the Causes of those Principles were not yet discovered: and therefore I scruple not to propose the Principles of Motion above-mentioned, they being of very great Extent, and leave their causes to be found out.[16]

Elsewhere, Newton goes further:

> And though every true Step made in this Philosophy brings us not immediately to the Knowledge of the first Cause, yet it brings us nearer to it, and on that account is to be highly valued.[17]

It is instructive in this connection to observe Maxwell's approach to a particularly recalcitrant problem—that of the molecular constitution of bodies; he discussed the problem in a lecture given to the London Chemical Society in 1875, titled "On the Dynamical Evidence of the Molecular Constitution of Bodies" (*SP* ii/418 ff.). In taking his bearings on the new problem, he begins with the Newtonian theory of gravitation as the example of all dynamical theory. We are baffled, however, in attempting to apply the method to bodies which are not, like the planets, individually visible to us. In the case of electricity, we have something approximating the astronomical problem, for there at least we can observe the electrified bodies:

14. Newton, *Opticks*, p. 376

15. *Ibid.*, p. 401

16. *Ibid.*, p. 402.

17. *Ibid.*, p. 370.

> [W]e can still observe the configuration and motion of electrified bodies, and thence, following the strict Newtonian path, deduce the forces with which they act on each other... (*SP* ii/419)

But there is a difficulty because there appears to be substance invisible to us which is relevant to the problem:

> [T]hese forces are found to depend on the distribution of what we call electricity. To form what Gauss called a "construirbar[e] Vorstellung" of the invisible process of electric action is the great desideratum in this part of science. (*SP* ii/419)

Action-at-a-distance theories, which is to say, dynamical theories in the manner of Cotes and the *Principia*, do not need to form a concept of an invisible process; but to form a "mental representation" of one (evidently, for Maxwell, an equivalent of Gauss's "Vorstellung") is, as we have seen, the ultimate goal of the *Treatise*.

Finally, in chemical problems, we deal with systems altogether invisible:

> In attempting the extension of dynamical methods to the explanation of chemical phenomena, we have to form an idea of the configuration and motion of a number of material systems, each of which is so small that it cannot be directly observed. We have, in fact, to determine, from the observed external actions of an unseen piece of machinery, its internal construction. (*SP* ii/419)

Note carefully that this is precisely what dynamical methods do *not* do in themselves: Lagrange's equations eliminate precisely those questions of "internal construction" about which Maxwell is curious. In terms of the image of the bell-ringers, (page 212, above) the motion of the ropes "is all that the men at the ropes can ever know." On the other hand, Maxwell looks upon the method of hypothesis with Newtonian distaste:

> The method which has been for the most part employed in conducting such inquiries is that of forming an hypothesis, and calculating what would happen if the hypothesis were true. If these results agree with the actual phenomena, the hypothesis is said to be verified, so long, at least, as some one else does not invent another hypothesis which agrees still better with the phenomena.
>
> The reason why so many of our physical theories have been built up by the method of hypothesis is that the speculators have not been provided with methods and terms sufficiently general to express the results of their induction in its early stages. (*SP* ii/419)

He concludes that the only recourse is an appropriate dynamical theory, skillfully employed to derive as much knowledge of the system as is warranted by the data:

> Of all hypotheses as to the constitution of bodies, that is surely the most warrantable which assumes no more than that they are material systems, and proposes *to deduce from the observed phenomena* just as

> much information about the conditions and connexions of the material system as these phenomena can legitimately furnish. (*SP* ii/420; emphasis added)

The following statement of method very closely parallels in form Newton's program in the *Principia*, applied by Maxwell now to a statistical aggregate of invisible bodies; first, an analysis of the forces acting:

> The next thing required is a dynamical method of studying a material system consisting of an immense number of particles, by forming an idea of their configuration and motion, and of the forces acting on the particles...

followed by a synthesis, demonstrating new phenomena:

> ...and deducing from the dynamical theory those phenomena which, though depending on the configuration and motion of the invisible particles, are capable of being observed in visible portions of the system. (*SP* ii/420)

The dynamical method in this case is that of Maxwell and Clausius for dealing with statistical aggregates of bodies.

I conclude that in this chemical problem, Maxwell takes the *Principia* as guide, both in formulating questions of principle and in designing his method of attack. In the same lecture, Maxwell formulates the task in the following words:

> To conduct the operations of science in a perfectly legitimate manner, by means of methodised experiment and strict demonstration, requires a strategic skill. (*SP* ii/420)

What conclusion can we draw from this concerning Maxwell's *Treatise*? I believe that it is shaped by the same Newtonian concepts. It is this demand, essentially Newtonian, for a "perfectly legitimate," strictly deductive method, which dictates the radical shift in the *Treatise* to an abstract dynamical theory. The parallel between the plan of the *Treatise* and that of the *Principia* is, I think, convincing. One can identify, in effect, a *Principia* within the five dynamical chapters of the *Treatise*.

Chapter V, which establishes the Lagrangian equations of motion, corresponds to Books I and II of the *Principia*; like them, it makes no reference whatever to existing phenomena, but speaks universally of motions in all possible worlds. Then, as the *Principia* turns at the beginning of Book III to the application of its mathematical methods to phenomena to answer the questions posed by the theory, so the *Treatise* in Chapter VI considers appropriate phenomena, obtained as Maxwell says "by means of methodised experiment," until the actually operative force law can be found out. This phase is completed in the first ten propositions of Book III of the *Principia* and in Chapter VI of the *Treatise*. The remainder of

Book III of the *Principia* builds, synthetically—that is, by demonstration, as Euclid builds the regular solids—the System of the World; while the remaining chapters of the dynamical theory in the *Treatise*, Chapters VII–IX, build the General Equations of the Electromagnetic Field.

Only after this work is completed does Maxwell go on to frame hypotheses. The discipline of the *Treatise* is learned from the *Principia*.

METAPHYSICS

If Maxwell takes his physics from Newton, it will not be surprising to learn that he takes his metaphysics from Kant, for in a sense Kant's problem was "How is the *Principia* possible?" The two men from whom Maxwell's philosophical position derives directly would seem to be Sir William Hamilton, Professor of Metaphysics and Logic at Edinburgh, and William Whewell, Master of Trinity College, Cambridge. Both held philosophical positions which were modifications of the philosophical theory of Kant. Inasmuch as we know much more about Maxwell's contact with Hamilton, and the marks of Hamilton's teaching are very evident in Maxwell's own statements, I shall discuss Hamilton's doctrine and its relation to Maxwell's thought first. Whewell must be discussed more briefly, although his effect upon Maxwell is also apparent.

Maxwell attended Hamilton's course of lectures during his first two years as a student at the University of Edinburgh (1847–49), and he was certainly much affected by Hamilton's teaching. The best testimony to this, apart from the evidence of Maxwell's own writing, is that of Lewis Campbell, who was often a sympathetic participant in Maxwell's speculations. Campbell tells us, in his *Life* of Maxwell:

> The lectures in Mental Philosophy, which were a prominent element in the Scottish University curriculum, interested him greatly; and from Sir William Hamilton especially he received an impulse which never lost its effect. Though only sixteen when he entered the Logic Class, he worked hard for it, as his letters show; and from the Class of Metaphysics, which he attended in the following year, his mind gained many lasting impressions. ... However strange it may appear that a born mathematician should have been thus influenced by the enemy of mathematics, the fact is indisputable that in his frequent excursions into the realm of speculative thought, the ideas received from Sir William Hamilton were his habitual vantage-ground. ... This is perhaps the most striking example of the effect produced by Sir William Hamilton on powerful young minds,—an effect which, unless the best metaphysicians of the subsequent age are mistaken, must have been out of all proportion to the independent value of his philosophy.[18]

18. Campbell, *Life*, p. 66.

Campbell suggests a number of concepts of Hamilton's which appear in Maxwell's work; we shall return to them after looking briefly at Hamilton's metaphysical system.[19]

It is most fortunate for a student of Maxwell that Hamilton's two-year sequence of lectures has come down to us, apparently almost precisely as Maxwell heard them, in the two volumes of the *Lectures on Metaphysics and Logic*. We shall here be concerned with the volume on metaphysics.[20]

> Consciousness is to the philosopher what the Bible is to the theologian. Both are revelations of the truth—and boh afford the truth to those who are content to receive it, as it ought to be received, in reverence and submission.[21]

For the philosopher whose work begins with this almost moral principle of Hamilton's, philosophy is the analysis of the phenomena of consciousness. The phenomena of consciousness must be the beginning of any knowledge, and it must be impossible for philosophy to penetrate beyond the evidence which they afford.

To have said this is to have accepted the challenge of Hume, for if there is to be any certain knowledge, it must in some way be drawn out of the flux of perception. Nothing within the series of past phenomena can guarantee, however, what the next phenomenon will be like. The classic answer to Hume is that of Kant, and essentially Kant's position is adopted by Hamilton: the possibility of knowledge—the "cause of philosophy," as Hamilton puts it—is in the conditions of our own minds. The laws of our very process of thought dictate that we shall be philosophers, that there must be a cause for every event. Only in these laws of our own minds can necessity arise. Like Kant, Hamilton sees space and time as modes of our own perception, and like Kant, he recognizes that in consequence our experience will have two sources: from phenomena we will derive the contingent, empirical content of perceptions while, independently, from an examination of the conditions of our own minds we will derive time and space as modes of our intuition, yielding the *a priori*, necessary sciences of arithmetic and geometry. In Hamilton's terminology, that work of mind which begins with phenomena and passes through stages which he designates successively perception, memory, representation, and imagination,

19. *Ibid.*, p. 70.

20. William Hamilton, *Lectures on Metaphysics and Logic*: *Volume I*, *Metaphysics*, ed. H. L. Mansel and J. Veitch (Boston: 1859). The original edition was Edinburgh, 1859–60. See also Hamilton's *Discussions on Philosophy and Literature* (New York: 1868); and H. L. Mansel, *The Philosophy of the Conditioned* (London, 1866), an answer to an attack by Mill.

21. Hamilton, *Lectures on Metaphysics*, p. 58.

culminates in "elaboration." This is the faculty of comparison, abstraction, judgement, and reasoning, which he relates to *Verstand* in Kant, and διάνοια in Plato. That work of mind which is based upon the conditions of its own operations he calls "reason," as opposed to "reasoning," and he relates it to *Vernunft* in Kant and νοῦς in Plato. These two separate beginnings for cognition lead to a twofold structure in philosophy which reappears in terms of dichotomies in every aspect of thought. I believe that ultimately it is this distinction that is at the root of Maxwell's division between "dynamical" and "physical" modes of scientific thought.

Hamilton moderates—perhaps compromises—the Kantian position in certain ways, with a note of pessimism at one end, and of optimism at the other. At the highest level, Hamilton emphasizes how little of a positive nature pure reason is able to yield. Causality is a *condition* of our knowledge, not as affording a positive insight, but as a consequence of sheer impotence: we insist upon an uncompromising law of causality, not because we have any knowledge of a necessary causal relation, but simply because we are powerless to conceive its alternative: we lack the power to conceive an absolute beginning or end of any existence, so we fall back upon insistence on a prior existence in every case, which we denote the "cause." The Law of Causality has no more content than this, and Hamilton scorns philosophers who attempt to draw positive knowledge from such a negative source. It should be added, on the other hand, that perception can add nothing to our idea of *necessary* cause; whether we find the required causes among phenomena or not, our conviction that every event must have a cause is unaltered, precisely because its foundation is in no way empirical.

This would seem the darkest scepticism, but it is matched by an act of optimism in regard to perception which had the effect of shifting the center of gravity of Hamilton's system away from reason and toward "fact." Hamilton insists on the immediate duality of all perception: we invariably find mixed in perception facts of two kinds, "internal facts" and "external facts." The former testify to an inner reality, the ego, and the latter to an external reality, the non-ego. This duality of perception is ignored, Hamilton complains, by philosophers who are determined to achieve unity and simplicity by the elimination of one or the other aspect of perception. To be more precise, no one (according to Hamilton) fails to recognize that consciousness testifies to these two realities: what is at issue is whether to honor this testimony. Hamilton accepts the testimony as valid for both the inner and the external object, and thereby commits himself to what he names Natural Dualism, or Natural Realism. Furthermore, he asserts as a guiding principle an extension of the creed of consciousness quoted earlier:

> ...I am ... bold enough to maintain, that consciousness affords not merely the only revelation and only criterion of philosophy, but that this revelation is naturally clear.[22]

It is perhaps not immediately evident what this might mean, but I think the following paragraph will suggest the force of the remark:

> [W]hat is meant by perceiving the material reality?
>
> In the first place, it does not mean that we perceive the material reality absolutely and in itself ... on the contrary, the total and real object of perception is the external object under relation to our sense and faculty of cognition. But though thus relative to us, the object is still no representation. ... It is the non-ego,—the non-ego modified, and relative, it may be, but still the non-ego. ... Suppose that the total object of consciousness in perception is =12; and suppose that the external reality contributes 6, the material sense 3, and the mind 3;—this may enable you to form some rude conjecture of the nature of the object of perception.[23]

Hamilton has in effect opened a window halfway on the *Dinge-an-sich* of Kant. Conciousness, then, "reveals" with a clarity not compatible with a strict understanding of Kant. Hamilton goes so far as to admit the assignment of primary qualities to external reality, one of these being *extension*. How can "extension" be attributed to external reality, while "space" is asserted to be a mere condition of our consciousness? Hamilton pauses for a long breath at this problem, and then takes a plunge which is perhaps more wishful than persuasive:

> To this difficulty, I see only one possible answer. It is this:—It cannot be denied that space, as a necessary notion, is native to the mind; but does it follow, that, because there is an *a priori* space, as a form of thought, we may not also have an empirical knowledge of extension, as an element of existence? The former, indeed, may be the only condition through which the latter is possible. It is true that, if we did not possess the general and necessary notion of space anterior to, or as the condition of, experience, from experience we should never obtain more than a generalized and contingent notion of space. But there seems to me no reason to deny, that because we have the one, we may not also have the other. If this be admitted, the whole difficulty is solved; and we may designate by the name of *extension* our empirical knowledge of space, and reserve the term *space* for space considered as a form or fundamental law of thought.[24]

In all cognition, for Hamilton, both the internal and external elements are present; never is there an absolute or "unconditioned" knowledge of either internal or external reality. But the testimony of both elements is

22. Hamilton, *Lectures on Metaphysics*, p. 185.

23. *Ibid.*, p. 357.

24. Hamilton, *Lectures on Metaphysics*, p. 346.

"naturally clear," and hence we have a relative or conditioned knowledge of both. This is reflected in the Hamiltonian Principle of Relativity, and is the origin of the name adopted by Hamilton for his position: the Philosophy of the Conditioned.

Although neither Campbell's *Life* nor the published correspondence makes it explicit, it seems quite clear that Maxwell followed a second master in metaphysics: William Whewell, who was Master of Trinity College over a long period which included Maxwell's years as a student and later fellow of Trinity. Whewell had before that time written his *History of the Inductive Sciences* (1837) and his *Philosophy of the Inductive Sciences* (1840), among other works, and since 1838 had been Professor of Moral Theology at Cambridge.[25] Quite possibly Maxwell attended Whewell's lectures on moral philosophy; he expressed an opinion of Whewell's *Morality*, the published version of the course, in 1851.[26] Though Maxwell does not allude directly to having read Whewell on the sciences, his 1856 essay, "On Analogies," leaves little doubt that he had been reading the *Philosophy of the Inductive Sciences*.[27] A passage in that essay parallels closely Whewell's discussion "Of the Idea of Cause" in Chapter II of Book III of his *Philosophy*; and a number of terms echo Whewell. This, I think, evidences that Maxwell had made acquaintance with Whewell's philosophy of science before he left Cambridge—which indeed would seem likely, as Maxwell was involved in discussions of many subjects of interest among his colleagues at Trinity. That Whewell's views left a lasting and favorable impression is testified by Maxwell's review of Todhunter's *Writings and Letters of William Whewell* (*SP* ii/528 ff., 1876), in which he forcefully draws out the central point of Whewell's philosophy of the sciences from Todhunter's rather heterogeneous survey. What is more telling, however, is Maxwell's attribution of the terms "Experiment of Illustration" and "Experiment of Research" to Whewell—terms which Maxwell had already used five years before in the famous statement of his philosophy of Experimental Physics at Cambridge (*SP* ii/241 ff., 1871). In that lecture, and in the equally well-known Address to the British Association of the previous year (*SP* ii/215 ff., 1870), Maxwell shaped some of his most earnest thoughts on the nature of science in Whewell's terms, or in terms which, as the 1876 review shows, Maxwell associated with Whewell.

Whewell and Hamilton were open and bitter intellectual enemies, although the difference between their philosophical positions does not

25. William Whewell, *History of the Inductive Sciences* (3 vols.; London: 1837); *Philosophy of the Inductive Sciences* (2nd. ed., 2 vols.; London: 1847); *Elements of Morality* (2 vols.; London: 1845)

26. Campbell, *Life*, pp. 110, 116.

27. Maxwell, "Analogies" (1856), in Campbell, *Life*, pp. 347 ff. Both Maxwell and Whewell participated in the Cambridge Philosophical Society; on May 7, 1855, Maxwell spoke on colors, and Whewell on Plato's dialectic (*Proc. of the Camb. Phil. Soc.* (1843–63), p. 149).

seem at this distance very significant.[28] Both were spokesmen for modified Kantian positions, and in each the modification was in the direction of discounting the role of pure reason and admitting a certain validity to the testimony of perception as to the nature of external reality. Hamilton stresses the negative role of the *conditions* of our understanding, while Whewell's discussions of what he terms the Fundamental Ideas, which indeed result from these conditions, gives greater importance to them.[29] This metaphysical difference had a seemingly disproportionate practical outcome in the question of the place of mathematics in the curriculum of liberal education, which was the focus of a rancorous debate between the two men.[30]

Actually, Whewell's philosophy can be reconciled with Hamilton's, or very nearly so, on the basis of Whewell's own remarks.[31] The argument is not really one of philosophy, though it is of great interest otherwise, and reflects a deep concern with the question of the relation of mathematics to what seemed to many, with Hamilton, more significant human concerns. I suspect that Maxwell is reflecting a certain perplexity at the apparently groundless dispute of Whewell and Hamilton when he writes in 1854 (the

28. Whewell's philosophy is reviewed by Curt J. Ducasse, "William Whewell's Philosophy of Scientific Discovery," Chapter IX of E. H. Madden, ed., *Theories of Scientific Method* (Seattle: University of Washington Press, 1960).

29. On "fundamental ideas" in Whewell, see Ducasse, in Madden, *op. cit.*, p. 184, and references to Whewell's writings given there. Compare Hamilton, *Lectures*, pp. 525 ff. Hamilton compresses his resentment of philosophies which convert the negative conditions into positive powers into an extreme statement: "[T]hat the imbecility of the human mind constitutes a great negative principle, to which sundry of the most important phenomena of intelligence may be referred, appears to me incontestable" (*op. cit.*, p. 526).

30. Whewell, *Thoughts on the Study of Mathematics as a Part of a Liberal Education* (Cambridge: 1835), answered by Hamilton in a review reprinted in *Discussions*, pp. 257 ff. The same volume also includes a reprint of a letter of Whewell's to the *Edinburgh Review*, replying to Hamilton's review, and Hamilton's reply to that letter in turn. The ill feeling involved is very little disguised. Writing to Forbes at Edinburgh, Whewell asks, "By the way, who is to succeed your professor of the Middle Ages, Sir William Hamilton?" (Isaac Todhunter, *William Whewell: an Account of his Writings*. 2 vols.; London: 1876, 2, p. 407). And in 1860, to the same: "Sir W. Hamilton, whose metaphysics I always thought as worthless as he was subtle and learned" (*loc. cit.*). On the important role of the dispute over the character of higher education in Scotland, see the excellent study by George E. Davie, *The Democratic Intellect: Scotland and her Universities in the Nineteenth Century* (Edinburgh: Aldine Press, 1961).

31. Whewell wrote: "We use the word *Ideas* ... to express that element, supplied by the mind itself, which must be combined with Sensation in order to produce Knowledge. For us, Ideas are not objects of Thought ... but rather Laws of Thought. Ideas are not synonymous with Notions, they are Principles which give to our Notions whatever they contain of truth." (*History of Scientific Ideas*, quoted by Ducasse, *op. cit.*, p. 186.) The emphasis may differ slightly, but this statement seems to me in essential agreement with Hamilton at the crucial point.

remark immediately concerns Berkeley):

> It is curious to see how speculators are led by their neglect of exact sciences to put themselves in opposition to them where they have not the slightest point of contact with their systems.[32]

For Maxwell, the perplexity must have been heightened by the fact that the Edinburgh teachers Forbes and Hamilton were arch rivals, while Forbes was a close friend of Whewell's.[33] My own judgement is that for the present purpose—which is to unlock Maxwell's concept of the status and nature of "dynamics" in the *Treatise*—whatever distinction there may be between the metaphysics of Whewell and Hamilton will not be important, and it will be a reasonable hypothesis to suppose tentatively that Maxwell was an adherent of both, borrowing from one or the other on occasion. The two philosophers of the Kantian tradition stand clearly opposed to empirical thinkers such as J. S. Mill.[34]

When Maxwell was called upon, in 1870, to speak as President of Section A (Mathematical and Physical) of the British Association, he evidently saw before him, as if embodied in the persons of the mathematicians and natural scientists who made up the membership of the organization, the two poles of human thought; and his Address concerns their relationship. Even within this one lecture he finds many ways to characterize the dichotomy between them, but repeatedly here and elsewhere he identifies the realm of pure thought with "mathematics," and that of fact and reality with the term "physical." The proper relation of these two orders of inquiry is not only a philosophical enigma which he regarded from the point of view of Hamilton and Whewell, but a constant practical problem for him, governing the strategy of his scientific work. His opening statement poses the dichotomy in terms which are essentially those of Hamilton:

32. Campbell, *Life*, p. 150.

33. "When two such unbending natures as those of Forbes and Sir William Hamilton came into contact, as they did more than once, the shock was a rough one, and the result not generally beneficial. A middle course, was ... an impossibility." (John Shairp, *Life and Letters of James David Forbes*, London: 1873, p. 134.) The growing intimacy of Forbes and Whewell is apparent in their extensive correspondence, part of which is to be found in Shairp, part in Todhunter. When Maxwell went to Cambridge, Forbes offered him a letter of introduction to Whewell (Campbell, *Life*, p. 100).

34. J. S. Mill, *Examination of Sir William Hamilton's Philosophy* (London: 1865), answered by Mansel in the *Philosophy of the Conditioned*, cited above. The following observation of Maxwell's in an 1856 letter is revealing: "I find I get fonder of metaphysics and less of calculation continually, and my metaphysics are fast settling into the rigid high style, that is about ten times as far *above* Whewell as Mill is *below* him." (Campbell, *Life*, p. 178). Above Whewell there would only be Hamilton, or Kant himself! This is consistent with observations about metaphysics in the 1856 Introductory Lecture at Aberdeen.

> [W]e are met as cultivators of mathematics and physics. In our daily work we are led up to questions of the same kind as those of metaphysics. ...
>
> As mathematicians, we perform certain mental operations on the symbols of number or of quantity. ...
>
> As students of Physics, we observe phenomena under varied circumstances, and endeavour to deduce the laws of their relations. (*SP* ii/216–17)

For Maxwell, these two phases of scientific work are rooted in a metaphysical difference between the internal realm of mind ("the mental operation of the mathematician") and the world of external reality ("the physical action of the molecules") (*SP* ii/216), in such a way that their origins are altogether independent and their relation highly problematical. Again, to quote the opening of the British Association Address:

> But who will lead me into that still more hidden and dimmer region where Thought weds Fact, where the mental operation of the mathematician and the physical action of the molecules are seen in their true relation? Does not the way to it pass through the very den of the metaphysician, strewed with the remains of former explorers, and abhorred by every man of science? It would indeed be a foolhardy adventure for me to take up the valuable time of the Section by leading you into those speculations which require, as we know, thousands of years even to shape themselves intelligibly. (*SP* ii/216)

Surely this is the problem of Hamilton and of Whewell, originating in the philosophy of Kant. The echoes of these two metaphysicians are so clear in the Address that it is easy to imagine that Maxwell had read his old notes in thinking through the problem afresh for this statement. For Hamilton, the point where "Thought weds Fact," in Maxwell's terms, is exactly that at which the fundamental Duality must be accepted as given: theories which attempt to account for the relation between mind and fact, Hamilton had said, "are unphilosophical, because they all attempt to establish something beyond the sphere of observation, and consequently, beyond the sphere of genuine philosophy. ... A contented ignorance is, indeed, wiser than a presumptuous knowledge; but this is a lesson which seems the last that philosophers are willing to learn."[35] This, then, is the "very den of the metaphysician" of which Maxwell speaks. Maxwell's passage is, again, very close to this of Whewell's, which comes at a point of transition in the *Philosophy of the Inductive Sciences*:

> In the case of the Mechanical Sciences, we have endeavoured to show, not only that Ideas are requisite in order to form into a science the Facts which nature offers to us, but that we can advance, almost or quite, to a complete identification of the Facts with the Ideas. In the

35. Hamilton, *Lectures on Metaphysics*, pp. 214–15.

> sciences to which we now proceed, we shall not seek to fill up *the chasm by which Facts and Ideas are separated*...[36]

Whether their meeting is at a "chasm" or in a "den," the relation between Fact and Thought is obscure, precisely because they have independent origins—the one in the phenomena of sensation, and the other in what Maxwell with Hamilton and Whewell regards as an *a priori* character of mathematical thought in phrases such as these: The "native penetrating power of our minds," "native modes of thought," or "the irrepressible secretions of their own minds"(*SP* ii/216, 419). Compare the term "native" in this sense to the quotation from Hamilton's *Lectures* on page 244, above: "space, as a necessary notion, is native to the mind."

Mathematics is not, however, a realm of *a priori knowledge*. I have pointed out that, for Hamilton, the conditions which give rise to necessary ideas are essentially negative (for example, in geometry, we are limited to Euclidean space by an incapacity to imagine four dimensions); and I cited earlier (page 246n, above) a passage from Whewell in which it is declared, in satisfactory accord with Hamilton, that "Ideas are not objects of Thought, but rather Laws of Thought." Mathematics is clearly not a realm which Maxwell finds attractive in its own right; for him, it is a realm of suspect symbols. This point of view of Maxwell's accords in some ways with his education. Hamilton, as we know, despised mathematics; and Forbes, Maxwell's sympathetic Professor of Natural Philosophy at Edinburgh, was not at ease with analytic mathematics.[37] When Maxwell entered Cambridge and, having passed the first examinations, began training for the Mathematical Tripos, he clung to methods that were personal, involving an insistent reinterpretation of the mathematical transformations.[38] From these days of drill at Cambridge, Maxwell learned a certain resentment of the rules of symbols and often speaks of them with irony.[39] In the British Association Address, he damns the realm of symbols as "serene," but "dry," "tenuous and pale," merely "abstract," the place of "unembodied symbols." (*SP* ii/216, 219, 220, 242). With Maxwell's term "unembodied symbols," compare Hamilton's characterization of the philosophers who ignore the data of consciousness as dealing with "so many empty spectres ... the heroes of Valhalla."[40] Often Maxwell remarks that the steps of an abstract mathematical argument are

36. Whewell, *History of Scientific Ideas: Being the First Part of the Philosophy of the Inductive Sciences* (2 vols.; London: 1858), *1*, p. 287.

37. Tait, "Forbes' Scientific Work," Chapter XIV of Shairp, *op. cit.*, p. 466.

38. Campbell, *Life*, p. 123. J. J. Thomson, "James Clerk Maxwell," in the memorial volume *James Clerk Maxwell* (Cambridge: Cambridge University Press, 1931), p. 11.

39. Campbell, *Life*, pp. 110, 123, and the verses on Cayley, p. 414.

40. Hamilton, *Lectures on Metaphysics*, p. 207.

difficult to remember: "...mere abstract terms are apt to fade entirely from the memory" (*SP* ii/242). There are, or may be, a few persons who are at home in such a region—compare the verses on Cayley[41]—but Maxwell certainly does not pretend to be among them. Now all this reflects not only Maxwell's personality (and perhaps that of his father as well, with his interest in practical affairs and processes), but Hamilton's view of the role of the mind. In weighing the meaning of a science whose roots are in the mind, such as dynamics, it is important to recall the precept concerning consciousness with which our review of Hamilton's doctrine opened (page 242, above). The mathematical sciences are "tenuous and pale" because they are very nearly emptied of conscious content.

With these remarks in mind, let us turn to Maxwell's discussion of the crucial concept, "matter." Much as Hamilton distinguished two senses of "space" (page 244, above), Maxwell separates two distinct meanings of "matter." On the one hand, insofar as it applies to real bodies as we know them through phenomena, it refers to "that unknown substratum, against which Berkeley directed his arguments. ... Real bodies may or may not have such a substratum, just as they may or may not have sensations." (*SP* ii/781). It is "nothing but an enchanted vase of Circe" and "*never* perceived by the senses."[42] "[W]e ... have to deal with something which claims the title of Matter..." (Hamilton had said: "Matter is the name of something unknown and inconceivable.").[43]

All this concerns matter in its first meaning, as a supposed substratum of sensations; as such, it is unknown and unknowable. (The Hamiltonian position is that we are neither able to know it, nor to put aside our conviction of its existence.) In its second, independent, sense, matter is the *quantity* "mass"; it is part of abstract dynamics, "something as perfectly intelligible as a straight line or a sphere" (*SP* ii/781). Indeed, a treatise on abstract dynamics has the same status for Maxwell as Euclid's treatise on geometry:

> Why, then, should we have a change of method when we move on from kinematics to abstract dynamics? Why should we find it more difficult to endow moving figures with mass than to endow stationary figures with motion? The bodies we deal with in abstract dynamics are just as completely known to us as the figures in Euclid. They have no properties whatever except those which we explicitly assign to them. (*SP* ii/779)

The last sentence of this quotation, very much in the spirit of Hamilton, takes away much of what might otherwise seem to have been granted. "They" [the bodies dealt with in abstract dynamics] are intelligible,

41. Campbell, *Life*, p. 414

42. Knott, *Life of Tait*, p. 414.

43. (*SP* ii/780). Hamilton, *op. cit.*, p. 97

precisely because they are so limited in content. The absence of properties is the price of intelligibility.

The fact that Maxwell here says we "assign" the properties to the bodies of abstract dynamics does not mean, as it might seem, that the science of dynamics is an arbitrary construction in any formalistic sense. At least this much of Kant has survived translation to Edinburgh and Cambridge: the Fundamental Ideas (using Whewell's terms) are *necessary*, as the fixed conditions of our minds. Maxwell calls the dynamical laws "the necessary laws of dynamical systems" (*SP* ii/781), and in this vein he says in the *Treatise*:

> The *fundamental dynamical idea of matter*, as capable by its motion of becoming the recipient of momentum and energy, is so *interwoven with our forms of thought* that, whenever we catch a glimpse of it in any part of nature, we feel that a path is before us leading, sooner or later, to the complete understanding of the subject. (*Tr* ii/197, emphasis added)

The definition implied here, that matter is "the recipient of momentum and energy," is close to the definitions expressed in *Matter and Motion* and elsewhere. There is nothing in it of "solidity" or other properties by which "matter" in the completely different and independent sense of a substrate of real bodies might be identified.

There are, then, as Maxwell tells his students of experimental physics at Cambridge, "two gateways" to knowledge; one via the "doctrines of science"; the other via "those elementary sensations which form the background of all our conscious thought" (*SP* ii/242). They are in principle independent, but if it is possible to "effect a junction in the citadel of the mind, the position they occupy becomes impregnable" (*SP* ii/246). To make this "junction" is to "wed Thought to Fact"; to effect a mystery. This may well be the key to both the content and the style of the *Treatise*: it is dedicated to achieving this union in the science of electromagnetism.

Such a radical view of the nature of knowledge has correspondingly radical implications for scientific method.[44] In the broadest view, Maxwell distinguishes three historic stages, corresponding in general to the stages of learning set forth by Hamilton. The first is the affective stage, in which the aspect of pleasure and pain in the phenomena attracts attention; the culmination of this is in a theory oriented toward the good (evidently, Aristotle is intended as the paradigm of this). The second phase is that of analysis, in which out of the confused, complex whole of sensation the investigator

44. Hamilton, *op. cit.*, Lecture VI, and pp. 247–77, 279–84, and 327 ff. These pages, among others, outline the process of learning as Hamilton views it. The point is that there must be a sharp break in the process, corresponding to the stage at which mind exercises a decisive function.

"pieces out" what is of primary importance, the process yielding a small selection of "clear and distinct" elements.[45] By this simplification and reduction, perceptions are narrowed to a form in which the mind can bring to bear upon them such ideas as those of substance, continuity, causality, and the ideas of mathematics. At one point, Maxwell compares this to the progress of a blind man, who strikes with his stick a few isolated objects—here, a few single and indivisible precepts. According to its own notions of continuity and probability, Maxwell says, the mind then fills in the sequence.[46]

The third, synthetic, stage begins with the intervention of mind, operating by laws reflecting its own constitution. Here the mind makes constructions of its own, which may take the form either of pure mathematical theory or of physical hypothesis. The purely mathematical method Maxwell sometimes calls the "direct" method, since the equations are based on the elements analysed from phenomena, without the intervention of a mediating hypothesis (*SP* i/155). It is in this sense that Ampère's theory is termed "direct" in the *Treatise* (*Tr* ii/158). Maxwell certainly favors the mathematical method rather than the hypothetical, and it is this method which has led to the success of the physical sciences. Hence the third stage is characteristically quantitative, and the transition from the second stage to the third is a shift from quality to quantity. Having already made our senses familiar with the phenomenon, we find out what is measurable; and we return, armed with a theory into which it fits, to measure it (*SP* ii/244). In the two scientific stages, the second and third, we recognize the broad pattern of the relationship between Faraday's work and Maxwell's—first the qualitative "piecing out" of first principles, and then "the intervention of mind," marking the transition from qualitative to quantitative modes, and from analysis to synthesis. Within Part iv of the *Treatise* we observe the same pattern again—from the first four chapters of exploration, to the recognition of first principles (momentum and energy) at the end of Chapter IV, and the beginning of the mathematical, deductive motion. Finally, within Chapter III itself we have the seen the same cycle—from the first synoptic review of

45. (*SP* ii/217). The criteria of "clarity and distinctness" of course are Descartes', and they are important in Whewell (Ducasse, *op. cit.*, p. 91). Maxwell must have thought of his terms "matter and motion" as Cartesian (cf. Hamilton, *op. cit.*, p. 51), and it is significant that he devotes a page of *Matter and Motion*, short as the work is, to an analysis of Descartes' error in respect to the former. The ideal of a complete and inherently intelligible explanation as the goal of mathematical physics is a Cartesian basis for Maxwell's thought in general. He seems to believe that Thomson's vortex atoms might finally achieve the Cartesian goal (*Encyclopædia Britannica* article "Atom" (1875) (*SP* ii/450, cf. 447).

46. (*SP* ii/777). It is interesting to compare Faraday's very different use of the image of the blind man: "I am obliged to feel my way by facts…" (page 18 above); and to recall the passage in the *Treatise* concerning the priority of the *whole* in perception (pages 177 ff., above).

the phenomena and concepts of electromagnetic induction, to their systematic exposition based upon the precise information supplied by the Felici induction balance (page 191, above). In Maxwell's view of our relation to nature, this rhythm is inherent and universal; having once attained a point of contact with nature, mind projects itself, legitimately or otherwise, back upon perceptions. Thus Maxwell speculated in an early essay:

> To inquire why these peculiarities of these fundamental ideas are so would require a most painful if not impossible act of self-exenteration; but to determine whether there is anything in Nature corresponding to them, or whether they are mere projections of our mental machinery on the surface of external things, is absolutely necessary to appease the cravings of intelligence.[47]

We shall return to his further formulation of this problem later, in connection with the concept of "analogy."

Thus Maxwell says of the contemporary scientific world, in which mathematical physics abounds, that "quantity has everywhere encroached upon quality," bringing "physical research under the influence of mathematical reasoning" (*SP* ii/217). There is danger of a careless reliance upon symbols, of operating with them as if they were in themselves meaningful; and of a distortion of our view of physical reality through a selection of "partial data" convenient to a certain mathematical synthesis prepared by the mind, picking from the phenomena those which are "most intelligible to us and most amenable to our calculation."[48]

To these two stages of inquiry—the qualitative and the quantitative—correspond two modes of experimenting: the Experiment of Illustration and the Experiment of Research. To most readers of Maxwell, these terms are familiar from the "Introductory Lecture on Experimental Physics" of October, 1871, in which Maxwell characterized the place of the new Cavendish Laboratory as an integral part of the liberal arts curriculum at the University of Cambridge (*SP* ii/241 ff.). Maxwell had used the same terms, however, since at least 1860, in his Introductory Lecture at King's College, London; and, as mentioned earlier, he himself ascribes them to Whewell in a late review of the latter's work.[49] It is indeed a distinction

47. Campbell, *Life*, p. 349.

48. (*SP* ii/228). The term "partial" is a key word for Maxwell, emphasizing the selection which the mind inevitably performs in its effort to explain the complex phenomena of consciousness (cf., for example, *SP* i/156). The term is to be found in the same sense in Faraday (Bence Jones, *Life of Faraday, 1*, p. 303). The problem is forcefully stated by Hamilton, *Lectures on Metaphysics*, Lecture IV, "The Causes of Philosophy."

49. (*SP* ii/242, 530; ULC Add.MSS.7655, V, h/7). Maxwell would seem to suggest that these two terms were common in Whewell's writing, but I have not found them in quite this form. The distinction they express is certainly a regular part of Whewell's doctrine. Cf. Whewell, *History of Scientific Ideas, 1*, p. 248, and Chapter VII, *passim*.

which gains special significance in the framework of Whewell's philosophy of science. Discussing Whewell's view, Maxwell says that an experiment of illustration was understood by Whewell as one which *suggests* truths to the mind, but "the doctrine when once fairly set before the mind is apprehended by it as strictly true, the accuracy of the doctrine being in no way dependent on the accuracy of observation of the result of the experiment" (*SP* ii/530).

In an early statement of the distinction, in the course of the 1860 lecture referred to above, Maxwell compared the role of the experiment of illustration to that of the diagram in a mathematical proof:

> We shall also have to distinguish between experiments of illustration, which, like the diagrams of Euclid, serve merely to direct the mind to the contemplation of the desired object, and the experiments of research, in which the thing sought is a quantity, whose value could not be discovered without experiment.[50]

The example is revealing as a paradigm of the two stages of learning to which the two categories of experiment correspond: the Euclidean diagram does not constitute a proof, but it has an intimate relation to the demonstration. We may "see" the principle of the proof in the figure, to the extent that the unfolding of the strict demonstration is a matter of course once the relations contained in the figure have been grasped. On the other hand, mathematical precision and the strict necessity binding the steps of a logical argument are displayed in the phase of demonstration, and not in the figure, which may itself be extremely crude and imprecise.

Maxwell, I think, finds Whewell employing this same distinction of two phases of learning as a theme in the interpretation of the history of science; he says of Whewell:

> He therefore regarded experiments on the laws of motion as illustrative experiments, meant to make us familiar with the general aspect of certain phenomena, and not as experiments of research from which the results are to be deduced by careful measurement and calculation. (*SP* ii/530)

And again:

> Thus, then, we are led by experiments which are not only liable to error, but which are to a certain extent erroneous in principle [Maxwell refers here to the use of weighing as a method of comparing inertial masses], to a statement which is universally acknowledged to be strictly true [the conservation of *mass* in chemical reactions]. Our conviction of its truth must therefore rest on some deeper foundation than the experiments which suggested it to our minds.

50. (ULC Add.MSS.7655, V, h/7). So Whatley, writing of arguments by Illustration, says that they correspond "to a geometrical demonstration by means of diagram." He uses "illustration" to translate Aristotle's "parable," a form of argument by "example" (Whatley, *Rhetoric*, p. 68).

> The belief in and the search for such foundations is, I think, the most characteristic feature of all Dr. Whewell's work. (*SP* ii/352)

The aim of the experiment of illustration is "to throw light upon some scientific idea so that the student may be enabled to grasp it" (*SP* ii/242). I think that Maxwell here understands the term "illustration" etymologically, as an experiment of *illumination*, shedding light on a pre-existing, but previously obscure, idea. It is the vehicle of the analytic phase of inquiry: "The circumstances of the experiment are so arranged that the phenomenon which we wish to observe or to exhibit is brought into prominence, instead of being obscured and entangled among other phenomena, as it is when it occurs in the ordinary course of nature" (*loc. cit.*). The concept of an illustrative experiment which brings intellectual clarity goes beyond, I think, the common notion of a "demonstration experiment" accompanying a lecture, even though experiments of illustration were a regular part of Maxwell's lectures. The function of the Experiment of Illustration is to trigger a transition from contingency to necessity, from empirical observation to an undoubtable basic concept. It is a special kind of physical situation, which has the shape of a Fundamental Idea.

Analogy may have a special role in bringing about this transition from the physical context to a mathematical concept at a higher level of abstraction. In the following paragraph on "illustration" (from Maxwell's "Address to the Mathematical and Physical Sections of the British Association" in 1870), we recognize a theme from Thomson's early, bold work on what Maxwell called "allegories":

> Now a truly scientific illustration is a method to enable the mind to grasp some conception or law in one branch of science, by placing before it a conception or a law in a different branch of science, and directing the mind to lay hold of that mathematical form which is common to the corresponding ideas in the two sciences, leaving out of account for the present the difference between the physical nature[s] of the real phenomena. (*SP* ii/219)

Thus the themes of "analogy" and "illustration" weave together in a way which we cannot trace in further detail, but in which both point toward the grasp of a truth of a different order from the empirical phenomena themselves.

To characterize the correspondence between a physical phenomenon and the scientific idea with which it is best grasped, Maxwell commonly uses a term which stems from the tradition of rhetoric: the idea is *appropriate*, he says, to the phenomenon:

> Experiments of illustration may be of very different kinds. Some may be adaptations of the commonest operations of ordinary life, others may be carefully arranged exhibitions of some phenomenon which occurs only under peculiar conditions. They all, however, agree in

> this, that their aim is to present some phenomenon to the senses of the student in such a way that he may associate with it the *appropriate* scientific idea. When he has grasped this idea, the experiment which illustrates it has served its purpose. (*SP* ii/243; emphasis added)

In rhetoric, propriety concerns the right relation of the word to the idea; it is the vehicle of clarity, which is the virtue of prose.[51] Here, the experiment of illustration simply supplants the word: it has the role of a symbol; it relates to the idea in the same way, and it is judged by the same criterion. That Maxwell applies the term "appropriate" to words as well as to experiments is shown in the following passage, again from the article on Whewell:

> To watch the first germ of an appropriate idea as it was developed either in his own mind or in the writings of the founders of the sciences, *to frame appropriate and scientific words* in which the idea might be expressed, and then to construct a treatise in which *the idea should be largely developed and the appropriate words copiously exemplified*—such seems to have been the natural channel of his intellectual activity in whatever direction it overflowed. (*SP* ii/529; emphasis added)

We have seen something of Whewell's role as a word-maker; it was he who suggested the term "sphondyloid" to Faraday (page 23, above). Maxwell frequently speaks in this same vein, concerning the role of the appropriate terms in suggesting ideas, or developing right scientific concepts.

Just as Maxwell intends "illustration" literally, as *illuminating*, so he intends "research" literally, as re-search: "In experimental researches, strictly so-called, the ultimate object is to measure something *which we have already seen*" (*SP* ii/243; emphasis added). Research is not primary discovery, but second search. It is the work of the quantitative phase, in which the mind supplies the mathematical forms under which the phenomena, already discovered, are to be scientifically viewed. This is the role which he ascribes to the Cavendish Laboratory. Surely the passage in which he makes this assertion has puzzled many readers, as it has the present writer, for today the "research" laboratory functions constantly, even primarily, to turn up new phenomena; the Cavendish is of course renowned for its role in pioneering work.

51. Aristotle, *Rhetoric*, III, ii, 2: "Among nouns and verbs, it is the *appropriate* ones [τὰ κύρια, L. *Propria*] that make things *clear* [σαφής, L. *clarus*]" (emphasis added). Cf. Richard Whately, *Elements of Rhetoric* (Boston: 1844), Part III, Chapter II, 1 (pp. 183 ff.).

I do not know to what extent Maxwell drew directly upon technical rhetoric; I believe he knew Whately's treatise, but he told Campbell in 1849 that he was not attending the course in rhetoric at Edinburgh and did not know Aristotle's *Rhetoric*—though he invites Campbell to tell him about it (Campbell, *Life*, p. 83). A general familiarity with such terms was part of a liberal education; note Hamilton's use of the term "appropriate" in describing induction from an experiment (Hamilton, *Lectures on Logic*, pp. 445–46). It is surely relevant that Campbell, Maxwell's early and intimate friend, became a student of, and later a collaborator with, Jowett, and an editor and translator of Greek classics in his own right.

But for Maxwell, the original illustrations can occur anywhere—in games and gymnastics (compare Maxwell's beloved "devil on two sticks," or Tait's study of the flight of a golf ball), in storms—"wherever there is matter in motion" (*loc. cit.*). They *may* occur in the physical laboratory, for "the riches of creation" are "unsearchable," and a re-search may lead to an unexpected further illumination. Neverthess, the search for initial discoveries, on the one hand, and research for mathematical precision, on the other, are in principle two distinct functions even if both are conducted in the same place. The Cavendish was conceived as a teaching laboratory, it must be remembered, in which students have a certain role in the genuine progress of science—but not as discoverers at the level of Faraday. The laboratory is to include a "course" of experiments, not an open-ended search:

> In the order of time, we should begin, in the Lecture Room, with a course of lectures on some branch of Physics aided by experiments of illustration, and conclude, in the Laboratory, with a course of experiments of research. (*SP* ii/242)

There is, admittedly, a certain ambivalence in the concepts of "illustration" and "research" which is accentuated by throwing together quotations from different contexts, as I have done here. Sometimes Maxwell speaks of the roles of these experiments *vis-a-vis* the student; at other times, he has in mind the natural philosopher. Thus, in speaking of the course at Cambridge, the experiments of illustration are regarded as lecture-demonstrations, though with the special significance Maxwell gives them in every case; at other times, as in Whewell's theory of the history of science, they are the initial illuminations which come to the pioneer discoverer. The distinction is not, however, as fundamental as it seems, for when teaching is at its best—when the teacher is Faraday, writing of his work in the *Experimental Researches*—the student vicariously relives the experience of the philosopher (compare page 176, above). This, I think, is a fundamental point for Maxwell: learning is the induction of the appropriate ideas in the mind of the student, and this process is essentially one, whoever the student may be. This is the significance of Faraday's style, in which the nascent idea is reported in its naked form: the birth is repeated in the mind of the reader. It *has* to be; for such is the nature of learning, as Maxwell understands it.[52]

I believe the source of the distinction between "experiments of illustration" and "experiments of research" is in Bacon, and though I do not know that Maxwell learned of it from Bacon's writing's directly, it seems to me that the parallel to the *Novum Organum* helps greatly in understanding Maxwell's use of these terms. This would be true even if the connection is principally through Whewell, for example.

52. This is, after all, a Socratic point. The curious aspect is that Maxwell believes that electromagnetism must be taught with the same indirection with which Socrates teaches justice.

In recasting the "Organon" (the traditional set of arts of the trivium—grammar, rhetoric, and logic, as contained in the treatises of Aristotle), Bacon concentrates on *inventio*, the art of discovery, which had always been regarded as the distinctive field of genius in the art of rhetoric. There, however, it was a question of the invention of arguments and speeches; Bacon in the *Novum Organum* is interested in the invention of arts and sciences. Invention—*inventio*—therefore both links the new organon to the old and marks the difference between the two. The invention of arts and sciences proceeds by experiment, and here Bacon distinguishes two aspects of experimentation, called the Hunt of Pan and the Interpretation of Nature. The Hunt of Pan proceeds from experiment to experiment; he calls it "learned experience" (*experientia literata*) to distinguish it from mere wandering among the phenomena, but it leads to "sagacity" (*sagacitas*) rather than to science (*scientia*). Because it does not lead to science, it lies outside the New Organon, which is confined to the true Interpretation of Nature. The Interpretation of Nature, on the other hand, is distinguished by the fact that it leads to the discovery of the true principles in nature—it does not lead from experiment to experiment, but from experiment to axiom, or from axiom to experiment.

The two modes of experimenting (the Hunt of Pan and the Interpretation of Nature) together make up the Art of Indication (*ars indicii*), of which Bacon says:

> This Art of Indication ... has two parts. For the indication either proceeds from one experiment to another; or else from experiments to axioms; which axioms themselves suggest new experiments. The one of these I will term Learned Experience, the other Interpretation of Nature, or the New Organon. But the former ... must hardly be esteemed an art or a part of philosophy, but rather a kind of sagacity. ... Nevertheless a man may proceed on his path in three ways: he may grope his way for himself in the dark; he may be led by the hand of another, without himself seeing anything; or lastly, he may get a light, and so direct his steps; in like manner when a man tries all kinds of experiments without order or method, this is but groping in the dark; but when he uses some direction and order in experimenting, it is as if he were led by the hand; and this is what I mean by Learned Experience. For the light itself, which was the third way, is to be sought from the Interpretation of Nature, or the New Organon.[53]

To be "led by the hand of another" is to move from experiment to experiment, the way of Learned Experience, while to "get a light" is to pursue the

53. James Spedding, Robert Leslie Ellis, and Douglas Heath, eds., *The Works of Francis Bacon* (14 vols.; London: 1858–74), *4*, p. 412. Note that the editors were all of Trinity College, Cambridge, and that the work was in progress at the time Maxwell was at Trinity College. This edition has been made available in a facsimile reprint (Stuttgart: Friedrich Fromann Verlag, 1963).

Interpretation of Nature, the New Organon. Now Bacon regularly distinguished *two kinds of experiment*:

> And it must ever be kept in mind (as I am continually urging) that Experiments of Light are even more to be sought after than Experiments of Fruit.[54]

> Our experiments we take care to be (as we have often said) either *experimenta fructifera* or *lucifera*; either of use or of discovery: for we hate impostures, and despise curiosities.[55]

> And for use; his lordship hath often in his mouth the two kinds of experiments, *experimenta fructifera* and *experimenta lucifera*: experiments of Use, and experiments of Light: and he reporteth himself, whether he were not a strange man, that should think that light hath no use.[56]

Maxwell, as I have suggested, intends the term "illustration" to mean "illumination," the bringing of light; hence I think Maxwell's "experiments of illustration" correspond to Bacon's *experimenta lucifera*: they are the real experiments of discovery and therefore belong most properly to the New Organon. They are the experiments which open first principles to our view. In relation to the student, they are artfully contrived instances, such as those Maxwell devised for lecture demonstrations, and those on which we have remarked in the opening chapters of Part iv of the *Treatise*, out of which ideas unfold. In relation to the pursuit of natural philosophy, they are such experiments as Faraday's, out of which new insights were born: the vision of the electrotonic state in the discovery of electromagnetic induction; the illumination of a line of force in the magnetic rotation of polarized light; the unity of conduction and induction in the melting of an electrolyte; the observation that bismuth moves not to or from a pole, but toward a region in which the field is weak; the discovery of the sphondyloid of power by means of the moving wire. They bring, in every case, new lights: the electrotonic state, the unity of conduction and induction, the existence of fields rather than poles, the equal importance of the sphondyloid compared to the material body of the magnet. All these are new visions, true intellectual "breakthroughs."

The term "research" likewise has an antecedent in Bacon's classifications. He lists five categories of experimenting as belonging to the Hunt of

54. Spedding, Ellis, and Heath, eds., *The Works of Francis Bacon, 4*, p. 421 (from the *Advancement of Learning*).

55. *Ibid., 4*, p. 501.

56. William Rawley (Bacon's chaplain), Preface to *Sylva Sylvarum*, in Bacon, *Works, 2*, p. 336. Whately takes the term "illustration" etymologically, as "bringing light," just as I here assume Maxwell does (Whately, *Rhetoric*, p. 70). The distinction between experiments of "light" and "fruit" is found in Whewell, *Philosophy, 2*, p. 247.

Pan: variation, production, translation, inversion, compulsion. "Production" means, literally, leading the same procedure further; and it is done, he says, either by *repetition* or *extension*:

> Production of experiment is of two kinds; repetition and extension; that is when the experiment is either *repeated*, or *urged to some effect more subtle*. As an instance of repetition: spirit of wine is made from wine by simple distillation, and is much more pungent and stronger than wine itself; will then spirit of wine, if it be itself distilled and clarified, proportionately exceed itself in strength?[57]

Compare Maxwell:

> In experiments of research we are supposed to be already familiar with the general aspect of the prominent phenomena exhibited but *we repeat the experiment* under carefully arranged conditions in order to measure certain quantities or *to detect the cause of variations* or *by varying the conditions* we endeavour to discover some new phenomena.[58]

"Variation," "repetition," and "urging to some effect more subtle" are on the whole common to Maxwell's "research" and to Bacon's "Hunt of Pan"—though admittedly, the element of measurement, which is of the essence for Maxwell, is lacking in Bacon. The overall distinction seems to be fundamentally the same, between *light*, which is the discovery of first principles, and *research*, which has rather to do with extension and refinement than with original discovery. Very generally, we can see in this the roles of Faraday and Maxwell, in relation to electromagnetism. Faraday, depite the great mass of "research" reported in his papers, is distinctively the one who brought new lights: virtually every insight in the *Treatise*, Maxwell insists, is originally attributable to Faraday. Maxwell's role is to rework and extend Faraday's beginnings, and especially to bring them under quantitative control. Again, this might be compared to the relation of the Royal Institution to Cambridge University: the former is a place of original discovery, the latter, of mathematical discipline and refined measurement. Maxwell probably did not think it appropriate for the Cavendish Laboratory to attempt the kind of original work Faraday had done at the Royal Institution. This, I think, helps us in turn to understand better Maxwell's understanding of the project of "translating" Faraday in a Cambridge *Treatise*: it is a task of intellectual refinement, based on lights already provided by the true intellectual pioneer.

Despite the title of his work, I think then that Faraday's real importance in these special terms was to bring "light," rather than to conduct "research." In the first lines of the following quotation from the Preface to

57. Bacon, *Advancement of Learning*, *Works*, *4*, p. 415.

58. Maxwell, "Methods of Physical Science," (H.512).

Bacon's *Great Instauration*, I think we can recognize the spirit of Maxwell's critique of Ampère; and in the remainder, a characterization of the special power of illustration, and of the contribution of the *Experimental Researches*:

> [A]ll industry in experimenting has begun with proposing to itself certain definite works to be accomplished, and has pursued them with premature and unseasonable eagerness; it has sought, I say, experiments of Fruit, not experiments of Light; not imitating the divine procedure, which in its first day's work created light only and assigned to it one entire day; on which day it produced no material work. ...
>
> But the universe to the eye of the human understanding is framed like a labyrinth, presenting as it does on every side so many ambiguities of way, such deceitful resemblances of objects and signs, natures so irregular in their lines, and so knotted and entangled. And then the way is still to be made by the uncertain light of the sense, sometimes shining out, sometimes clouded over, through the woods of experience and particulars. ... Our steps must be guided by a clue ... a more perfect use and application of the human mind and intellect [must] be introduced.[59]

Ampère is not pursuing "works" in the sense of making gold out of lead; but he is pursuing a repetition and extension of Newton's work, with "premature and unseasonable eagerness," and he is closed to new light. Faraday, by contrast, is keen to find a *clue*, and thereby to move to "a more perfect use and application of the human mind." This, I think, is precisely the way Maxwell understands the contrast of the two men: Ampère's experiments are essentially experiments of research, fitting nature to a mold long before established; Faraday's experiments are experiments of illustration, performed in order to get a clue, a new insight. Maxwell, in the "dynamical theory," is now working in the style of Ampère—but he is repeating not the old and inappropriate paradigm of Newton, but the new light which Faraday has discovered, the concept of a connected mechanical system in space; and he is bringing to bear on it, not the old mechanics of actions between point-centers, but a new theoretical method: Lagrange's abstract mechanics of a generalized connected system.

Maxwell's best known, and most significant, method for maneuvering between Thought and Fact is the device he sometimes calls "scientific metaphor," sometimes "physical analogy" (page 47, above). Like the term *appropriate*, *scientific metaphor* is a term conceived by Maxwell in the context of a rhetorical tradition; he defines it as "the figure of speech of thought by which we *transfer* [μεταφερεῖν] the language and ideas of a familiar science to one with which we are less acquainted..." (*SP* ii/227).[60] The transfer is

59. Bacon, Preface to the *Great Instauration*, *Works*, *4*, p. 17

60. Aristotle, *Rhetoric*, III, x, 7 ff.; *Poetics*, xxi–xxii. Whately, *Rhetoric*, Part III, Chapter II, 3 (pp. 189 ff.).

made by virtue of an analogy or proportion, a likeness in form, between two branches which in their real, physical nature are altogether different. All abstract terms, Maxwell says, are thus metaphorical, and as an example of scientific metaphor, he offers the generalization of the terms of Newtonian dynamics into those of the Lagrangian scheme (*SP* ii/227). The extension of the same terms into the realm of electrical phenomena will constitute a still greater stretch of analogy. The work of the mind in generalizing, then, is always by virtue of analogy or metaphor, and this rhetorical term becomes a further way of viewing the relation of mind to physical reality. Analogy makes it possible to reach from the realm of mathematics toward phenomena, and thus to lend color to the pale and lifeless symbolic expressions. Again a rhetorical term is useful: analogy brings *vividness*, which is crucial to the productive life of the mind.[61]

It was by means of this method of physical analogy that Maxwell hoped to solve the methodological enigma which he posed for himself in "Faraday's Lines": how to avoid commitment either to a purely mathematical theory (the style of Ampère), or to a physical hypothesis (the style of Faraday)?

> We must therefore discover some method of investigation which allows the mind at every step to lay hold of a clear physical conception, without being committed to any theory founded on the physical science from which that conception is borrowed, so that it is neither drawn aside from the subject in pursuit of analytical subtleties, nor carried beyond the truth by a favorite hypothesis. (*SP* i/156)

Maxwell's resolution of the problem is this: "In order to obtain physical ideas without adopting a physical theory we must make ourselves familiar with the existence of physical analogies" (*ibid.*). The paper itself serves to illustrate this new method. As we have seen, there is no suggestion that the ideal fluid proposed there is really present in electrical phenomena; by a merely formal similarity, it serves to illustrate only the mathematical idea which the two sciences have in common.

What Maxwell hopes to show in his dynamical theory of electromagnetism is, in these terms, that the phenomena of electromagnetism can be illustrated in the behavior of visible mechanical systems, and thereby be brought under the forms of thought, or fundamental ideas, of dynamics—and hence rendered intelligible. He applies the dynamical terms to electromagnetism by analogy.

What does this procedure then tell us of the reality of electrical systems? Does it imply that electrical currents and familiar mechanical systems in fact have "matter" in common? Maxwell raised this basic question of the

61. Aristotle, *Rhetoric*, III, x, 11; *Poetics*, xvii. Whately, *Rhetoric*, Part III, Chapter ii, "Of Energy." The term "energy" stems from Aristotle's phrase ἐνεργοῦντα σημαίνει, "signify things *in act*"; words which do this also "place things before the eyes" (πρὸ ὀμμάτων)—i.e., make them vivid. Thus "energy" and "vividness" are allied terms in rhetoric.

significance of such analogies in an early essay written at the time he was a fellow of Trinity College. This is the essay, "Analogies," printed in Campbell's *Life* of Maxwell. There the basic relation between the concept of analogy and the metaphysical inheritance of Kant is spelled out: the one great analogy, the principle of all others, is the analogy of Thought and Fact. Thought deals in formal relations; whenever the opportunity arises, by a necessity of our natures we project these relations upon phenomena, thereby imposing analogies. Is this then no more than a "mere projection of our mental machinery on the surface of external things"?[62] That may be the case, but it is also possible that the formal patterns found by mind correspond to an underlying unity of structure in reality. The success of dynamical analogies may be evidence that the phenomena of nature are really various instances of matter in motion:

> Now the question of the reality of analogies in nature derives most of its interest from its application to the opinion, that all the phenomena of nature, being varieties of motion, can only differ in complexity, and therefore the only way of studying nature, is to master the fundamental laws of motion first, and then examine what kinds of complication of these laws must be studied in order to obtain true views of the universe. If this theory be true, we must look for indications of the fundamental laws throughout the whole range of science. ... In this case, of course, the resemblances between the laws of different classes of phenomena should hardly be called analogies, as they are only transformed identities.[63]

The alternative is this:

> If, on the other hand, we start from the study of the laws of thought (the abstract, logical laws...), then these apparent analogies become merely repetions by reflexion of certain necessary modes of action to which our minds are subject.[64]

The question of ultimate interest to Maxwell, both in the essay "Analogies" and more generally, is that of the relation of these dynamical laws to moral freedom. We cannot pursue that problem here, nor need we in order to follow Maxwell's thinking about the meaning of the term "dynamical theory" in relation to electromagnetism. It is interesting to note in passing, though, that "to show to what length analogy will carry the speculations of men" Maxwell suggests in the essay the notion of the *propagation of consequences (with finite velocity) through a moral medium*—many years before his announcement of his theory of the propagation of effects through an electromagnetic medium![65]

62. Campbell, *Life*, p. 349

63. Campbell, *Life*, pp. 352–53

64. *Loc. cit.*

65. Campbell, *Life*, p. 353.

To summarize Maxwell's metaphysical position in relation to his own work as a scientist: for Maxwell, "science" deals properly with relations, and with quantity. These represent the contribution of the mind, through the operation of the necessary laws of its own constitution. Science begins in the serene, pale realm in which unity is so inviting, and seems so possible. We encounter Reality through perception, whose testimony of an external existence Maxwell, as a Hamiltonian Natural Realist, accepts. This Reality is complex, diverse, and colored by all the differences and affective qualities that make up the vividness of existence. All of Maxwell's strategy in the fabrication of mental representations, analogies, illustrations, or scientific metaphors concerns the one effort to throw bridges across the chasm dividing Thought and Fact.

Science, then, has *a priori* beginnings, and pure dynamics is an *a priori* science, though it may have taken many illustrative instances to awaken in men an awareness, a "grasp," of the appropriate Fundamental Ideas. But this *a priori* science is not knowledge: as *necessary*, it is only a reflection of our own condition. In this sense, Maxwell's Chapter V, on the laws of motion, is a mirror of ourselves, the science of our *concept* of a connected system. That such thought may "wed fact"—that is, do more than merely project itself on the surface of our perceptions, and make any junction with the reality of which they give testimony—is the Hamiltonian optimism superimposed, perhaps vainly, on the Kantian division of mind from the *Dinge-an-sich*.

It is in the spirit of this problem, I think, that we must read the following statement, with which Maxwell draws together the argument of his essay "Analogies":

> I have been somewhat diffuse and confused on the subject of moral law, in order to show to what length analogy will carry the speculations of men. Whenever they see a relation between two things they know well, and think they see there must be a similar relation between things less known, they reason from the one to the other. This supposes that although pairs of things may differ widely from each other, the *relation* in the one pair may be the same as that in the other. Now, as in a scientific point of view the *relation* is the most important thing to know, a knowledge of the one thing leads us a long way towards a knowledge of the other. If all that we know is *relation*, and if all the relations of one pair of things correspond to those of another pair, it will be difficult to distinguish the one pair from the other. ... Such mistakes can hardly occur except in mathematical and physical analogies. ... Perhaps the 'book,' as it has been called, of nature is regularly paged; if so, no doubt the introductory parts will explain those that follow, and the methods taught in the first chapters will be taken for granted and used as illustrations in the more advanced parts of the course; but if it is not a 'book' at all, but a

> *magazine*, nothing is more foolish [than] to suppose that one part can throw light on another.[66]

The "dynamical theory of electromagnetism" is, in this sense, a venture in reading nature as a book.

THE MEANING OF DYNAMICAL TERMS

Before turning in the next chapter to Maxwell's application of dynamics to electromagnetism, it will be well to collect here in summary the meaning, for Maxwell, of certain basic dynamical terms. We have seen that their meaning presents a problem, in view of the vast stretch which they have undergone since their Newtonian definitions, and of the complexity of the very notion of "meaning" itself within the philosophical framework in which Maxwell thinks.

From the metaphysical point of view, and as we have seen in the case of "matter" and "space," a term has dual and in principle independent meanings: first, that belonging to pure dynamics; and second, that properly applicable to external reality, the realm of physics. The meanings which we shall review here belong to pure dynamics; Maxwell's *Matter and Motion* and Chapter V of the *Treatise* will serve as our principal guides. The relation between these meanings and anything which can be predicated of the external world arises by virtue of that great analogy between Thought and Fact which has been the subject of the preceding section.

Matter and Motion is formulated consistently in terms of a strict relativity of space, with a door left open to the possibility of a relativity of time as well.[67] Thus the basic term upon which others are based is *configuration*:

66. Campbell, *Life*, p. 354.

67. *Matter and Motion*, pp. 31–32. See the remarkable section, "Relativity of Dynamical Knowledge," *ibid.*, pp. 80–81: "Our whole progress up to this point may be described as a gradual development of the doctrine of relativity of all physical phenomena."

Maxwell's formulation here would seem to put the concept of a single, uniform time-scale for all systems on a purely empirical basis. "The statement that equal intervals of time are those during which equal displacements occur in any such independently moving system, is therefore equivalent to the assertion that the comparison of intervals of time leads to the same result whether we use the first system of two bodies or the second system as our timepiece." Maxwell understands this to be an empirical, not an *a priori* assertion, as suggested in the early essay "Analogies": there he asserts that the correspondence between space as a condition of our minds is a *real* analogy producing an objective truth, but the correspondence between our idea of time and an order of events is a *mere* analogy—i.e., the single time-scale which we perceive is not necessarily binding upon nature (Campbell, *Life*, p. 350). As is well known, Maxwell was much interested in questions of relativity in relation to the optical ether; it is part of his deliberate program of "generalization" of dynamical concepts. See Maxwell, "On an Experiment to Determine Whether the Motion of the Earth Influences the Refraction of Light" (April, 1864) (H.227) [*continued on next page*]

> When a material system is considered with respect to the relative position of its parts, the assemblage of relative positions is called the Configuration of the system.
>
> The configuration of material systems may be represented in models, plans, or diagrams. The model or diagram is supposed to resemble the material system only in form, not necessarily in any other respect.[68]

These are diagrams without indication of any coordinate system, the coordinate framework being left relative to the observer. Maxwell says of his diagrams, very much as he said earlier of physical analogies, that they represent "exactly what we can know about the motion and no more."[69]

The term *force* is probably the most difficult to purge of the physical, phenomenal connotations which have no place in abstract dynamics. There must be no suggestion of effort, resistance, or struggle. In dynamics, Maxwell liked to repeat, bodies all move "like the blessed gods"; the "Manichean doctrine of the depravity of matter" has no place there.[70] With these aids withdrawn, it is difficult to extract a definition of force from the pages of *Matter and Motion*, beyond the statement that (in the guise of *stress*, which Maxwell introduces as a higher understanding of *mutual* force) it is "one aspect of the mutual action between bodies."[71] *Which* aspect cannot, apparently, be said in any useful way; as I have pointed out, force seems to be taken as an intuitively primary term in Chapter V of the

---

I gather that this was rejected by Stokes, with a request that Maxwell write a section on alternative hypotheses, which he declined to do (Larmor, *Memoir of Stokes*, *2*, pp. 23–25); there had been earlier correspondence on the subject of the velocity of light in media (*ibid.*, *2*, p. 6). See also the famous letter of Maxwell's to D. P. Todd, published after Maxwell's death by Stokes: "On a Possible Method of Detecting a Motion of the Solar System through the Luminiferous Ether," *Proceedings of the Royal Society*, *30* (188), pp. 108–110 (H.734), and a reply from John Couch Adams, director of the Cambridge Observatory, to an inquiry of Maxwell's concerning an ether-drift experiment based on the moons of Jupiter (ULC Add.MSS.7655, II/195).

68. *Matter and Motion*, pp. 2–3.

69. *Ibid.*, p. 21.

70. Maxwell wrote, of the replacement of the vector "force" by scalar methods, in reference to Tait's lecture on "Force" (Address to the British Association, 1876):

> The Universe is free from pole to pole,
> Free from all forces.
> Rejoice! ye stars—like blessed gods ye roll
> On in your courses.

(Campbell, *Life*, p. 418). Tait wrote later (1882) of his own effort to formulate a mechanics without placing the concept "force" in a central position: "I tried to find some simple mode of getting rid of what I find Maxwell has called Personation" (Knott, *Life of Tait*, p. 234). On the "Manichean doctrine," see (*SP* ii/780).

71. *Matter and Motion*, p. 80.

Treatise (page 215, above).[72] The root idea which yields the concept of force is this:

> The new idea appropriate to dynamics [by contrast with kinematics] is that the motions of bodies are not independent of each other, but that, under certain conditions, dynamical transactions take place between two bodies, whereby the motions of both bodies are affected. (*SP* ii/780)

This passage, from the 1879 Thomson and Tait review, is followed by a discussion which places mass and momentum logically before force (as does the *Treatise on Natural Philosophy* under review). But no matter how the relevant terms are introduced, the one "new idea" is that of the *interdependence* of motions—*not* of an innate resistance to motion:[73]

> Is it a fact that "matter" has any power, either innate or acquired, of resisting external influences? Does not every force which acts on a body always produce exactly that change in motion of the body by which its value, as a force, is reckoned? Is a cup of tea to be accused of having an innate power of resisting the sweetening influence of sugar, because it persistently refuse to turn sweet unless sugar is actually put into it? (*SP* ii/779)

The formal likeness which is reproduced in the diagram or model is precisely what Maxwell has elsewhere called analogy; the configuration catches just what is essential, from a formal point of view, about the relative positions of the parts of the system. This is, in turn, what is represented by any complete set of generalized coordinates (page 208, above). The system of bells, in the example quoted earlier, is fully represented whether by the angular positions of the bells themselves, or by any other sufficient set of independent measurements taken at other points in the mechanical system. "Motion," in turn, becomes *change of configuration*—hence the "motion" of the title of Maxwell's book is not in general linear, local motion, but any change of form.

The reader who finds the above discussion of force unduly obscure and indirect should consider that we are here attempting to find a definition

72. Essentially, the "aspect" of mutual action which distinguishes "force" is that of *cause*. Both dynamics and kinematics deal with motion; we distinguish the former from the latter only in that it deals with interatctions from the point of view of cause; *Matter and Motion*, p. 26. Compare "Analogies": "When the objects are mechanical, or are considered in a mechanical point of view, the causes are still more strictly defined, and are called *forces*" (Campbell, *Life*, p. 350).

73. The notion that "force" does not mean a resistance would seem to be flatly contradicted by the following passage, and I admit the difficulty: "Work is the act of producing a change of configuration in opposition to a force which resists that change" (*Matter and Motion*, p. 54). Maxwell is committed in other passages, however, to the position that matter does not really "resist" acceleration. I think, then, that the sentence quoted here represents a tentative statement, which he could not stand by.

in words of a concept which arises initially in the symbolism of *mathematics* alone—following Maxwell in the effort which he makes consistently in all his work, but heroically in *Matter and Motion*, to clothe the symbols of mathematical physics in words to awaken ideas—and that it is by no means simply Newtonian "force" that is in question. Metaphor has operated to carry the term outside itself, and a new idea is in birth. This is at least Maxwell's conviction: that new dynamical ideas, meaningful and intelligible, will shape themselves around the symbolical transformations of Lagrange. For Maxwell, of course, they are "new" only in the sense that we are only now led to recognize them.

Energy is simply another aspect of the interactions of motions. Historically, in our developing understanding of the relation between motions, energy has come to take priority over force. Maxwell sketches the sequence as beginning with "force" in Newton, then emphasizing "stress" as the mutuality of force becomes better understood, and focusing finally upon "energy," as the law of the conservation of energy becomes known. If we first identify force as a measure of the interaction of motions, *work* will be defined as the product of the force and the change of configuration, and *energy* as the capacity for doing work. Conversely, "work" is the transference of energy from one system to another.[74] The primacy of this new concept reflects its universality: "the doctrine of the Conservation of Energy is the one generalized statement which is found to be consistent with fact, not in one physical science only, but in all."[75] Maxwell distinguishes, as always, between the "doctrine" as "a deduction from observation and experiment," on the one hand, and as "a scientific or science-producing doctrine" on the other. In the latter capacity, he says "it is always acquiring additional credibility from the constantly increasing number of deductions which have been drawn from it, and which are found in all cases to be verified by experiment."[76] I must admit that if we had this passage alone as a statement of Maxwell's view of the role of a fundamental doctrine, we could easily interpret it as a merely empirical hypothesis supposted by many observations. But in the context of all we have seen of Maxwell's views, it is rather to be understood as a Fundamental Idea which we are only beginning to grasp in its inherent clarity. There is a great difference. For in the latter case, the Conservation of Energy is a principle inherently intelligible, and carrying a necessity which we can come to appreciate. That is the point of a book like *Matter and Motion*: to expose the intelligibility of a latent Fundamental Idea.

For Maxwell, kinetic energy is thus intelligible; he ends a discussion of kinetic energy with the remark:

74. *Matter and Motion*, p. 56.

75. *Ibid.*, p. 55.

76. *Loc. cit.*

> We can now *see the appropriateness* of the name *kinetic energy*, which we have hitherto used merely as a name to denote the product $\frac{1}{2}MV^2$. For the energy of a body has been defined as the capacity which it has of doing work, ... The *kinetic* energy of a body is the energy it has in virtue of being in *motion*.[77]

On the other hand, potential energy and elasticity are *not* intelligible:

> The potential energy of a material system is the capacity which it has of doing work depending on other circumstances than the motion of the system. In other words, potential energy is that energy which is not kinetic.
>
> *In the theoretical material system which we build up in our imagination from the fundamental ideas of matter and motion*, there are no other conditions present except the configuration and motion of the different masses of which the system is composed. Hence in such a system the circumstances upon which the energy must depend are motion and configuration only, so that, as the kinetic energy depends on the motion, the potential energy must depend on the configuration. ...
>
> But when all is done, the nature of the connexion between configuration and force remains as mysterious as ever. *We can only admit the fact*, and if we call all such phenomena phenomena of elasticity, we may find it very convenient to classify them in this way, provided we remember that *by the use of the word* elasticity *we do not profess to explain the cause* of the connexion between configuration and energy.[78]

This reduces to the problem of action-at-a-distance, as Maxwell points out directly. I think it is clear that Maxwell was willing to accept the term "potential energy" as a convenience of classification, as he suggests above, but that he ultimately requires the reduction of every case of potential energy to kinetic energy. Such reduction is not a practical necessity: it is an importunate demand of mind. The objection to potential energy is that it does not lie within the scope of the *fundamental ideas*, or the conditions of our own minds—and hence must remain unintelligible.

The ultimate account of any phenomenon will be, then, phrased in terms of energy:

> A complete knowledge of the mode in which energy of a material system varies when the configuration and motion of the system are made to vary is mathematically equivalent to a knowledge of all the dynamical properties of the system.[79]

Then we may, if we wish, dispense with force, to make way for this more fundamental idea:

77. *Matter and Motion*, p. 60; first emphasis added.

78. *Ibid.*, pp. 65–66; emphasis added.

79. *Matter and Motion*, p. 68.

> But see! Tait writes in lucid symbols clear
> One small equation;
> And force becomes of Energy a mere
> Space-variation.[80]

What, then, is matter? In Maxwell's dynamics, the concept of matter is deadlocked with that of energy. There is nothing here of the school of Lucretius, Gassendi, or Newton: nowhere is *matter*, as a dynamical concept, related to hard bodies or a solid substrate. Maxwell will only give us this dual statement:

> Hence, as we have said, we are acquainted with matter only as that which may have energy communicated to it from other matter, and which may, in its turn, communicate energy to other matter.
>
> Energy, on the other hand, we know only as that which in all natural phenomena is continually passing from one portion of matter to another.[81]

He then gives us this "test of a material substance": "Energy cannot exist except in connexion with matter." Then, where there is energy there is "matter," by our very understanding of the terms. There can be no doubt, then, for Maxwell, that interplanetary space—being full of radiant energy—is by the same token full of matter. This does not mean ponderable stuff; it means rather what it says, that we have a case of communication of energy, or a case of interrelated motions.

We now return to Maxwell's proposal to "examine the consequences of the assumption that the phenomena of the electric current are those of a moving system" (*Tr* ii/198). I submit that we could not responsibly have done so without the digression of this section. Maxwell is asserting in effect that electromagnetic phenomena are those of moving material systems, but in the sense only of the new dynamics, related to the old only by a long, metaphorical, reach.

80. Campbell, *Life*, p. 418.

81. *Matter and Motion*, p. 89.

Chapter 6. Application of Dynamical Theory to Electromagnetism

JUDGEMENT OF THE TERMS IN LAGRANGE'S EQUATIONS OF MOTION

One of the difficulties in the way of translating Faraday's *Experimental Researches* into scientifically acceptable form is the apparent imprecision of his expressions. This is sometimes a question of the use of terms: words such as "power," "force," "action," and "tendency" are used as if without any consciousness of the fact that they have strict technical definitions in a systematic science. Sometimes, on the other hand, Faraday's seeming imprecision represents a deliberate effort to avoid premature commitment to an overly-definite hypothesis, as in the case of his reluctance to speak definitely of the electric "current," substituting rather the baffling phrase "axis of power." Either way, this largeness of Faraday's style is an important aspect of his mode of approach to nature; it is related to the freedom which both Tyndall and Maxwell recognized as a consequence of his innocence of mathematical physics.[1] A just translation of Faraday would have to preserve this openness of view, retaining at the same time the almost inflexible commitment to certain overall beliefs relating, for example, to conservation and continuity.

None of the interpreters of Faraday whom we discussed in Chapter 2 offered any real equivalent to Faraday's method in this respect. Mossotti's caloric-ether interpretation of electrostatics, for instance, had already moved to a specific hypothesis, which, as we saw, not only failed to eliminate action-at-a-distance, but was too limited to reflect the range of thought which Faraday himself carefully preserved. It was a particular hypothesis, and hence a "partial view," rather than a mathematical version of Faraday's own thought. On the other hand, a purely mathematical interpretation, such as that of van Rees, left out Faraday's conviction of the physical reality of the system of lines of force, and thus substituted mathematical symbols for the world which Faraday wished to grasp philosophically. Possibly the nearest approach to the liberality of Faraday's thought was the free use of analogy in the writings of Thomson. But what was really demanded was a

1. Compare pages 45–46 above. Maxwell said of Faraday's reluctance to speak of the "electric current" in the usual terms: "...Faraday, who constantly endeavoured to *emancipate his mind* from the influence of those suggestions which the words 'electric current' and 'electric fluid' are too apt to carry with them, speaks of the electric current as 'something progressive, and not a mere arrangement' (*Tr* ii/211; emphasis added).

mathematical theory that would preserve exactly the balance of physical significance and openness of view of the *Experimental Researches* themselves.

Maxwell, I believe, found this mathematical theory in Lagrange's equations of motion of a connected system. First it was necessary, as we saw in Chapter 4, to restructure the account of Lagrange's theory so as to translate it into the language of dynamics. But we saw also that the effect of this is to generalize the concepts of dynamics so that in detail they are almost unrecognizable, while in their broadest significance they express those fundamental ideas in terms of which the physical world can rightly be known. This generalized mechanics, Maxwell found, was alone adequate to the translation of Faraday. I do not mean that he phrased his results in quite these terms. But the "dynamical theory of electromagnetism," the systematic theory which Maxwell unfolds in the *Treatise*, is the work that he offered as the translation of Faraday's ideas.

This makes good sense, for we have found that the sphondyloids of power, or systems of lines of force, whatever else they may be, can be interpreted as loci of energy; and if they are loci of energy, they necessarily constitute a *material* system in the very general sense which Maxwell has discovered in that term, and expressed in *Matter and Motion*. If we know that they form a material system, then we know that Lagrange's equations describe them, and the problem of theorizing about the sphondyloids or systems of lines becomes that of bringing Lagrange's equations to bear upon them. To use Newton's equations would demand too much of us—we would need to know specific places, paths, velocities and forces—and we would have to commit ourselves to a "partial" choice out of the range of possibilities. By sticking to Lagrange, we save the freedom of thought which Faraday's language and generalized mechanics have in common.

No two opposites could contrast more severely than Lagrange, with "no figures" in his work, and Faraday, with his head full of images. Yet Maxwell has seen that they are born for each other—that together, they can lead us away from the literal materialism of the seventeenth century into the much more generous view of nature that is demanded by a physics of imponderable and unbounded fields.

Having thus seen that Faraday's "system" is at the same time Lagrange's, Maxwell approaches the construction of his own electromagnetic theory with a clear initial vision of the shape it must take. He does not begin with a collection of basic empirical results and seek a merely complete and convenient set of equations which will save these appearances. Maxwell knows at the outset that his theory must take the form of the equations of motion of a moving material system; these, as we have seen, are Lagrange's equations of motion, which in Maxwell's view simply explicate mathematically our *a priori* concept of matter in motion. *A priori*, Maxwell's equations are merely a special case of Lagrange's equations.

Therefore, Maxwell's program for a "dynamical" approach to electromagnetism must be this: beginning with Lagrange's equations of motion, identify the generalized coordinates and velocities which characterize an electromagnetic system, and then determine by experiment which of the possible coefficients are actually operative in this particular science, and what relationships exist among the coefficients and the coordinates. Lagrange's equations, thus related to electromagnetism and sifted of inoperative terms, will be the basic equations of electromagnetism. At the same time, they will characterize in broad strokes a particular form of connected mechanical system.

Maxwell first identifies as one set of generalized coordinates the geometrical positions of the conductors, and as an independent set of generalized velocities, the electric currents which flow in them. The full set of generalized coordinates will then be these geometrical positions, together with the time-integrals of the currents. It would belong to a physical theory, not to a dynamical theory, to inquire further about the nature of these electric currents: they are assumed to be *generalized*—not necessarily Newtonian—velocities. In particular, it is by no means assumed that any sort of fluid is moving in the wire.

The construction of the electromagnetic theory becomes a great process of selection, with each step guided by empirical results deliberately called upon as required. The experimental results are then strictly "crucial" in determining in each case a choice between foreknown theoretical alternatives. The first mathematical step is to write the Lagrangian function for the electromagnetic system, at first admitting all possible terms. The system itself is initially thought of as containing any number of current-carrying conductors of any shapes, and placed in any relation to one another, together with the fields throughout space to which they give rise. At the beginning, electrostatics is left out of account, and the conductors are supposed placed in a vacuum. (We shall be concerned later with the problems which arise when electrostatics is integrated with this theory of electromagnetism; pages 300 ff., below.)

Under these assumptions, Maxwell is confident in assuming that the energy of the system will be entirely kinetic: "The electric current cannot be conceived except as a kinetic phenomenon" (*Tr* ii/211). He therefore writes the total energy of the system in an expression of the form:[2]

$$T = \tfrac{1}{2}\sum_{ij} P_{ij}\dot{q}_i\dot{q}_j\,.$$

The coefficients P_{ij} weigh the contributions of the various currents and

2. (*Tr* ii/213). In what follows, Maxwell uses the representation $T_{\dot{q}}$ (in the notation of Chapter 4 above), since the relevant electrical quantity, the current, has the nature of a velocity (*Tr* ii/211).

motions to the kinetic energy. Maxwell's notation distinguishes immediately between the two types of generalized velocity q_i: writing $\dot{x}_i$ for the geometrical velocities of the conductors and $\dot{y}_j$ for the currents, Maxwell recognizes three terms which contribute to the total kinetic energy (*Tr* ii/214):

$$T_m = \tfrac{1}{2}\sum_{ij} P^m_{ij}\dot{x}_i\dot{x}_j \qquad \left[P^m_{ij} \leftrightarrow \left(x_i\, x_j\right)\right]$$

$$T_e = \tfrac{1}{2}\sum_{ij} P^e_{ij}\dot{y}_i\dot{y}_j \qquad \left[P^e_{ij} \leftrightarrow \left(y_i\, y_j\right)\right]$$

$$T_{me} = \tfrac{1}{2}\sum_{ij} P^{me}_{ij}\dot{x}_i\dot{y}_j\,. \qquad \left[P^{me}_{ij} \leftrightarrow \left(x_i\, y_j\right)\right]$$

(Here, and in what follows, I shall consistently write P^m_{ij} for the inertial coefficient which Maxwell designates $(x_i\, x_j)$, etc.) T_m denotes the energy due to the motions of the conductors without regard to any current flow through them, and is hence the "mechanical" energy of the conductors in the ordinary sense. Correspondingly, T_e denotes the energy due to current flow without regard to any motions of the conductors, and is hence purely electrical kinetic energy. Lastly, T_{me} denotes that part of the kinetic energy which depends on both motions, and is zero when either is absent. In the most general case, P_{ij} might be a function of both the coordinates x_i and y_j, although not of the velocities.[3]

From this beginning, there now follows a marvelous sorting-process that must have few parallels in the history of physics; for out of this abstract universal schema, Maxwell selects the actual laws of the electromagnetic world. The nearest parallel may well be Newton's procedure in the "System of the World," in which our gravitational universe is sorted out of the possible worlds of the *Principia*. Many of the empirical touchstones which Maxwell will need have already been discussed in Chapters I–IV of Part iv of the *Treatise*, but there are other experiments which Maxwell is led to propose because they are demanded by the theory. As alternatives are eliminated empirically, the sorting becomes a rapid reduction, which proceeds in the following sequence of giant-steps.

(i) *In electromagnetism*, $P_{ij} = P_{ij}(x_1, \ldots x_n)$ *only*. This is to say that the generalized coordinates y_i have nothing to do with the values of the inertial coefficients P_{ij}. That this is the case follows immediately from experiment, for the electromagnetic behavior of the system has nothing to do with the length of time during which the currents have been flowing, while of course

3. In measuring the current in the ith conductor with the single quantity $\dot{y}_i$, Maxwell assumes the "equation of continuity" of current, as he himself points out (*Tr* ii/214). This is not ultimately a limitation on the theory, as he claims that this principle is never violated in nature; there is never an "accumulation of charge."

during this time the "coordinate" $y_i = \int \dot{y}_i \, dt$ has been steadily growing or diminishing if the currents have been undirectional (*Tr* ii/214).

(ii) *Since the terms* T_m *have nothing to do with the currents, they represent exclusively mechanical effects in the ordinary sense and may be ignored in an electromagnetic theory.*

The following terms remain, then, for consideration:

$$T = T_e + T_{me}$$

where

$$T_e = \tfrac{1}{2}\sum_{ij} P_{ij}^{e}(x_k)\,\dot{y}_i\,\dot{y}_j$$

$$T_{me} = \tfrac{1}{2}\sum_{ij} P_{ij}^{me}(x_k)\,\dot{x}_i\,\dot{y}_j\,.$$

(Here and in what follows, the notation (x_k) means $(x_1, x_2, \ldots)$.)

To pass judgement on the remaining inertial coefficients, we consider the sorts of force to which they would give rise through the Lagrangian equations, and then search for these forces in nature. Such forces will be of two types: forces exerted on the conductors when the coordinates x_i are varied, and the generalized forces, the "electromotive forces," which result from variation of the y_i coordinates. We consider the forces on the conductors first.

In this first category, forces of three types will be candidates for inclusion; I shall here designate them f_α, f_β, and f_γ, though Maxwell's notation is slightly different.[4] Consider first the forces due to the purely electrical terms T_e in the expression for the total kinetic energy T. We have for the force on the ith *mechanical* coordinate, due to the *electrical* part of the kinetic energy:

$$f_i^e = \frac{d}{dt}\frac{\partial T_e}{\partial \dot{x}_i} - \frac{\partial T_e}{\partial x_i}.$$

Let us term this force f_α. Now since

$$T_e = \tfrac{1}{2}\sum_{ij} P_{ij}^{e}(x_k)\,\dot{y}_i\,\dot{y}_j\,,$$

T_e has no dependence on $\dot{x}_i$; hence

$$f_\alpha = -\frac{\partial T_e}{\partial x_i} = -\tfrac{1}{2}\sum_{ij}\frac{\partial P_{ij}^{e}}{\partial x_i}\,\dot{y}_i\,\dot{y}_j\,.$$

Since each term in f_α is multiplied by the product of two currents, the force f_α will not be altered by reversing all currents in the system; this is the symptom by which we may recognize it. In fact, we know that this is

4. $[f_\alpha \leftrightarrow x_e']$, $[f_\beta \leftrightarrow x_{me}'$ (first term)$]$, $[f_\gamma \leftrightarrow x_{me}'$ (second term)$]$ (*Tr* ii/215–16).

characteristic of the *Ampèrian force* between currents with which we are already familiar, and we may now so identify f_α.

In calculating the *mechanical* forces due to the *electromechanical* energy terms T_{me} we get two terms, both of which remain candidates for inclusion:

$$f_i^{me} = \frac{d}{dt}\frac{\partial T_{me}}{\partial \dot{x}_i} - \frac{\partial T_{me}}{\partial x_i}.$$

Let us identify the two terms in this expression separately as force f_β and force f_γ, each a force on the ith coordinate:

$$f_\beta = \frac{d}{dt}\frac{\partial T_{me}}{\partial \dot{x}_i}$$

$$f_\gamma = -\frac{\partial T_{me}}{\partial x_i}.$$

(iii) Maxwell's third major step is now to show that *there is no force of type* f_β. Since we have

$$T_{me} = \tfrac{1}{2}\sum_{ij} P_{ij}^{me}(x_k)\dot{x}_i\,\dot{y}_j\,,$$

then, differentiating,

$$\frac{\partial T_{me}}{\partial \dot{x}_i} = \tfrac{1}{2}\sum_{j} P_{ij}^{me}(x_k)\dot{y}_j\,;$$

so that

$$f_\beta = \frac{d}{dt}\frac{\partial T_{me}}{\partial \dot{x}_i} = \tfrac{1}{2}\sum_{j} P_{ij}^{me}(x_k)\ddot{y}_j\,.$$

Hence f_β will be a *mechanical* force proportional to the *acceleration* of electricity, that is, the time-rate of change of current. Maxwell sketches (Figure 25) a simple apparatus by which to search for such a force, which would require in effect that the current have a momentum in the ordinary sense, and that this momentum be imparted to the conductor through interaction with the current when that current is varied. He concludes that "no such phenomenon has yet been observed" (*Tr* ii/217), but to be sure, he must weigh carefully the tokens by which it would be known: it would be a mechanical effect dependent not on the strength of the current, but on its variation, and (because in the equation above the force is proportional to the acceleration of a single variable y_j) it would be a force whose direction would reverse with the reversal of all the currents in the system. Such a description fits no known phenomenon.

I have not found any indication in the published literature or in the Cambridge University manuscripts that Maxwell actually performed experiments with the apparatus sketched in Figure 25, though the device is so simple, and the question so fundamental, that I am inclined to suppose that he did.[5] The result would of course have been negative, as indeed he asserts ("no such phenomenon has been observed"); for, as we have since learned to say, the "ratio of mass to charge of the electron" happens to be a small number. A positive result in what was acknowledged to be essentially the experiment of Figure 25 was first achieved by S. J. Barnett in 1931; the apparatus indicated by a few lines in Maxwell's drawing had become very complex and delicate.[6]

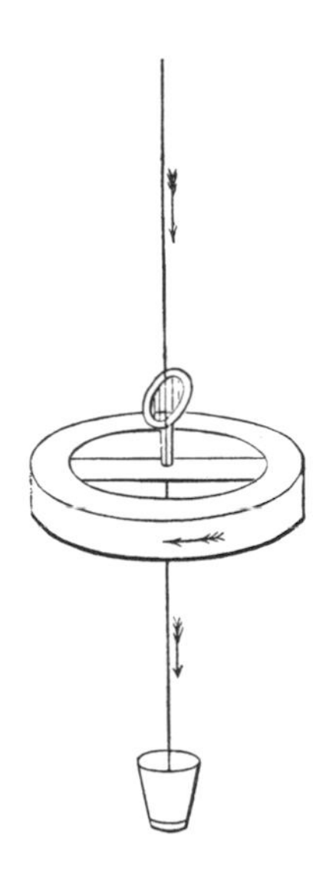

Figure 25 (*Treatise*, vol. II, Figure 33)

What indeed was at stake, in Maxwell's view, in the empirical judgment of the force f_β? The experiment is certainly *not* deciding the question whether the electric current involves the motion of a material system; for, given Maxwell's definition of a material system (page 270, above), this has already been guaranteed by the fact that energy is associated with the electric current. Clearly Maxwell understands the electric current to *be* the motion of a material system in some sense; the question subject to empirical test can only concern the nature and spatial location of this moving system. Maxwell has been careful not to assume that the so-called electric "current" is actually "a current of a material substance" (*Tr* ii/218); *the matter and the motion may be anywhere, and not necessarily in the conductor*—nor need they be linear even if the conductor is. Furthermore, having boldly enlarged the concept

5. Maxwell's wording of the description, by contrast with some others, is admittedly quite hypothetical. He says "We therefore *take* a circular coil..." and "We shall *suppose* the current..." *etc.* (*Tr* ii/216). He surely does not consider he has performed the experiment definitively, and there is no evidence that he tried it at all. Compare Heinrich Hertz, "Versuche zur Feststellung einer oberen Grenze für die kinetische Energie der elektromotorischen Strömung," *Annalen der Physik, 10* (1880), pp. 414 ff.

6. S. J. Barnett, "A new electron-inertia effect and the determination of m/e for the free electrons in Cu," *Philosophical Magazine, 12* (1931), pp. 349 ff. The elaborate apparatus used in this experiment was not constructed for the purpose, but rather to investigate the "Einstein-de Hass effect" (sometimes called the "Richardson-Einstein-de Hass effect"—see the reference to Richardson's work at page 283n below, also G. Joos, *Theoretical Physics*, p. 464). The Einstein-de Hass effect is the mechanical reaction on the body of a material whose state of magnetization is altered; it is due to the angular momentum imparted to the atomic electrons, and the necessity that the total angular momentum of the system be conserved. It is thus very closely related to the force f_β, and indeed Maxwell himself relates the apparatus depicted in Figure 25 to the angular momentum of molecular magnetizing currents; in a letter to Thomson dated December 10, 1861, he writes: "I think it more probable that a coil of conducting wire might be found to have a slight shock tending to turn it about its axis when the electricity is let on or cut off, or that a piece of iron magnetized by a helix might have a similar impulse" (Larmor, *Origins*, p. 36).

of "matter" in his discussion of the foundations of dynamics, Maxwell is now able to draw a possible distinction between matter in general and "*ordinary matter*"; in evaluating the absence of the force we have labeled f_β he writes:

> In fact, if electrical motions were in any way comparable with the motions of ordinary matter, terms of the form T_{me} would exist, and their existence would be manifested by the mechanical force X_{me} [f_β]. (*Tr* ii/218)

Reversing this sentence, we may conclude that since forces f_β do not occur, electrical motions (which are nonetheless motions of matter in the larger sense) are *not* motions of *ordinary* matter, or comparable with them. The negative empirical result would force Maxwell to consider the electric current as a motion of a form of matter that is not "ordinary." In distinguishing two classes of coordinates of the dynamical system he is investigating, Maxwell has opened the possibility that though the electrical variable is coupled inertially with other electrical variables, and the mechanical with other mechanical variables, the two are not inertially coupled to each other: the electrical momentum of a suddenly arrested current cannot be transferred as mechanical momentum to the conductor, as we would expect if the current were thought of as a flow of ordinary matter along the length of the conductor, nor to any other mechanical variable of the system. The absence of any terms of the form f_β would show that an electrical acceleration does not have as consequence *any* sort of mechanical force upon any conductor of the whole system—though we might have expected such remote forces to arise if the motion associated with the current were a motion of "ordinary matter," but occurring elsewhere than in the conductor through which the current is said to flow. There does remain the formal possibility of currents of ordinary matter flowing always in equal and opposite pairs (Fechner's hypothesis), in such a way that the total momentum is forever zero. Perhaps foreseeing difficulties which he will discuss in the final chapter of the *Treatise*, Maxwell does not linger over this hypothesis, though he admits it as a possible alternative.[7]

It should be emphasized that Maxwell does not regard the evidence before him as final in respect to f_β; but insofar as it does rule out such inertial reactions between "ordinary matter" and electricity, Maxwell must admit matter and forces of a second and distinct kind. This *second matter* is inertial and bears real energy, but it does not interact inertially with ordinary matter, or yield ordinary mechanical forces.

Does a positive result in Maxwell's experiment, as has since been achieved by Barnett, but was much earlier assured by the "discovery of the electron," fundamentally alter this conclusion? Maxwell asserted:

7. Whittaker, *History*, p. 201. Fechner's hypothesis is discussed briefly at (*Tr* ii/481–82).

> If any action of this kind were discovered, we should be able to regard one of the so-called kinds of electricity, either the positive or the negative kind, as a real substance, and we should be able to describe the electric current as a true motion of this substance in a particular direction. (*Tr* ii/218)

Though an "action of this kind" would seem to have been discovered, I do not think it is what Maxwell meant: quite apart from the particular character of the "electron," Maxwell did not envision here a "coupling" of charge and mass, but an *identification* of the electric current with a material system of a certain sort. Perhaps this distinction is difficult to trace; but it would mean, I think, that the same current would in all circumstances geometrically alike show the same mechanical-inertial effects. The mass/charge ratio would be intrinsic and hence universal throughout nature; it could not vary with the species of "charged particle." Maxwell is looking for an inertial effect of the current *as such*, and not for the arbitrary coupling of charge to mass which appeared with the discovery of the electron.

(iv) *Similarly, there is no force of type* f_γ. This force, which was defined on page 276, above, will be given by:

$$f_\gamma = -\frac{\partial T_{me}}{\partial x_i};$$

and since

$$T_{me} = \tfrac{1}{2}\sum_{rs} P_{rs}^{me}(x_k)\dot{x}_r\dot{y}_s,$$

we have for f_γ:

$$f_\gamma = -\tfrac{1}{2}\sum_{rs}\dot{x}_r\dot{y}_s\frac{\partial}{\partial x_i}P_{rs}^{me}(x_k).$$

This would be a force exerted mechanically on a conductor due to the presence of currents and motions in other conductors; no pair of conductors could contribute to the force on the kth conductor unless there was a current $\dot{y}_j$ in one and a mechanical velocity $\dot{x}_i$ of the other, and unless their mutual inertial coefficient (product of momentum) P_{ij}^{me} was a function of the position of the ith conductor in question.

As in the case of f_β which we have just considered, there are many conceivable ways in which a force of this general form might manifest itself; but Maxwell takes, as a likely case that can be put to experimental test, a kind of electromagnetic gyroscope, which is a piece of apparatus he tells us he had constructed and used in 1861. The nature of the action corresponding to the force f_γ that is at issue will be made clearer if we follow Maxwell's analysis of his device.

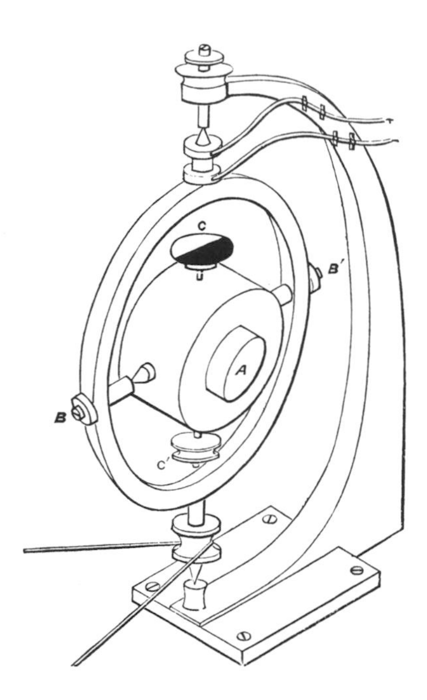

Figure 26. Maxwell's electromagnetic top (*Treatise,* vol. II, Figure 34).

Maxwell's drawing of the electromagnetic top (*Tr* ii/219) is reproduced as Figure 26.[8] Its geometry may be represented as in Figure 27 (showing the top, for pictorial convenience, in a quite different position). Here *A* denotes the axis of the coil, *BB′* the horizontal axis, and *C* the axis through the red-and-green disc, which presents the appearance of a red circle when the angle θ is positive, green when negative. Once the top is spinning with θ constant in an equilibrium position, any change in this equilibrium will be indicated by a change in the apparent color of this disc.

8. Maxwell's electromagnetic gyroscope is still extant at the Cavendish Laboratory. A photograph of it is reproduced in R. L. Smith-Rose, *James Clerk Maxwell* (London: Longmans, Green, and Company, 1948), facing p. 11. A replica is in the posession of the Division of Electricity of the Museum of History and Technology in Washington, DC. The experiment is mentioned in the same letter to Thomson (December 10, 1861) that was quoted on page 277n, above:

> I think that molecules *of iron* are set in motion by the cells and revolve the opposite way so as to produce a very great energy of rotation but an angular momentum in the opposite direction to that of the vortices.
>
> I find that unless the diameter of the vortices is sensible, no result is likely to be obtained by making a magnet revolve about an axis perpendicular to the magnetic axis. I have tried it but have not yet got rid of the effects of terrestrial magnetism which are very strong on a powerful electromagnet which I use. (Larmor, *Origins*, p. 36)

This correspondence makes it clear, if it were not so otherwise, that the experiments discussed in this chapter of the *Treatise* were conceived by Maxwell in connection with his

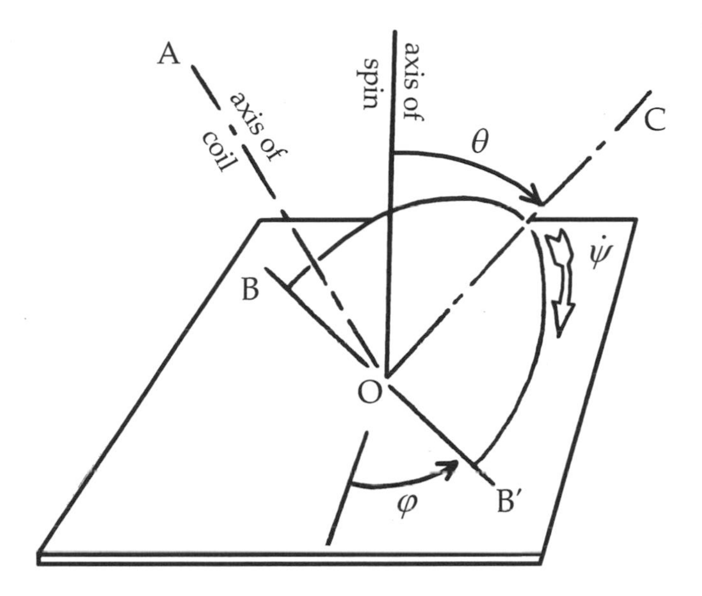

Figure 27. Geometry of Maxwell's electromagnetic top.

The equations of motion of the top may be written as follows. If ψ be taken as the variable whose time-derivative is the current in the coil, then the three independent generalized coordinates of the system are:

$$x_1 = \varphi \quad x_2 = \theta \quad y_3 = \psi,$$

with ψ measured *relative to the coil.* Maxwell writes as A, B, and C the three moments of inertia about the axes denoted by the same letters; resolving the angular velocity $\dot{\varphi}$ about the vertical axis into component angular velocities about axes A and C, we may write the total mechanical kinetic energy T_m as:

$$T_m = \frac{1}{2}\sum_{ij} P_{ij}^m \dot{x}_i \dot{x}_j = \frac{1}{2}\Big[\big(A\sin^2\theta + C\cos^2\theta\big)\dot{\varphi}^2 + B\dot{\theta}^2\Big].$$

Let us now write the kinetic energy of the current. If we assume that the "current" indeed flows as inertial matter around the coil, we must write its angular velocity as $(\dot{\psi} + \dot{\varphi}\sin\theta)$, the electrical motion being added to that

speculations leading up to Parts III and IV of his "Physical Lines of Force" (*Philosophical Magazine, 23* (Jan. & Feb., 1862), pp. 12 ff., 85 ff.) (*SP* i/492, 503 ff.). He expected that in ferromagnetism the angular velocities would be much greater than in ordinary electromagnetism, as the iron particles themselves participated in the motion (*SP* i/507–08). The discussion of the same experiments in Chapter VI of Part iv of the *Treatise* represents, then, something of a retreat, doubtless occasioned by the negative results of these experiments; what he had hoped could become a physical theory becomes here a mere question of formal terms in an abstract dynamical theory. By the same token, the question has become a more fundamental one.

of the conductor in which it flows. Then its kinetic energy will be given as a term of the form:

$$\begin{aligned} T &= \tfrac{1}{2}E\left(\dot{\psi}+\dot{\varphi}\sin\theta\right)^2 \\ &= \tfrac{1}{2}E\left(\dot{\psi}^2+2\dot{\psi}\dot{\varphi}\sin\theta+\dot{\varphi}^2\sin^2\theta\right), \end{aligned}$$

in which E denotes the moment of inertia of the current. We see that the assumption that the electrical motion and the mechanical motion of the coil add together has yielded a cross-term of the type T_{me}, so that we have altogether the following terms in the kinetic energy:

$$\begin{aligned} T_m &= \tfrac{1}{2}\left(P_{11}^m\dot{x}_1^2+P_{22}^m\dot{x}_2^2\right) \\ T_{me} &= \tfrac{1}{2}P_{13}^{me}\dot{x}_1\dot{y}_3 \\ T_e &= \tfrac{1}{2}P_{33}^e\dot{y}_3^2, \end{aligned}$$

where

$$\begin{aligned} P_{11}^m &= (A+E)\sin^2\theta+C\cos\theta \\ P_{22}^m &= B \\ P_{13}^{me} &= E\sin\theta \\ P_{33}^e &= E. \end{aligned}$$

Forces of the type we are investigating arise from the kinetic energy T_{me}, and hence in this case from the inertial term P_{13}^{me}; since it is a function of the position coordinate θ, we will have a generalized force (a torque) tending to rotate the top about the θ-axis, if and only if we were correct in regarding the current as flowing in the conductor with an angular velocity to which that of the conductor itself adds. This torque will be:

$$\begin{aligned} f_\gamma &= -\frac{\partial T_{me}}{\partial\theta} \\ &= -\frac{\partial}{\partial\theta}\left(\frac{1}{2}P_{13}^{me}\dot{x}_1\dot{y}_3\right) \\ &= -\frac{1}{2}\dot{x}_1\dot{y}_3\frac{\partial P_{13}^{me}}{\partial\theta} \\ &= (E\cos\theta)\dot{x}_1\dot{y}_3 \\ &= (E\cos\theta)\dot{\varphi}\dot{\psi}. \end{aligned}$$

When a large current is flowing through the coil, the top is spinning rapidly about the vertical axis, and the axis A is vertical, so that θ is zero and therefore $\cos\theta$ is a maximum, this interaction between the current and the mechanical motion should be best revealed. However, f_γ is not the *total* torque about θ. Maxwell shows how the instrument may be adjusted so that θ will take a stable value with the current off; when the current is then switched on, the term f_γ, if it is present, should disturb the equilibrium, as shown by a change in color of the indicator disc.[9] Maxwell experienced difficulties arising from interaction with the earth's magnetic field, but with that qualification he reports a negative result.

By going one step further with the experiment, Maxwell brought it to bear upon an additional question. Initially, it was really only a second approach to the question whether a current "flows" in a conductor; but Maxwell then added an iron core to the coil to form a strong electromagnet. If Ampère's hypothesis were correct, currents should then be circulating in the iron with the same sense as the original current about the same axis, and thus, should greatly increase the electrical angular momentum. The experiment becoms a test of Ampère's hypothesis.

Maxwell's results remained negative even with the iron core in place, and of course the angular momentum of a magnet is in fact extremely difficult to detect. It was demonstrated in 1915 by the same experimenter, S. J. Barnett, who later demonstrated the inertial effect of a current in the experiment to which we have already referred.[10] In the present experiment Barnett used no current other than the hypothetical Ampèrean currents of a permanent magnet; the magnet, suspended and spinning about a vertical axis, was equivalent to Maxwell's electromagnetic top. The torque f_γ was observed, not in a bodily motion of the magnet, but in a change in orientation of the circulating currents within the magnet, which was detected with a fluxmeter as a change in the magnetization. Barnett attributes the concept of this experiment, as he did the 1931 experiment discussed above, to Maxwell, referring to the section of the *Treatise* which we are examining

9. The equilibrium condition is obtained by calculating the total torque, using the complete kinetic energy T, and then setting the total torque equal to zero. The result is (*Tr* ii/220):

$$\sin\theta = \frac{E\eta}{(C-A)\dot{\varphi}} \qquad [\eta \leftrightarrow \gamma]$$

where η is equal to $(\dot{\varphi}\sin\theta + \dot{\psi})$ and depends upon the strength of the current. This shows that we have equilibrium for zero current when θ is zero. The magnet revolves "about an axis perpendicular to the magnetic axis," which is the way Maxwell used it, as shown by the letter quoted at page 280n, above.

10. S. J. Barnett, "Magnetization by Rotation," *Physical Review, 6* (1915), pp. 239 ff. Barnett says that the experiment occurred to him in 1909, in connection with speculations about the earth's magnetism; could it be accounted for, he wondered, by the tendency of initially randomly oriented circulating currents to align themselves gyrostatically with the rotation of the earth? (That is, in our analysis, by the torque f_γ.) The same question, he says, had occurred to Sir Arthur Schuster in 1912 (*ibid.*, p. 240). He recognizes the effect he has produced as the counterpart of that sought in 1908 by O. W. Richardson, who attempted to demonstrate the torque of reaction upon a piece of iron due to a change in its magnetization. This experiment, which Richardson describes as being in progress at Princeton in 1908 but not yet successful, is exactly that suggested at the end of Maxwell's letter to Thomson, quoted at page 280n, above. O. W. Richardson, "A Mechanical Effect Accompanying Magnetization," *Physical Review, 26* (1908), p. 248.

here. Again it must be pointed out, however, that the significance of Barnett's positive result is not quite what Maxwell intended, for it is the mass of the electrons, not an inertia of the current *as such,* which Barnett demonstrated.

Of the three thinkable mechanical actions of currents, only the first, f_α, reveals itself to Maxwell as existing in the actual world. Thus the concept of a generalized mechanical system including all possible interactions has reduced, in its application to the forces of currents upon conductors, to a system including two categories of moving elements in which any inertial interaction between the two elements is absent. Let us now turn to the question of the action of currents upon currents, and derive the appropriate expressions from the same set of kinetic energy terms. These will then be tested one-by-one for existence.

Let us denote the generalized forces acting on the electrical coordinates as g_i^e and, as before, consider first the forces arising from the kinetic-energy terms T_e. We have:

$$\begin{aligned} g_i^e &= \frac{d}{dt}\frac{\partial T_e}{\partial \dot{y}_i} = \frac{1}{2}\frac{d}{dt}\sum_j P_{ij}^e(x_k)\dot{y}_j \\ &= \frac{1}{2}\sum_j P_{ij}^e(x_k)\ddot{y}_j\,. \end{aligned}$$

This is, then, a generalized force on the current in the ith conductor, a force that is zero when the currents are constant, but directly proportional to the time rate of change of the currents when they vary. Out of the Lagrangian mists emerges the familiar figure of Faraday, for these g_i^e are Faraday's *electromotive forces* of self- and mutual-induction: the coefficients P_{ii}^e (moments of inertia—the case $i=j$) measure the self-inductances of the system's circuits individually, and P_{ij}^e (products of inertia—the case $i\neq j$) measure the mutual inductances.

It might be well to pause briefly to consider the term "electro-motive force." It is, I think, sometimes objected that "electro-motive force" is something of a misnomer, since its units are those of work per unit charge, rather than units of force. The answer to this in the context of the dynamical theory is evident. For in the dynamical theory, "current" is a generalized velocity, and hence the quantity of electricity which "passes" in a period of time is a generalized displacement. But the product of a generalized force and a generalized displacement is the corresponding quantity of work. Hence the term "electro-motive force" is used precisely in Maxwell's dynamical theory to denote the *generalized force* acting on that *generalized coordinate* whose time-derivative is the current.

It remains to consider the forces g_i^{me} arising from the kinetic energy terms T_{me}:

$$\begin{aligned} g_i^{me} &= \frac{d}{dt}\frac{\partial T_{me}}{\partial \dot{y}_i} - \frac{\partial T_{me}}{\partial y_i} \\ &= \frac{1}{2}\frac{d}{dt}\sum_j P_{ij}^{me}(x_k)\dot{x}_i \\ &= \frac{1}{2}\sum_j P_{ij}^{me}(x_k)\ddot{x}_i . \end{aligned}$$

This would be an electromotive force resulting from acceleration of the conductor. It should arise if there is in the conductor an electric fluid which in any sense exhibits inertia. Note that in this case the action is not of the electricity upon the conductor, but of the electricity upon other currents which may be used to detect the presence of the electro-motive force, as for example in accelerating current flow through the coil of a detecting galvanometer. It was of this force that Maxwell wrote to Tait, in the verse "Answer to Tait," concerning Rowland's spinning disc. Maxwell proposed that an electromotive-force experiment might be made with a spinning disc by bringing it to a sudden stop and observing the electromotive force due to the deceleration:

> Not while the coil is spinning sleeps
> On her smooth axle swift and steady,
> But when against the stops she sweeps,
> To watch the light spot then be ready,
> That you may learn from its deflexion
> The electric current's true direction.
>
> It may be that it does not move,
> Or moves, but for some other reason,
> Then let it be your boast to prove
> (Though some may think it out of season
> And worthy of a fossil Druid)
> That there is no electric fluid.[11]

(The "light spot" is that of a mirror galvanometer used to detect the electromotive force g_i^{me}.) This was written in 1877, about seventeen years after Maxwell had first described possible experiments to detect the accelerative electromotive force. There is no evidence, according to Tait's biographer, that Tait ever undertook the experiment.[12] Maxwell emphasizes in the *Treatise* that this force g_i^{me} should be detected much more easily than mechanical forces previously evaluated, since "few scientific observations can be made with greater precision than that which determines the existence or non-existence of a current by means of a galvanometer" (*Tr* ii/221). Maxwell closes this Chapter VI with a reminder to the reader that he has described

11. Knott, *Life of Tait*, p. 95.

12. *Loc. cit.*

methods for seeking three different forces arising from the kinetic energy terms T_{me}: these are the forces we have called f_β and f_γ, and the accelerative electromotive force just discussed:

> We have thus three methods of detecting the existence of the terms of the form T_{me}, none of which have hitherto led to any positive result. I have pointed them out with the greater care because it appears to me important that we should attain the greatest amount of certitude within our reach on a point bearing so strongly on the true theory of electricity. (*Tr* ii/222)

This is clearly an invitation to the reader to undertake a further experimental search for the missing forces.

Like the other forces, the accelerative electromotive force actually observed is a consequence of the coupling of charge to mass in the electron. Electron-inertia effects have been exhibited in a number of fascinating experiments in the twentieth century. Potentials have been produced by, in effect, centrifuging electrons, and also by shaking conductors so as to cause them to move relatively to the electron population in them. These last experiments were facilitated by shaking the conductors at frequencies resonant with the measuring equipment.[13]

Despite its power as a suggestive instrument of search, Maxwell's dynamical theory has caught no new forces in its meshes—though this is not a fault, but a consequence of the proportions of nature. Here, in a sense, is one of those points at which the apparently arbitrary value of a fundamental dimension of the physical world, rather than any want of scientific imagination, sets limits to the development of theory. If the ratio of mass to charge of the electron were considerably greater than it is, Maxwell might have detected forces that would seem to correspond to terms of his dynamical equations. In actuality, however, these would not have been the terms *which he had in mind*. Detecting electron-inertia would have been the discovery of a New World, not the continent toward which he was steering. The forces Maxwell was searching for would have implied that the electric current was at least involved with, if it were not identical to, the motion of

13. R. C. Tolman and T. D. Stewart, "The Electromotive Force Produced by the Acceleration of Metals," *Physical Review, 8* (1916) pp. 97 ff. The long and highly interesting history of experiments on this problem is sketched in this article, with abundant references to the literature. Experiments involving ions rather than electrons succeeded first; one researcher produced a potential in 1882 by dropping a tube of electrolytic solution into a box of sand (*ibid.*, p. 97). Lodge is said to have reported negative results in the Maxwell experiment (*ibid.*, p. 99). Tolman and Stewart reported definitive results in centrifuging electrons with improved equipment in 1917 (*Physical Review, 9* (1917) pp. 164 ff.). Results with an oscillating apparatus are reported by Tolman, Karrer, and Guernsey in *Physical Review, 21* (1923) pp. 525 ff., and there is a further report by Tolman and Mott-Smith in *Physical Review, 28* (1926) pp. 794 ff. See also Whittaker, *History*, ii, p. 243, which is where my attention was first called to these experiments.

a fluid coupled to the conductor—a fluid which exhibited its presence by pushing and pulling lengthwise on the conductors through which it flowed. This would have formed a beginning-point for the design of "a complete dynamical theory of electricity":

> A knowledge of these things would amount to at least the beginnings of a complete dynamical theory of electricity, in which we should regard electrical action, not, as in this treatise, as a phenomenon due to an unknown cause, subject only to the general laws of dynamics, but as the result of known motions of known portions of matter, in which not only the total effects and final results, but the whole intermediate mechanism and details of the motion, are taken as the objects of study. (*Tr* ii/218)

In the absence of these sought-for terms, Maxwell remains on open ocean, without clues. What he is able to do in these circumstances, we shall see below. Before going on, however, it may be prudent to collect the results of the reduction process we have just reviewed. From the equations of motion of a possible electromagnetic system, only two have proved actual: the electromagnetic force between conductors, and the electromotive force acting upon currents. In dynamical theory guise, they appeared as generalized forces derived from those terms of the kinetic energy which involved electrical velocities only:

$$T_e = \tfrac{1}{2}\sum_{rs} P^e_{rs}\,\dot{y}_r\dot{y}_s\,.$$

The first force arises from the dependence of the inertial coefficients upon the geometrical coordinates:

$$f^e_i = -\tfrac{1}{2}\sum_{rs}\frac{\partial P_{rs}}{\partial x_i}\,\dot{y}_r\dot{y}_s\,.$$

The electromotive force arises from acceleration of the electrical variables:

$$g^e_i = \tfrac{1}{2}\sum_{r} P_{ri}\,(x_k)\,\ddot{y}_r\,.$$

This might seem small reward for large labor, but we have not heard the last of the dynamical theory.

Maxwell pauses, in Chapter VII, to reformulate the results he has attained, using terms appropriate to electric circuits. At the same time, he shows how the theory will apply to the simplest case of two circuits inductively coupled. The coefficients of the terms in the expression for kinetic energy, which appeared in the dynamical theory as moments of inertia and products of inertia, become now coefficients of self-induction and mutual induction "when we wish to avoid the language of the dynamical theory" (*Tr* ii/223–24). (Maxwell speaks two languages, much as Homer could name things in the language of men and the language of gods: the dynamical theory is an Olympian view of electromagnetism.) The kinetic energy T_e, to which the total kinetic energy relevant to electromagnetism has reduced, Maxwell now names the "Electrokinetic Energy." The generalized forces on the currents are, as we have seen, identified as the "electromotive forces," and the mechanical momentum associated with a given current $\dot{y}_i$ will be termed its "electrokinetic momentum." If the symbols L_i and M_{ij} denoted the coefficients of self- and mutual-induction (the coefficients P_{ii} and P_{ij} above), then we have:

$$T = \tfrac{1}{2}\left(\sum_i L_i \dot{y}_i^{\,2} + \sum_{ij} M_{ij}\dot{y}_i\dot{y}_j\right)$$

$$= \tfrac{1}{2}L_1\dot{y}_1^{\,2} + \tfrac{1}{2}L_2\dot{y}_2^{\,2} + \cdots + M_{12}\dot{y}_1\dot{y}_2 + \cdots$$

$$P_i = \frac{\partial T}{\partial \dot{y}_i} = L_i\dot{y}_i + M_{1i}\dot{y}_1 + M_{2i}\dot{y}_2 + \cdots.$$

We have seen that the momentum associated with the ith current is in no sense "in" the ith circuit. It rather measures the effectiveness of a change in the ith current in contributing to the total energy of the system; and it depends in general on all the other currents in the system, and the coupling of each to the ith circuit. We may well revert to the image of the bell-ropes (page 212, above): the ropes may be as light as we please; the momentum with which each is associated is not "in" that particular rope but distributed in the system in a way which we may not know, and the motions of all the ropes contribute to the momentum of each through connections which may remain obscure. We are dealing in these equations with what Maxwell has called "total effects and final results" (page 287, above). He will go on in Chapter X to speculate, in terms of a physical theory of the electromagnetic field, about the possible nature of these connections.

It will not be necessary to examine Maxwell's application of the theory to the case of two circuits in detail, as it is set out in a way that is quite straightforward in the *Treatise*, and the unfolding of the theory is not at stake (*Tr* ii/226–27). It is important to notice, however, the role of the coefficient of mutual induction of two circuits. We know that the geometrical position of each circuit is specified by a set of coordinates x_k; if the circuits are regarded as rigid and moving without rotation, three coordinates (which we may take as rectangular Cartesian coordinates) suffice to determine the position of each. The force corresponding to each such coordinate will be given by the corresponding partial derivative of the total energy T, and in the case of rigid circuits the self-induction coefficients will be constant as the circuits are moved. The expression for the force becomes very simple:

$$f_i = -\frac{\partial T}{\partial x_i} = -\left(\frac{1}{2}\dot{y}_i^{\,2}\frac{\partial L}{\partial x_i} + \dot{y}_i\dot{y}_2\frac{\partial M}{\partial x_i} + \frac{1}{2}\dot{y}_2^{\,2}\frac{\partial N}{\partial x_i}\right)$$

where

$$\frac{\partial L}{\partial x_i} = \frac{\partial N}{\partial x_i} = 0.$$

(Here we follow Maxwell in writing L for L_1 and N for L_2.) Then

$$f_i = -\dot{y}_1\dot{y}_2\frac{\partial M}{\partial x_i}\,.$$

The vector force on a conductor at point $P(x, y, z)$ will then be:

$$\mathbf{f} = -\dot{y}_1\dot{y}_2\nabla_P M\,,$$

from which we see that the coefficient of mutual induction, multiplied by the currents, functions here as a mechanical potential.[14] This is the same potential as we met in Maxwell's earlier chapters (pages 133 and 192, above), but this "potential" now emerges from the dynamical theory as arising entirely from the operation of kinetic energies. It is a measure of the interactions arising within a system of "matter and motion."

THE CONSTRUCTION OF THE ELECTROMAGNETIC FIELD

Thus far, the development of Maxwell's dynamical theory has proved correct and clear, but disappointingly conservative. Although it has invited three new fundamental experiments, their outcome has been negative and the work has yielded nothing new for electromagnetic theory. Maxwell has, however, prepared in the dynamical theory a theoretical instrument for which he has ambitious purposes, and through the negative results of the experiments designed to detect the inertia of the electric "fluids," he has prepared us to look elsewhere than in the conductors for the electromagnetic "system" in question. With the beginning of Chapter VIII, attention shifts from the conductors to the space which surrounds them, and each further step in the development of the theory brings us closer to a delineation of the electromagnetic *field* as a connected system of matter in motion. At the same time, the argument departs from the firm ground of unquestioned laws of motion and direct experimental evidence and becomes an imaginative—and questionable—construction.

We are asked first to consider two conductors, of which the first is fixed and rigid and carries constant current. The electrokinetic momentum referred to the second conductor may be written:

14. This equation for **f** expresses a force applied to the system to move the conductor against the electromagnetic force. Maxwell, writing the expression with a positive sign, indicates rather the electromagnetic force itself (*Tr* ii/227).

$$p_2 = M_{12}(x_k)\dot{y}_1 + L_2(x_k)\dot{y}_2,$$

where the coordinates x_k measure the positions of the two coils; only those belonging to the second conductor will be variable. Let us consider just that part of p_2 which measures the effect of coupling to $\dot{y}_1$. Since we shall reason extensively about this term, it will be convenient to denote it simply p. Then:

$$p = M_{12}(x_k)\dot{y}_1.$$

Now p is a quantity associated with the second circuit as a whole in its relation to the first circuit; to every position and configuration of the second circuit, a definite value of p will correspond. Note that p has nothing at all to do with the current in the second circuit, and would indeed be unchanged if $\dot{y}_2$ were made zero, or if the second conductor were not present. Its value is then a function of the "circuit" considered as a *geometrical path*:

> [W]e are not now considering a *current*, the parts of which may, and indeed do, act on one another, but a mere *circuit*, that is a closed curve along which a current *may* flow, and this is a purely geometrical figure, the parts of which cannot be conceived to have any physical action on each other. (*Tr* ii/230)

What was introduced as an examination of two circuits carrying currents has quite suddenly become a study of one current, $\dot{y}_1$, and the *space around it*, in which the second circuit is a mere exploratory path. The quantity whose value we trace in this region is the momentum of $\dot{y}_1$ referred to the exploratory circuit as a whole. The next step will be to refer this momentum not to the whole path, but to individual points in the space about $\dot{y}_1$. Maxwell argues:

> Since the quantity p depends on the form and position of the circuit, we may suppose that each portion of the circuit contributes something to the value of p, and that the part contributed by each portion of the circuit depends on the form and position of that portion only, and not on the position of other parts of the circuit. (*Tr* ii/230)

Since the quantity p in question does not depend on current flow in the second circuit it cannot—so Maxwell reasons—depend on any interaction of the parts of that circuit with one another. Hence the effect of any part of the circuit must be independent of the location of other parts of the circuit, dependent only on its own position and orientation. Maxwell denotes this momentum-per-unit-length of the secondary circuit J, and thus writes:

$$p = \oint \mathrm{J}\, ds.$$

J depends, in some way we do not initially know, upon the orientation of the element of the path at the point. Maxwell now shows carefully (*Tr* ii/230–32) that the values of J for various orientations of the element of the circuit must be related as the components of a single vector. The reasoning draws upon elements of the argument at the basis of Ampère's theory, which Maxwell has employed in his Chapter II; it culminates in the demonstration that a crooked line segment must contribute the same amount to p as the straight line segment between the same end-points; hence the contributions to p add as the components of a vector. (Maxwell himself compares this theoretical result to Ampère's second experiment (*Tr* ii/160, 206).) This permits us to write p as the line integral of a vector that is a function of position only:

$$p = \oint \mathbf{A}\cdot\mathbf{ds}$$
$$J = A_s .$$

This is perhaps the most important single step in the *Treatise*, for Maxwell has now introduced the vector which he believes catches the essence of Faraday's insight into the nature of the electromagnetic field. This vector **A** enters the dynamical theory as the measure of a dynamical property at a point P in the space surrounding a current $\dot{y}_1$ flowing in a given conductor. Specifically, it is the electrokinetic momentum referred to unit length of a geometrical path **ds** placed at the point P, and oriented so that this momentum is a maximum. In saying this, it is important to keep in mind the bell-ringers: the Lagrangian analysis does not ascribe the momentum to the point, or to the line segment locally. The path segment **ds** is an exploratory bell-pull, and we have simply referred to it a "momentum" which may in fact be of any sort, and locally distributed in any way in the space. What we have learned is that the coordinates of one point P are so coupled to those of a circuit elsewhere, S, that if a conductor did pass through P and a current were flowing through S, we would have an (electrical) inertial reaction at P to any change in the current in S. The quantity A is proportional to the current $\dot{y}_1$ flowing in the circuit S; but aside from this factor of proportionality, it reflects the geometry of the situation. Maxwell does not do this, but it may be useful to point out that if we divided out the magnitude of the current $\dot{y}_1$ and thus defined a vector $\mathbf{A}'$, which is independent of the current, we would have the following relations:

$$\mathbf{A}' = \frac{1}{\dot{y}_1}\mathbf{A}$$
$$p = \oint \mathbf{A}\cdot\mathbf{ds} = \dot{y}_1 \oint \mathbf{A}'\cdot\mathbf{ds} .$$

But also

$$p = \dot{y}_1 M_{12} ,$$

therefore

$$M_{12} = \oint \mathbf{A}'\cdot\mathbf{ds} .$$

Then **A′** would be simply a way of representing the coupling coefficient P_{12} or M_{12} in terms of a distribution over space. We have understood from the beginning that the coupling coefficients would be functions of the geometry alone (page 133, above), and we see that **A′** is a measure of the geometrical configuration as well.

Maxwell defines the vector **A** in the following way:

> The vector **A** represents in direction and magnitude the time-integral of the electromotive intensity which a particle placed at the point (x, y, z) would experience if the primary current were suddenly stopped. We shall therefore call it the Electrokinetic Momentum *at the point* (x, y, z). (*Tr* ii/232)

The term "electromotive intensity" has not previously been defined in the dynamical context; the definition which is to follow (at *Tr* ii/240) is that the electromotive intensity $\mathscr{E}$ is the electromotive force per unit length along a path. (We shall return to a discussion of this "definition" at the point at which it is formally introduced, page 298 below.) With this definition of electromotive intensity, the "time-integral of the electromotive intensity" becomes "the time-integral of the electromotive force per unit length." The time-integral of the electromotive force is, however, the electrical *impulse*, and since

$$f_i = \frac{d}{dt}\frac{\partial T}{\partial \dot{y}_i} - \frac{\partial T}{\partial y_i} = \frac{d}{dt}\frac{\partial T}{\partial \dot{y}_i} = \frac{dp_i}{dt}$$

(since T is not a function of y_i), this is the change of electrokinetic momentum. Maxwell's definition of **A** above might therefore be modified to read "the electrokinetic momentum per unit path length which a particle placed at the point would experience if the primary current were suddenly stopped."

The same device that Maxwell has used often throughout the *Treatise*, and which we have discussed earlier (pages 129 and 194)—which is indeed one of the principal figures in his mathematical rhetoric—can be used to change the aspect of this theory without altering its mathematical content: the line integral may be replaced by a surface integral. We have only to introduce a vector $\mathbf{B} = \text{curl } \mathbf{A}$, for then

$$\oint \mathbf{A} \cdot d\mathbf{s} = \oint \mathbf{B} \cdot d\mathbf{S}.$$

B must be a vector such that div $\mathbf{B} = 0$, for div curl $\mathbf{A} = 0$ identically; but otherwise we have as yet no reason to recognize in **B** a familiar physical quantity. Maxwell calls it simply "a new vector." We might, indeed, quite properly describe **B** as "the electrokinetic momentum per unit area due to current $\dot{y}_1$ in circuit S, referred to point P." The flux of the vector **B** through the surface,

$$\varphi = \int_{\mathbf{S}} \mathbf{B} \cdot \mathbf{dS}.$$

emerges as the *electrokinetic momentum of the enclosing circuit*, and it is in this guise that Faraday's lines of force make their way into the dynamical theory.

The physical significance of the quantities **A** and **B** is to be revealed by varying the coordinates: as in the case of the belfry, we pull on one of the ropes. Here we might move some part of the secondary circuit, or vary the current $\dot{y}_1$. Maxwell's strategy is to choose a mathematical method that is shaped to suggest the experimental procedures used by Faraday as nearly as possible. He has in mind a great return, by which the dynamical theory can be brought into accord with the physical method of the "moving wire." Maxwell will thereby be imitating Faraday's physical method mathematically.

As we recall, Faraday used, as an instrument for exploration of the magnetic lines of force, an exploring wire connected to a galvanometer (page 22, above). Motion across the lines of force yielded maximum induction; motion parallel to the lines yielded none at all. To Faraday, the moving wire demonstrated the existence of a state of the space through which it was moved:

> [T]his current, having its full and equivalent relation to the magnetic force, can hardly be conceived of as having its entire foundation in the mere fact of motion. The motion of an external body, otherwise physically indifferent, and having no relation to the magnet, could not beget a physical relation such as that which the moving wire presents. There must, I think, be a previous state, a state of tension or a static state, as regards the wire, which, when motion is superadded, produces the dynamic state or current of electricity. This state is sufficient to constitute and give a physical existence to the lines of magnetic force. (*XR* iii/422)

It is this state, the "electrotonic state," so real to Faraday, that Maxwell proposes to translate into dynamical terms (compare pages 14 ff., above). Maxwell has recognized the possibility that what Faraday had seen as a "state of tension or a static state," and for which a scholar had supplied the term τόνος, might be a state of motion; so that the shock which Faraday had taken as marking the release of a tension might be an inertial reaction to the acceleration of a moving system.

Maxwell makes the program which he now plans to pursue very clear. He must demonstrate that the new vector **B** is indeed the magnetic induction already familiar in other contexts: Faraday's lines of magnetic induction will then be seen to spring, miraculously, out of the mathematical theory:

> In the present investigation we propose to deduce the properties of this vector from the dynamical principles stated in the last chapter, with as few appeals to experiment as possible.

> In identifying this vector, which has appeared as the result of a mathematical investigation, with the magnetic induction, the properties of which we learned from experiments on magnets, we do not depart from this method, for we introduce no new fact into the theory, *we only give a name* to a mathematical quantity, and the propriety of so doing is to be judged by the agreement of the relations of the mathematical quantity with those of the physical quantity indicated by the name. (*Tr* ii/234; emphasis added)[15]

This *naming* is of course the pattern we have already followed in identifying the forces on conductors and currents which arose from the terms of the kinetic energy by way of the Lagrangian equations. It is the pattern we know from a much earlier Creation Story: as the creatures appear, they require to be given appropriate names. To test what Maxwell calls the "propriety" of the procedure, he observes the behavior of each new quantity. In the case of the vector **B**, he explicitly models the mathematical procedure after Faraday's physics:

> We have next to deduce from dynamical principles the expressions for the electromagnetic force acting on a conductor carrying an electric current through the magnetic field, and for the electromotive force acting on the electricity within a body moving in the magnetic field. *The mathematical method which we shall adopt may be compared with the experimental method used by Faraday* ... in exploring the field by means of a wire. ... What we have now to do is to determine the effect on the value of p, the electrokinetic momentum of the secondary circuit, due to given alterations in the form of that circuit. (*Tr* ii/235; emphasis added)

To accomplish this, Maxwell begins for simplicity with a rectangular circuit of which three sides are fixed while the fourth moves parallel to itself: the mathematical surrogate of the moving wire is a purely geometrical path (Figure 28). If **B** is assumed uniform over the enclosed surface and at an angle η to the normal $\hat{\mathbf{n}}$, the electrokinetic momentum p_i of the circuit i is

$$p_i = \mathbf{B} \cos \eta \, S_i \,,$$

where S_i is the area of the loop. If the sliding member is shifted so that it sweeps out the area ΔS_i while covering the distance AA′ in unit time, the increment in p_i will be

$$\Delta p_i = \mathbf{B} \cos \eta \, \Delta S_i .$$

In this simple case the *Faraday force* will be:

$$E_i = \frac{\Delta p_i}{\Delta t} = \mathbf{B} \cos \eta \frac{\Delta S_i}{\Delta t} .$$

For example, we may assume with Maxwell that the field through the rectangular loop is due to current flow $\dot{y}_1$ in a single primary circuit at some

15. Compare the agreement of the mathematical and physical realms by *relation* here, with the role of relations in Maxwell's essay "Analogies" (Campbell, *Life*, pp. 347 ff.).

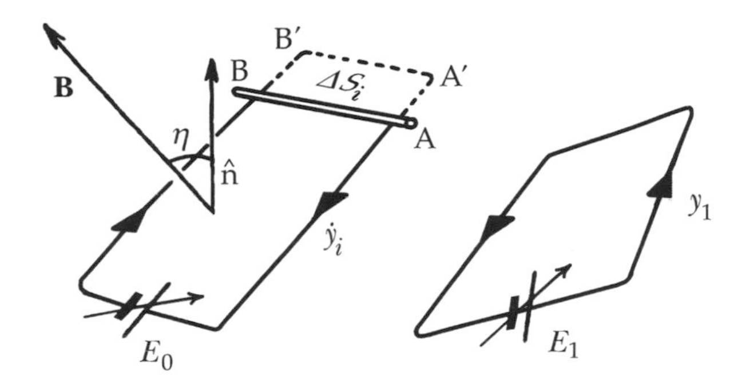

Figure 28. Magnetic induction in the Dynamical Theory.

distance, and that both currents are held constant. (The two adjustable cells E_0 and E_1 have been included in Figure 28 in order to achieve this constancy of the currents as the geometrical relation of the circuits is altered.) Then the electromagnetic force on the conductor, which we may call the *Ampère force*, will be

$$f = \dot{y}_i \dot{y}_1 \frac{\Delta M}{\Delta x}.$$

The Faraday force is an electromotive force induced in the loop; it is independent of current flow $\dot{y}_i$, whereas there will be no Ampère force unless currents flow in both circuits.

The expression for the Ampère force can be transformed into terms of the lines of magnetic induction by considering the energy involved. With the currents held constant (a process which of course involves work on the part of the adjustable cells), the work done by the electromagnetic force on the moving segment will be:

$$f\Delta x = \dot{y}_i \dot{y}_1 \Delta M,$$

$$M = \frac{1}{\dot{y}_i}\oint \mathbf{A}\cdot d\mathbf{s} = \frac{1}{\dot{y}_i}\int \mathbf{B}\cdot d\mathbf{S} = \frac{\varphi}{\dot{y}_i}.$$

Hence over the course of the process,

$$\Delta M = \frac{\Delta\varphi}{\dot{y}_i}$$

$$f\Delta x = \dot{y}_i \Delta\varphi.$$

Or, if we write $\Delta\varphi = B\cos\eta\,\Delta S = B\cos\eta\, l\Delta x$, then

$$f = \dot{y}_i B\, l \cos\eta\, .$$

Thus both the force on the current (the electromotive force) and the force on the conductor can be characterized in terms of Faraday's lines of magnetic inducion. The dynamical theory has deflected from its natural course to introduce the vector **B**, and thus to make connection both with Faraday's usual mode of expression, and with the formulation Maxwell has

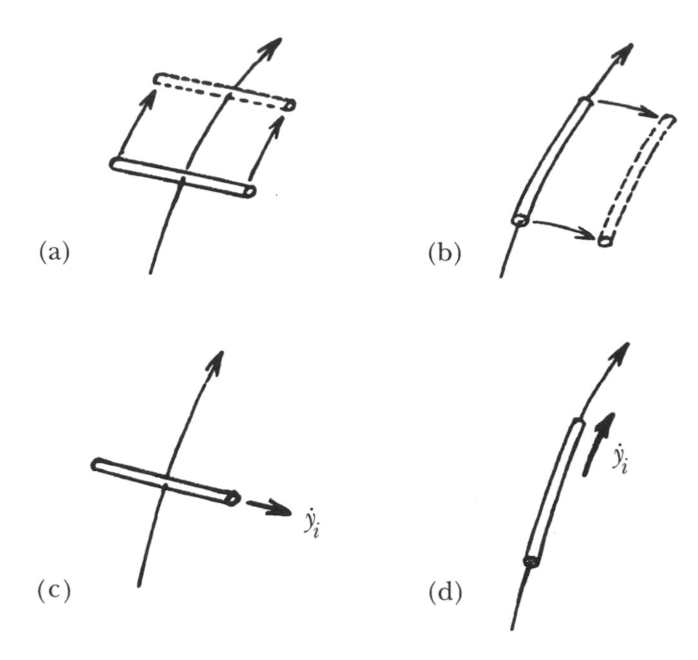

Figure 29. Four definitions of magnetic induction.

given to electromagnetic theory in Chapters I and III of Part iv of the *Treatise*. Maxwell will return to the development of this theory in terms of vector **A**, after a brief consideration of the possible definitions of **B**.

It is now possible to draw more than one definition of a *line of magnetic induction* from this discussion of the moving segment. In fact Maxwell does this in four ways (Figure 29). A line of magnetic induction is a line such that no electromotive force is induced (a) in a segment which moves along it parallel to itself, or (b) in a conductor which lies in it and is displaced parallel to itself; it is equally a line (c) such that a conductor which carries current and is free to move only along the line will experience no tendency to do so, or (d) such that a conductor carrying current and lying along the line will experience no force. Note that the first two of these four definitions do not require the physical presence of a conductor; the electromotive force will exist in a moving line segment.

These last two definitions are equivalent to those given by Ampère for his *directrix*, which, as we have seen, is the same as Faraday's line of force (page 158, above). It is interesting that Faraday's lines enter the dynamical theory very much as they do Ampère's—defined negatively, and as merely geometrical directrices. For Maxwell, I believe, this is a symptom that **B** is not really the significant quantity in the dynamical theory. He returns from this to a further discussion of the electrokinetic momentum, **A**, and thus to what he feels is Faraday's deeper and truer insight, the electrotonic state at the circuit itself.

The next step must be to shift from the special case of the stationary rectangular exploring loop to general expressions for the Faraday and Ampère forces in any case. We must let **A** vary as a function of time, and we must permit the exploring circuit to be in motion. We have found this expression for the Faraday electromotive force around a closed loop:

$$E = -\frac{dp}{dt} = -\frac{d}{dt}\oint \mathbf{A}\cdot\mathbf{ds} = -\oint \frac{\mathbf{dA}}{dt}\cdot\mathbf{ds},$$

where now $\mathbf{A} = \mathbf{A}(x, y, z, t)$ and, since the loop may be in motion, $x = x(t)$, $y = y(t)$, $z = z(t)$. The resulting differentiation, though straightforward, is quite extensive because of the dependence of **A** on four variables. The differentiation yields:

$$E = \oint \left[\mathbf{v}\times(\nabla\times\mathbf{A}) - \dot{\mathbf{A}}\right]\cdot\mathbf{ds}.$$

This equation is then simplified by expressing it in terms of **B**, though its significance is still in terms of the electrokinetic momentum at the circuit:

$$E = \oint \left[(\mathbf{v}\times\mathbf{B}) - \dot{\mathbf{A}}\right]\cdot\mathbf{ds}.$$

The resulting expression is not, however, quite complete. For we may add to the integrand—without changing the value of the integral around the closed loop—the gradient of any function ψ. Maxwell introduces the function ψ as a merely possible term:

> The terms involving the new quantity ψ are introduced for the sake of giving generality to the expressions for ... the integrand. They disappear from the integral when extended round the closed circuit. The quantity ψ is therefore indeterminate as far as regards the problem now before us, in which the electromotive force round the circuit is to be determined. We shall find, however, that when we know all the circumstances of the problem, we can assign a definite value to ψ, and that it represents, according to a certain definition, the *electrical potential* at the point (x, y, z). (*Tr* ii/239)

In this way—certainly most obliquely—a junction is initiated between static electricity and electromagnetism, between Parts i and iv of the *Treatise*. The unity of the work depends on the success of this hesitant graft.[16]

One of the rhetorical devices on which Maxwell's field concept relies most heavily is that which we might call "detachment of the integrand." It consists simply in supposing that the integrand of a definite integral has significance outside the integral; as a mathematical move it is perfectly

16. Duhem objects to the identification of the added term as the gradient of the electrostatic potential: "La fonction est absolument indéterminée et l'on ne saurait logiquement souscrire à l'affirmation" (*Les théories électriques*, p. 163). As I have indicated in the text, the identification is better regarded as rhetorical than logical at this point.

legitimate, of course—so long as conclusions are ultimately drawn only for the definite integral. Physically, it can change a world-view.

Here, Maxwell writes the bracketed integrand separately, and defines it as the *electromotive intensity* at the point $P(x, y, z)$ on the loop. We have then, by definition:

$$\mathscr{E} = (\mathbf{v}\times\mathbf{B}) - \dot{\mathbf{A}} - \nabla\psi$$

with

$$E = \oint \mathscr{E} \bullet \mathbf{ds}.$$

$\mathscr{E}$ will be the electromotive force *per unit length* induced in a circuit that is placed at some point at which the magnetic induction is **B** and the electrokinetic momentum is varying at the rate **A**, the circuit itself meanwhile moving as a whole with the velocity **v**. The term "intensity" refers to this distribution over the length of the circuit.

It is important, I think, to reflect on this device of detachment of the integrand. Its power is evident: a moment before, we were speaking of quantities p_i and E_i which were defined as belonging to the circuit i as a whole. E_i was the generalized force on the integral current associated with the entire circuit. Although Maxwell has cautioned us not to think of this as a fluid (which might of course be partitioned over the circuit), we now introduce an intensive quantity contributing to the action on this whole circuit and yet associated with each element of the path. It may indeed be difficult to supply any physical understanding of this intensive quantity without thinking of a fluid; but so long as our physical image is restricted to the two currents in the original pair of circuits, and to physical measurements made upon such currents (employing instruments such as Faraday's galvanometer) no harm will be done. As a mathematical move, it is unobjectionable, and all physical measurements are made upon the integrated quantity.

But Maxwell is gradually weaning our thoughts away from closed circuits acting upon one another at a distance, and directing us instead to think in terms of dynamical properties referred to points of space. We must recognize this detachment of the integrand in the expression for electromotive force, not as a mere deduction from the theory which precedes, but as a stage in theoretical construction, an imaginative step into a new realm of thought, which is unknown and must remain suspect. He has not shown that the separated $\mathscr{E}$ has any counterpart in physical phenomena. He has certainly not shown that the $\mathscr{E}$ thus defined is the same as the $\mathscr{E}$ of electrostatics. He has, rather, given a rhetorical twist to the theory—and thereby pointed out a direction in which our thought may move in its search of the possible realms of physics. He has taken an admissable step in the mathematical theory, which may or may not prove to have any counterpart in the physical world.

The preceding has concerned the Faraday electromotive force, and yielded the suggestion that the generalized force E acting on the electrical coordinate, the integral current, can be understood as the sum of contributions distributed over the length of the circuit. Maxwell now continues by making the analogous transition for the Ampère force. Earlier, this force was written:

$$f = \dot{y}_1 \dot{y}_2 \nabla M = \dot{y}_2 \nabla \oint \mathbf{A} \cdot \mathbf{ds}.$$

Maxwell now supposes the secondary circuit to be completely flexible, and he allows any continuous distribution whatever of displacements of parts of the circuit, each displacement of a segment **ds** being denoted $\boldsymbol{\delta}\mathbf{r}$. If we now detach the integrand of the above equation, and suppose that it represents a force per unit length per unit current in the secondary circuit, we may write the total work for the arbitrary distribution of the loop as

$$dW = \dot{y}_2 \oint \boldsymbol{\delta}\mathbf{r} \cdot \nabla(\mathbf{A} \cdot \mathbf{ds}).$$

which becomes, by straightforward transformations,

$$dW = -\dot{y}_2 \oint (\mathbf{B} \times \mathbf{ds}) \cdot \boldsymbol{\delta}\mathbf{r} + \dot{y}_2 \oint \nabla(\mathbf{A} \cdot \boldsymbol{\delta}\mathbf{r}) \cdot \mathbf{ds}.$$

The second term on the right, having the form

$$\oint \nabla u(\mathbf{r}) \cdot \mathbf{dr},$$

is identically zero. Then since the remaining expression for the work done in the arbitrary distortion of the whole loop is

$$dW = -\dot{y}_2 \oint_{\mathbf{s}} (\mathbf{B} \times \mathbf{ds}) \cdot \boldsymbol{\delta}\mathbf{r},$$

the integrand may be interpreted as representing a force on an element of the conductor given by

$$\mathbf{df} = -\dot{y}_2 (\mathbf{B} \times \mathbf{ds}).$$

We have made two moves that are mathematically analogous: in the cases of both the electromotive force on the currents, and the electromagnetic force on the conductor, we began with an expression for the forces referred to the entire circuit, and passed from this force on the whole to a specific expression for the force referred to unit length of the path. In the former case, there is no obvious way to test the result physically; in the latter, as we have already seen, the result is familiar and was tested empirically by Ampère (page 132, above).

We are very far from having reproduced in the dynamical theory all of Ampère's theory. The related quantities M, **A**, and **B** have not yet been evaluated at all in terms of the configurations of the circuits; and in these quantities, Ampère's law of force between currents is still latent. To evaluate these quantities will require an investigation of a different kind, one in which new empirical information must enter in order to specify the type of connections that exist within the mechanical system we are describing. In other words, we have been concerned thus far with the presence or absence of terms, and with mathematically alternative expressions for the terms which were found by experiment to exist. Now we need further empirical information as to the specific forms of these terms. The new information is essentially the same as that furnished by Ampère's experiments. Descending from a view of the assumed system as a whole, Maxwell has now arrived at those questions of detail from which Ampère began.

COMPLETION OF THE DYNAMICAL THEORY OF THE FIELD

Up to this point, one may say that Maxwell's theory is only loosely a dynamical one; in many ways it might still satisfy advocates of action-at-a-distance. But in Chapter IX of Part iv, in a series of moves that come very quickly, Maxwell will convert his dynamically-leaning theory into a strictly dynamical theory of the electromagnetic *field*, one in which action-at-a-distance is explicitly replaced by processes in a distributed medium.

So far, Maxwell's theory has been remarkably parallel to Ampère's, deriving Faraday induction in the spirit of strictly formal mathematical inquiry. The principal difference—admittedly an extremely important one—has been that Maxwell's theory is based on Lagrange's equations and the concept of energy, while Ampère's is based on the *Principia* and the concept of force. Maxwell's approach has implied a progression downward from the concept of the system of the whole to the behavior of the part, while Ampère's implied a synthesis, beginning with forces acting on the least parts and building up from these.

Maxwell has indeed shaped his mathematics with an eye toward the field view. He has introduced quantities that are distributed over the paths in question, while making it clear that the "circuits" in the case of Faraday induction are only geometrical paths, and do not necessarily imply the presence of conductors. This has led him to refer a quantity of electrokinetic momentum per unit of path length—the momentum vector—to *a point in space*, and so to regard the Faraday and Ampère forces as arising locally from the relations of elements of those paths with this distributed momentum vector. But Maxwell has yet to take an altogether new step—a step into the field view, and into the unknown. With that step he will complete his dynamical theory.

The dynamical theory until now has had nothing to do with electrostatics, except for the one possible term which Maxwell introduced into the equation for electromotive intensity, the term $\nabla\psi$. Mathematically, this was no more than a formal recognition of the fact that a gradient of a potential would integrate out around the closed path in question, and so might be present—though rhetorically it was clear to Maxwell and the reader that this was to be the electromotive intensity. One of the several decisive moves of this chapter is to make this union explicit:

> We have now determined the relations of the principal quantities concerned in the phenomena discovered by Oersted, Ampère, and Faraday. To connect these with the phenomena described in the former parts of this treatise, some additional relations are necessary. (*Tr* ii/252)

A connection between current electricity and the electric charge had, of course, been established by others long before, and Maxwell introduces it here without delaying for proofs.[17] We do need, however, to maintain our bookkeeping scrupulously in order with respect to the *definitions* of the quantities involved. In the dynamical theory, $\mathscr{E}$ has appeared as the electromotive force per unit length of path, the electromotive force E having been introduced as the generalized force on the electric variable, which we really know in the dynamical theory only as the integral current. In the electrostatic theory, indeed, $\mathscr{E}$ was equated to an "electromotive force" per unit length of path, but it was *not so defined*. Rather, the electromotive intensity was defined as:

> the force which would be exerted on a small body charged with the unit of positive electricity, if it were placed ... [at the point] without disturbing the actual distribution of electricity. (*Tr* i/75)

The electromotive force was then defined in terms of the intensity:

> The Electromotive force along a given arc AP of a curve is numerically measured by the work which would be done by the electric intensity of a unit of positive electricity carried along the curve from A, the beginning, to P, the end of the arc. (*Tr* i/76)

We have thus met two definitions of "electromotive force," derived from phenomena which are completely different. The first has to do with what are called electric "currents" associated with entire closed paths: the electric coordinate thus belongs to the path as a whole, and is obtained indirectly as the time integral of the measured current. We have no empirical support, to say the least, from the experiments of Maxwell's Chapter VI of Part iv for any assumption of a fluid substance moving locally in the conductor; and neither Maxwell nor Faraday would wish to become committed to such a preconception. So we must think simply of a generalized coordinate y_i not specified in location, but merely associated with the path traced by the circuit in question. An electromotive force is associated with this

17. Whittaker, *History*, pp. 92, 175. Maxwell discusses the relation of electromotive force and static potential at a number of points in the *Treatise*; see for example Chapters II and III of Part ii, and especially (*Tr* i/362).

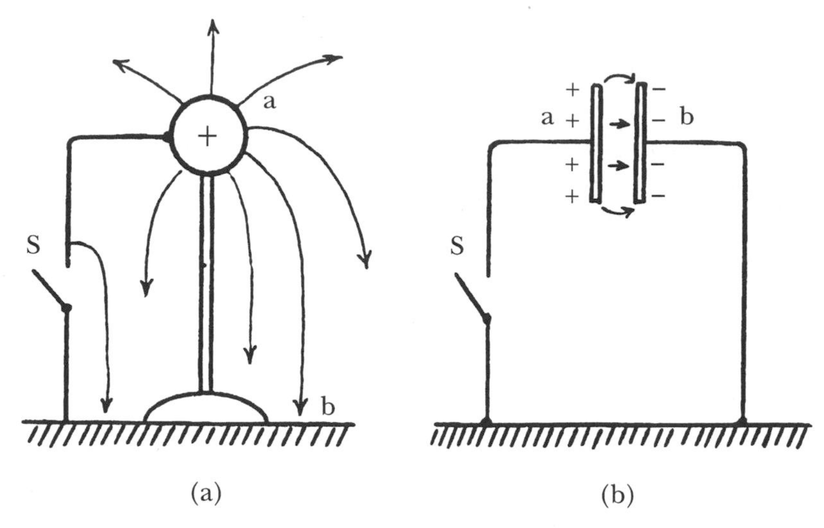

Figure 30. The parallel-plate condenser as a paradigm for electrostatics.

path as a whole, and is simply the generalized force derived from the system energy as a whole when referred to variations of this coordinate alone.

The other definition grew out of electrostatics, in which the observations concerned "electrified bodies" and the observed attractions and repulsions between pairs of them arranged in various configurations. What were observed were mechanical forces between ponderable bodies in virtue of their being electrified. The electromotive force associated with a particular path is measured by the amount of work done by this mechanical force on a body charged to unit amount and carried over the path in question.

In the dynamical theory, the electromotive force acts essentially on an entire closed current. In electrostatics, the electromotive force is always zero over the same closed path.

Maxwell, I believe, achieved a precarious reconciliation of electrostatics and electrodynamics in the dynamical theory. Maxwell's work on electrostatics is complex and difficult to interpret; over the years, Maxwell's papers are inconsistent with themselves. I think it is evident that Maxwell was groping for a unified view of electrostatics and electromagnetism, and that he failed to leave a single coherent account of the whole—as would surely have been an objective of the second edition of the *Treatise*. It has been impossible to include a review of the electrostatics in the *Treatise* in the present study. However, Maxwell's concept of "electric displacement," and the difficulties in which he becomes involved, have been analyzed by Bromberg[18] All I hope to do at this point is to indicate the reconstruction of electrostatics toward which I believe Maxwell to have been making his

18. Joan Bromberg, *Maxwell's Concept of Electric Displacement* (Ph.D. dissertation, University of Wisconsin, 1967).

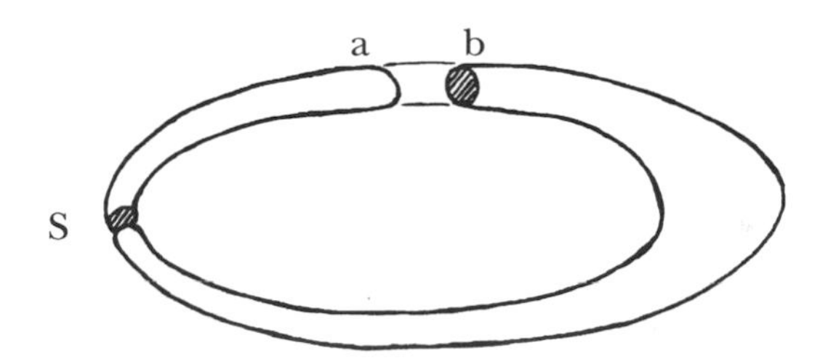

Figure 31. Solenoidal displacement.

way, and by which he was able to glimpse a unified dynamical theory. I offer this rather as a suggestion to readers of Maxwell, than as an interpretation which could be documented with precision. Nevertheless, Maxwell has left clues to the coherent form he wished ultimately to give his theory—not only in the *Treatise*, but especially in the *Elementary Treatise*, which is perhaps a better guide in this respect than is the second edition of the *Treatise* itself. Electrostatics and electromagnetism can be related through the key term "displacement," and through the principle which Maxwell affirms consistently in the *Treatise*: that all currents form closed circuits. For convenience, let us agree that all electrostatic situations are conceptually equivalent to the parallel-plate condenser, so that we need only discuss this one paradigm. Figure 30 suggests this equivalence; it follows from the principle that all lines of force terminate on equal and opposite charges, so that however isolated a charged body may seem to be, there is always, somewhere, a counterpart to the second plate of the condenser.

Looking then at Figure 30(b), I propose that we think of the (quasi static) process by which the condenser might be charged. Current flows, we commonly say, "in" the conductor. Of course we remind ourselves that this locution merely represents a process which may be going on anywhere—perhaps, indeed, everywhere—in the surrounding field. Call this current, as in the dynamical theory, $\dot{y}$. The *electric displacement* is the current integrated over any period of time Δt:[19]

$$\Delta y = \int_{t_1}^{t_1+\Delta t} \dot{y}\, dt .$$

Note that this is *not* the "displacement" vector **D**. The principle that all currents are closed requires that we view this displacement around the entire circuit—including the space between the condenser plates—as solenoidal, while **D** is of course not solenoidal. We may then represent the circuit of Figure 30(b) (and hence, equally, any configuration such as that of Figure 30(a)), by the tube of displacement of Figure 31.

19. The "integral current" (*Integralstrom*) is an important working concept in Franz Neumann's theory: "Die mathematischen Gesetze der inducierten elektrischen Ströme" (1845), *Gesammelte Werke*, *3*, p. 261. Compare page 191n, above.

We can express this solenoidal character of the electric displacement symbolically be writing the current in terms of a current density ι. Here ι is the *true* current density, comprising both conduction and displacement currents in the dielectric. Then we have the following relations:

$$\dot{y} = \int \iota \cdot \mathbf{dS}$$

$$\nabla \cdot \iota = 0.$$

The difference between electrostatics and current electricity is no longer very great. Maxwell accepts Faraday's insight, based on the experiment of the frozen dielectric, (*XR* i/420), and summarized in Faraday's principle that

> insulation and conduction depend only upon the same molecular action of the dielectrics concerned; are only extreme degrees of *one common condition* or effect; and in any sufficient mathematical theory of electricity must be taken as cases of the same kind. (*XR* i/418)

Maxwell says:

> Faraday made great use of this analogy between electrostatic phenomena and those of the electric current, or, as he expressed it, between induction in dielectrics and conduction in conductors, and he proved that, in many cases, induction and conduction are associated phenomena.[20]

He goes on to formulate his own view:

> But we may arrive at a more perfect mental representation of induction by comparing it, not with the instantaneous state of a current, but with the small displacements of a medium of invariable density [i.e., an incompressible medium, like that of "Faraday's Lines"].[21]

The following statement establishes the general ground for a unified treatment of electrostatics and electric circuits:

> That whatever electricity may be, and whatever we may understand by the movement of electricity, the phenomenon which we have called electric displacement is a movement of electricity in the same sense as the transference of a definite quantity of electricity through a wire is a movement of electricity, the only difference being that in the dielectric there is a force which we have called electric elasticity which acts against the electric displacement, and forces the electricity back when the electromotive force is removed; whereas in the conducting wire the electric elasticity is continually giving way, so that a current of true conduction is set up. …
>
> That in every case the motion of electricity is subject to the same condition as that of an incompressible fluid, namely, that at every instant as much must flow out of any given closed surface as flows into it.

20. Maxwell, *Elementary Treatise*, p. 55.

21. *Loc. cit.*

> It follows from this that every electric current must form a closed circuit. (*Tr* i/69–70)

We may speak, then, of the current $\dot{y}$ and the resulting diplacement Δy equally in the case of currents and electrostatics. In the case of a dielectric medium, the flow is blocked by an electromotive force which increases in proportion to the accumulating displacement; it may be regarded as a force on the current as a whole, just as in the case of electromagnetism. It is expressed as the path integral of the electromotive intensity through the medium:

$$E = \int_{\mathrm{a}}^{\mathrm{b}} \mathscr{E} \cdot \mathbf{ds}.$$

The relation between the electromotive intensity and the displacement is expressed as a relation between two vectors $\mathscr{E}$ and **D**. The latter simply represents the electric displacement in an insulating medium—which is to say, *blocked* displacement, or electric (not mechanical) *strain* in the dielectric. In the dielectric, then, after time Δt,

$$\mathbf{D} = \iota \Delta t \quad \text{and} \quad \nabla \cdot \mathbf{D} = 0.$$

In a conductor, current flow gives rise to displacement, but it does not give rise to *strain*; therefore in a conductor the *displacement-vector* (as distinguished from the actual *displacement*) is zero. Then at an interface such as *a* in Figure 31,

$$\mathbf{D} = \iota \Delta t = \sigma \hat{\mathbf{n}},$$

where $\hat{\mathbf{n}}$ is a unit vector normal to the interface surface and σ is the total displacement after time t.

Strictly, there is no "true charge" for Maxwell; there is only displacement and, in dielectrics, the resulting strain and electromotive stress. These are measured by the vectors **D** and $\mathscr{E}$, respectively. The relation between them is a function of the dielectric medium, and is measured by an inverse-elastic constant K, as

$$\mathscr{E} = \frac{4\pi}{K}\mathbf{D}.$$ [22]

What we think of as "charge," and what Maxwell continues to designate "true charge," is then simply the discontinuity of stress produced by the continuity of displacement at a boundary between a conductor and the vacuum. "Charge" is the terminus of a state of stress in the medium, just as Faraday had proposed in Series XI:

22. (*Tr* i/142–45). I have written the relation for an isotropic medium; in general, as Maxwell makes clear, the relation must be expressed by a tensor, and it may also be a function of the position in a non-homogeneous medium, and of the time (*Tr* i/145). Although the stress is in the direction opposite to the displacement, Maxwell does not write a negative sign here because $\mathscr{E}$ has been defined (by a convention hardly appropriate to his theory) as the force on a free charged body at the point. Such a body would be moved in the direction of the displacement, whereas the displacement is blocked by the elastic counter-force.

> According to this theory, all charge is the residual effect of the polarization of the dielectric. The polarization exists throughout the interior of the substance, but is neutralized by the juxtaposition of oppositely charged parts, so that it is only at the surface of the dielectric that the effects of the charge become apparent. (*Tr* i/167)

It is not necessary to linger over difficulties of the term "polarization," so long a problem in the discussion (compare page 87, above). We need only think of a single displacement, continuous throughout conductors and insulators alike, along with a resulting stress in a sense opposing the displacement and having divergence only at a surface of discontinuity in the medium.

What we call with Maxwell "true charge" appears at the interface of a perfect conductor and the vacuum. The latter is simply the "standard medium," the universal ether (*Tr* i/145). For this medium, $K = 1$ by definition, so

$$\mathscr{E} = 4\pi\sigma\hat{\mathbf{n}}.$$

But "apparent charges" arise in the same way, at points of discontinuity of K.

In employing this "representation" of electrostatics, it is essential constantly to bear in mind that it is no more than that. Though we diagram the displacement as being "in" the wire, and the electromotive intensity as a vector acting on it, the meaning remains as generalized as in the Lagrangian theory of electromagnetism:

> In speaking of the resultant electric intensity at a point, we do not necessarily imply that any force is actually exerted there, but only that if an electrified body were placed there it would be acted on by a force $\mathscr{E}e$, where e is the charge of the body. (*Tr* i/75) [$\mathscr{E} \leftrightarrow$ R]

We are reminded here as well of the distinction, always fundamental for Maxwell, that the electromotive force is properly a force on the matter of the electric current, not on the ponderable matter of the conductor or the electrified body. Electromotive force acts on the electrified *body* only incidentally, through whatever interaction there is between the matter which is displaced in the field as the electric current flows and the matter of ponderable bodies.

With this synopsis of electrostatics, we are ready, I think, to proceed with the dynamical theory of electromagnetism—having seen how Maxwell hopes to deal in every case with phenomena of solenoidal currents of electric displacement. The conclusion must be that the quantity $\mathscr{E}$ expressed by the equation

$$\mathscr{E} = (\mathbf{v}\times\mathbf{B}) - \dot{\mathbf{A}} - \nabla\psi$$

is the generalized force referred to the electric generalized coordinate, per unit of path length. This understanding will apply to the static term just as well as to the terms representing Faraday induction. The fact that the electrostatic term $-\nabla\psi$ is derivable from a potential simply means that no generalized force acts on the whole current to set it in motion—that is, that electrostatics *is* static, a case of equilibrium. (When the switches S of Figures 30 and 31 are closed, this equilibrium is destroyed and the electromotive force of the strained dielectric medium becomes a driving force on the current, giving rise immediately to kinetic effects; under these circumstances one can hardly speak of it as having a potential.) It appears to me that we must write in the general case:

$$\mathscr{E} = (\mathbf{v} \times \mathbf{B}) - \dot{\mathbf{A}} - \frac{4\pi}{K}\mathbf{D}.$$

Maxwell inserts at this point a most interesting extension of the investigation, entitled "On the Modification of the Equations of Electromotive Intensity when the Axes to which they are referred are moving in Space" (*Tr* ii/241). He allows a general motion of coordinate system, rotation as well as translation, and determines that the electromotive intensity with respect to the coordinate system moving with velocity **u** will be

$$\mathscr{E}' = (\mathbf{v}' \times \mathbf{B}') - \dot{\mathbf{A}}' - \nabla\psi + \nabla(\mathbf{u} \cdot \mathbf{A}'),$$

where the primed quantities refer to values which would be measured by an observer who made all his measurements in the moving system. We see that $\mathscr{E}'$ is related to the observed $\mathbf{A}'$, $\mathbf{v}'$, and $\mathbf{B}'$ just as in a system at rest, except for a term which is the gradient of a scalar. This will of course disappear on integration around a closed circuit. Maxwell concludes from this:

> In all phenomena, therefore, relating to closed circuits and the currents in them, it is indifferent whether the axes to which we refer the system be at rest or in motion. (*Tr* ii/243)

But all currents, and all displacements, are for Maxwell in closed circuits—hence this is a very sweeping result. It seems unlikely that Maxwell would have asserted it without drawing the conclusion that electromagnetic experiments are indifferent to motion relative to an ether in which, he has assumed, they are transmitted. Optical experiments, as a later chapter of the *Treatise* will show, are also electromagnetic ones; so it seems at least possible that Maxwell has inferred the impossibility of detection of absolute motion, whether by electromagnetic or optical observations, as a mathematical consequence of his theory. If so, this would lend additional point to his well-known letter to Todd proposing an effort to detect any such motion by astronomical measurements.[23] This speculation is not, however, borne out by any remark of Maxwell's of which I am aware; for example, in the article "Ether" (1878) (*SP* ii/763 ff.), where one might expect it.

23. Compare page 265n, above.

Maxwell's "dynamical theory" is indeed an abstract structure of thought. The notion of quantities of "electricity" residing on electrified bodies has been withdrawn; nor do we, in dynamical theory, think of fluids coursing through conducting circuits. Nothing has taken the place of these concrete images. We have instead only the unknown distribution in space of a generalized coordinate representing the electric variable in the connected dynamical system; "currents" are its generalized velocity, and "electric displacement" is its quantity. The complexity of the electromagnetic field view is reflected in the fact that the values of the coordinate y_i and its velocity $\dot{y}_i$, though referred to some particular circuit C_i, or to some particular point P at which the system has been sampled, do not tell us simply the state of affairs at that point, but are only symptoms of the state of affairs *everywhere.*

One hopes that such a union of dynamical theory with electrostatics is to be fruitful; one additional empirical principle will allow us to calculate the progeny:

> We ... observe the action between currents and magnets, and we find that a current acts on a magnet in a manner apparently the same as another magnet would act if its strength, form, and position were properly adjusted, and that the magnet acts on the current in the same way as another current. *These observations need not be supposed to be accompanied by actual measurements of the forces.* They are not therefore to be considered as furnishing numerical data, but are useful only in suggesting questions for our consideration.
>
> The question these observations suggest is, whether the magnetic field produced by electric currents, as it is similar to that produced by permanent magnets in many respects, resembles it also in being related to a potential?
>
> The evidence that an electric current produces, in the space surrounding it, magnetic effects precisely the same as those produced by a magnetic shell bounded by the circuit, has been stated in Arts. 482–485. (*Tr* ii/249–250; emphasis added)

We see here more fully than before the conflux of great sequences of the *Treatise,* with an experiment of illustration identifying the effects of currents and magnetic shells once again at the point of nexus, "suggesting" the unity of view. The magnetic shell theory of Ampère and Poisson, which we have already seen applied to electromagnetism in Part I of this study, is brought to bear on dynamical theory, but now it applies to processes in the dielectric medium as well, where we know that we have phenomena of currents again, and again all currents are closed. Universally, the principle Maxwell derived earlier for magnetic shells and currents in closed conducting circuits applies to the motions of the material system which constitutes the electromagnetic field: all its motions make closed paths of current, and each such path is equivalent to a magnetic shell. If the current in the shell is designated i, then we may write for it (page 138, above):

$$\oint \mathbf{H}\cdot d\mathbf{s} = 4\pi i.$$

What Maxwell has really proposed, in asserting his broad understanding of electric displacement, is that in any electromagnetic process space must be regarded as filled with currents. Any point at which what we call *charge* accumulates will, in the theory, be a source or sink of *dielectric* displacement and will arise only as a result of temporary current flow. To adapt the theory of circuits to this prospect of a continuum of displacement currents in space, it is necessary to replace currents supposed flowing in extended, distinct circuits by a continuum of current density. At the same time, the exploring circuit must be shrunk to the perimeter of a differential area in the field, thus constructing an intensive quantity which measures the value per unit area, as well as the maximal orientation, of the loop integral *at every point* in the field.

The mathematical method of expressing such a measure of the field, the quantity $\nabla\times\mathbf{H}$, was borrowed—as the whole solenoidal representation was borrowed—from the theory of fluids. This had been worked out, as we have seen, by Stokes and many others, and was used by Maxwell in his first electromagnetic paper (see pages 101ff., above). Maxwell is therefore able to take the step in the *Treatise* very readily; and he shifts, virtually without comment, from a theory appropriate to extended circuits acting upon one another at a distance to a strictly field representation. We recognize the result as one of "Maxwell's equations"—a statement valid everywhere in the universe if we continue to understand ι as denoting the *true* current density (page 304 above).

$$\nabla\times\mathbf{H} = 4\pi\iota.$$

It would be well to relate this statement specifically to the dynamical theory as we first met it. In terms of the generalized velocity $\dot{y}$, we would have in any tube of cross-sectional area S:

$$\dot{y} = \int_S \iota\cdot d\mathbf{S}.$$

The total displacement in any time interval Δt would then be:

$$\Delta y = \int \dot{y}\,dt = \int_t\int_S \iota\,dt\cdot d\mathbf{S}.$$

What Maxwell has done here, I suspect, is not so much to introduce a technique of calculation for fields, as to suggest a way of thinking about the details of the connected mechanical system. The elementary area in terms of which "curl" is reckoned is a generalized representation of an element of the machine; if the curl is not zero, then we know that a tiny Faraday rotator would run. Quite possibly Maxwell intends us to have that illustrative mechanism in mind and to recognize, in this equation of the field, evidence that in some sense there will be motors distributed throughout the mechanical system.

Taking this equation for curl **H** as universally valid, we can make a revealing calculation (*Tr* ii/251). Take the divergence of both sides, to obtain

$$\operatorname{div}\operatorname{curl}\mathbf{H} = 4\pi \operatorname{div}\boldsymbol{\iota}.$$

But identically,

$$\operatorname{div}\operatorname{curl}\mathbf{H} = 0.$$

We conclude that div $\boldsymbol{\iota} = 0$: all currents are closed; or, as we have already remarked, displacement is in every case solenoidal. Maxwell says of this result:

> [It] indicates that the current ... is subject to the condition of motion of an incompressible fluid, and that it must necessarily flow in closed circuits.
>
> This equation is true only if we take [$\boldsymbol{\iota}$] as that electric flow which is due to the variation of electric displacement as well as true conduction.
>
> We have very little experimental evidence relating to the direct electromagnetic action of currents due to the variation of electric displacement in dielectrics, but the extreme difficulty of reconciling the laws of electromagnetism with the existence of electric currents which are not closed is one reason among many why we must admit the existence of transient currents due to the variation of displacement. Their importance will be seen when we come to the electromagnetic theory of light. (*Tr* ii/251–52)

Grassmann, Clausius, and Helmholtz had all called attention to the need for work on open circuits, but the principal experiments were done after publication of the *Treatise*.[24] The mathematical deduction of course is not decisive—it perhaps only gives us back what we fed in, for it was derived from the equivalence of circuits to magnetic shells, an equivalence which has only been shown to obtain in cases in which the currents form closed loops. What Maxwell is saying in effect is that the unity of view which he has been able to achieve, combining the insights of Faraday with the mathematical techniques of the Continental theorists, is only compatible with the assumption that displacement is solenoidal. It also happens, I believe, to be a result very close to Maxwell's own intuition that intelligibility in nature can be found only with motion in the plenum of Descartes.

24. Hermann Grassmann, "Neue Theorie der Elektrodynamik," *Annalen der Physik, 64* (1845), pp. 1 ff.; Rudolf Clausius, "Ueber die von Gauss angeregte neue Aufassung der elektrodynamischen Ersheinungen," (1868), translated in *Philosophical Magazine, 37* (1869), pp. 445 ff.; Hermann Helmholtz, "Ueber die Bewegunsgleichungen der Elektrodynamik" (1870), *Wissenshaftliche Abhandlungen, 1*, pp. 545 ff., especially 547, 563. The above papers call attention to the need for experiments with open circuits. Such experiments were carried out not long after the publication of the *Treatise*, particularly by Helmholtz and by one of his students, N. Schiller; J. J. Thomson, *Report*, pp. 142–148, and Whittaker, *History*, pp. 305–06.

This brings the dynamical theory to completion. The quantities and relations needed to characterize the system of matter in motion are now in hand. From the equations already given, we can fill the gap left earlier when we introduced the quantity **A** without showing how to calculate its value (or other quantities derived from it) from a given set of currents.

In the "indirect method," discussed in Chapter 3 above, there were three steps; they culminated in an equation in which the field had cancelled out, so that there was nothing to distinguish the result from an ordinary action-at-a-distance equation except our recollection of how the result had been obtained (page 133, above). Much the same thing happens here again. Maxwell shows us that the familiar expression for the vector potential *at a distance r* from a current can be obtained from the equations of the dynamical theory, and he thereby teaches us to recognize the "electromagnetic momentum at a point" as the vector potential **A**. In other words, the dynamical theory has taught us to understand what the "vector potential" really is.

The derivation is straightforward. It proceeds from the equation for curl **H**, substituting $\mathbf{B} = \text{curl}\,\mathbf{A}$ for $\mu\mathbf{H}$, and then expanding the vector triple product. Maxwell then simply shows by trial that the equation

$$\mathbf{A} = \mu \int \frac{\iota}{r} d\tau, \quad \text{where } d\tau = dxdydz,$$

(*Tr* ii/257) satisfies the differential equation he has obtained.

The result is out of place in the context of the dynamical theory; like Coulomb's law, which Maxwell does not derive from his equations, although he might, it has become misleading—for actions do not take place at the distance *r*. Actions are propagated, as Maxwell will show in a later chapter of the *Treatise*, through the "connected mechanical system" with a finite and definite velocity, so that an equation of this sort can only be interpreted as a steady-state solution applicable to processes which do not vary too rapidly. It is disappointing, indeed, that Maxwell does not point out, as he might, the transformation which must take place in our interpretation of an equation such as this, when it is derived as a consequence of equations of the field. It is, curiously, just the time required to achieve this steady-state which Faraday was envisioning, when he wrote to Maxwell of his experiments in 1857 "on the *time* of the magnetic action, or rather on the time required for the assumption of the electrotonic state, round a wire carrying a current."[25]

Maxwell concludes the dynamical theory with a résumé of the field equations he has developed, expressed in the quaternion notation. Let us follow Maxwell in collecting the equations of the dynamical theory, at the same time reviewing briefly what the dynamical theory does—and above all, what it does not—assert.

25. Campbell, *Life*, pp. 200–201.

RÉSUMÉ OF THE DYNAMICAL THEORY OF THE ELECTROMAGNETIC FIELD

In reviewing Maxwell's version of "Maxwell's equations," we may be struck both by the fact that there are so many of them, and by the inherent sequence in which they unfold. As Maxwell counts them (he letters them serially), there are twelve, but if we add certain important corollaries which he lists but does not count, there are still more. He has not made any effort to reduce the number—though he knows this might be done—because they are not so much quantitative relations as they are vehicles of ideas. Not logical economy, but rather explicit expression, is to be sought:

> These may be regarded as the principal relations among the quantities we have been considering. They may be combined so as to eliminate some of these quantities, but our object at present is not to obtain compactness in the mathematical formulae, but to express every relation of which we have any knowledge. To eliminate a quantity which expresses a useful idea would be rather a loss than a gain in this stage of our inquiry. (*Tr* ii/254)

The progression of equations has a natural beginning in the principles of the dynamical theory; the first three or four equations are deductions from that basic foundation. The next two equations express the mechanical principles from which the electromagnetic equations derive; the first pertains to the matter of the electric current, and the second to ponderable matter. Each determines the force which acts in its realm; the first equation, then, determines an electromotive force, the second, a familiar Newtonian mechanical force:

$$E = -\frac{dp}{dt}$$

$$\mathbf{f} = \dot{y}_1 \, \dot{y}_2 \, \nabla M .$$

E and p are respectively the electromotive force and the electromagnetic momentum belonging to a complete circuit; **f** and M are respectively the mechanical force and electrical product of inertia relating two whole circuits carrying currents $\dot{y}_1$ and $\dot{y}_2$. Although these are the primary electromagnetic quantities, they do not appear explicitly in any of the twelve equations; they are represented by intensive quantities which pertain to points in the field rather than finite circuits. The first of these intensive quantities is **A**, the electromagnetic momentum per unit length of a circuit, or, as Maxwell is willing to say, "the electromagnetic momentum at a point" (*Tr* ii/257); we understand it as an intensive measure of the momentum of the circuit:

$$p = \int \mathbf{A} \cdot d\mathbf{s} .$$

The first of the equations that Maxwell counts as one of his "principal relations" defines the quantity **B** in terms of **A**; it is essential in the flow of thought that it be read from left to right:

$$\mathbf{B} = \nabla \times \mathbf{A}. \tag{A}$$

The solenoidal condition for magnetic induction follows as an immediate corollary; Maxwell does not identify it with a letter, but of course we recognize it as one of the contemporary set of "Maxwell's equations" and may identify it as (A′):

$$\nabla \cdot \mathbf{B} = 0. \tag{A′}$$

These equations serve to tell us what **B** is, dynamically—namely, it is an alternative intensive measure of electromagnetic momentum, in terms of a flux through an area rather than an intensity along a line—electromagnetic momentum per unit area. Like **A**, its ultimate meaning is as a measure of p, the quantity with which the theory begins.

Maxwell's second and third equations constitute a pair, expressing the two forces with which the dynamical theory is concerned: the electromotive force and the mechanical force on a circuit. We have seen how they follow from the corresponding basic mechanical equations, but it is important to recall that $\mathscr{E}$ and **f** are here both intensive quantities; they represent, respectively, the electromotive force per unit length, and the mechanical force per unit volume:

$$\mathscr{E} = (\mathbf{v} \times \mathbf{B}) - \dot{\mathbf{A}} - \nabla\psi \tag{B}$$

$$\mathbf{f} = (\iota \times \mathbf{B}) \tag{C}$$

They are mathematical consequences of the original mechanical idea, and as Maxwell emphasizes, no additional empirical information has gone into them. The significance of $\mathscr{E}$ is then in its integrated form, as contributing to the generalized force which gives rise to the current flow in the whole circuit. The ψ term therefore has really no effect, as we have seen. Although Maxwell does not include it in his final summary of the equations, we should not forget the generalization of equation (B) which he has given in the case of a moving coordinate system (page 307, above):

$$\mathscr{E}' = (\mathbf{v}' \times \mathbf{B}') - \dot{\mathbf{A}}' - \nabla\psi + \nabla(\mathbf{u} \cdot \mathbf{A}'), \tag{B′}$$

These three equations constitute, in effect, the *a priori* group; Maxwell said of them:

> Up to this point of our investigation we have deduced everything from purely dynamical considerations, without any reference to quantitative experiments in electricity or magnetism. The only use we have made of experimental knowledge is to recognize in the abstract quantities deduced from the theory, the concrete quantities discovered by experiment, and to denote them by names which indicate their physical relations rather than their mathematical generation. (*Tr* ii/249)

It is true that Maxwell included a fourth equation in this first set, namely the relation:

$$\mathbf{B} = \mathbf{H} + 4\pi\,\mathbf{I}, \qquad \text{(D)}$$

where **I** is Maxwell's intensity of magnetization—the contribution of a magnetic material to the resulting magnetic field (*Tr* ii/25). This equation seems to me, however, to fall in much more naturally with the later equations in which magnetic and dielectric materials are brought into consideration. Maxwell feels obliged to introduce it here because he needs to employ the vector **H**, and this must first be related to **B**; still, the inclusion of equation (D) here clouds the argument in a most unfortunate way. Strictly, Maxwell is not convinced that two different vectors, **B** and **H**, are needed; he rather inclines to admit Ampère's hypothesis and to deny that there are any magnets as distinguished from currents (*Tr* ii/254). If so, **B** is certainly primary, and **H** is unnecessary. There is even a further awkwardness, in that Maxwell does not here remind us of the transition in the role of **B** (or **H**) from representing the *electromagnetic momentum per unit area* (its true meaning in the dynamical theory) to a measure of *mechanical force on a "pole."* The link is readily supplied, however; we need only follow Ampère in integrating equation (C) to find the force on the near end of a half-infinite solenoid.

The next principal step is to introduce the independent idea that two circuits are equivalent in their interaction to two magnetic shells having the circuits as boundaries. We have noted that magnets do not necessarily enter Maxwell's theory as independent physical entities, but only as aggregates of currents. Nonetheless, the magnetic shell is important here as a device of thought: the *mathematics* of shells is clear, so the identification of currents with shells makes the mathematics of currents clear immediately. This gives us the following equation, which incorporates the insight that currents behave as shells would, together with the quantitative definition of unit current in electromagnetic units:

$$4\pi\iota = \nabla \times \mathbf{H}. \qquad \text{(E)}$$

As we shall see, Maxwell demonstrates that this equation together with our understanding of the meaning of **H** in terms of **A** permits the prediction of the electromagnetic effects of any given configuration of currents.

From this, Maxwell derives as a corollary a relation which is probably as important to his theory as any other, but which he does not himself identify as one of the fundamental equations; it is the solenoidal condition for true current:

$$\nabla \cdot \iota = 0. \qquad \text{(E}'\text{)}$$

Since all of the electromagnetic phenomena are assumed to depend on this true current ι (*Tr* ii/253), it is the current of equations (C) and (E) as well, and there is no need to distinguish between conduction and displacement components.

The remaining equations deal in a broad sense with the properties of materials. When the electromotive force acts on the matter of the electric current, either of two responses is possible, depending on the ponderable matter present at the point at which we are making the observations. We may have either continuous, solenoidal current flow, or blocked, solenoidal current flow together with a counterbalancing electromotive force. The blocked flow results in a dielectric displacement **D**, which is related to the generalized force by the following relation, in which the dielectric inductive capacity K measures the tendency of the ponderable medium to yield and so permit displacement of the material of the current:

$$\mathbf{D} = \frac{1}{4\pi} K\mathscr{E}. \tag{F}$$

If the medium yields continuously, we have the limiting case of continuous current flow with conductivity C:[26]

$$\iota_c = C\mathscr{E}, \tag{G}$$

where ι_c denotes conduction current density. The identity of the displacement in the two cases is immediately asserted by equation (H), and more explicitly spelled out by equation (I):

$$\iota = \iota_c + \dot{\mathbf{D}}, \tag{H}$$

$$\iota = \left(C + \frac{1}{4\pi} K \frac{d}{dt} \right) \mathscr{E}. \tag{I}$$

In terms of the dynamical theory, we are really speaking here of an unexplained but very simple set of relations between two kinds of matter—notwithstanding that experimentally, as we saw, the two matters appear to be inertially distinct.

These last four equations have established our understanding of electric displacement and the dielectric medium. It is impressive that the electric charge has never entered into our thinking; the electric fluid has been guaranteed to be solenoidal, so that no accumulation is possible. The electric charge is therefore defined, as a derivative concept, in the next to last of Maxwell's equations. He calls it "the volume-density of free electricity" or ρ (*Tr* ii/254), and we see immediately that *there is no free electricity*, but only the terminus of a dielectric displacement at the interface with a conductor:

$$\rho = \nabla \cdot \mathbf{D} \tag{J}$$

or, for the corresponding surface-density σ,

$$\sigma = \mathbf{D}_n - \mathbf{D}'_n. \tag{K}$$

Strictly speaking, equations (J) and (K) should be set equal to zero, reflecting in full Faraday's insight that there is no accumulation of "charge"

26. The question of the properties of dielectrics and conductors seems closely related to Maxwell's review of the "Constitution of Bodies" (1877), in which he considers all degrees of elasticity or yielding on the part of a continuous medium (*SP* ii/616 ff.).

on an electrified surface. Measured values of σ and ρ represent discontinuities, not in **D**, but in the elasticity of the medium, represented by its specific inductive capacity *K*.

One other concept remains to be reinterpreted in the theory; that is μ, the coefficient of magnetic permeability. If Maxwell is right in accepting Ampère, μ is not a new theoretical quantity, but a statistical measure of the behavior of the aggregate of permanent currents in a magnetic material. Maxwell has included it, as if with some reluctance, "in order ... to be able to make use of the electrostatic or of the electromagnetic system of measurement at pleasure" (*Tr* ii/255). He therefore writes:

$$\mathbf{B} = \mu \mathbf{H}. \tag{L}$$

The true theoretical quantity is **B**. Although μ and **H**, like σ and ρ considered previously, are measurable quantities, they are actually artefacts of fluid-based thinking and thus have a decidedly subordinate place in Maxwell's theory.

Maxwell's equations, as Maxwell teaches us to read them, form a chain of descent from the dynamical theory. Step-by-step, intelligibility is thereby lent to the familiar concepts of electromagnetism. These concepts, to be sure, are still not perfectly clear; but they are arranged to bring to the subject all the light Maxwell has been able to summom. We can see in them signs of a transition to new and sharper forms, but already they go far toward accomplishing Maxwell's purpose. They embody Faraday's principal insights: the electotonic state and its relation to lines of magnetic induction (equation A); the relation of the electrotonic state and magnetic induction to the induction of currents (equation B); the Faraday motor (equation E); the unity of induction and conduction (equation H); the understanding of electric charge as merely the terminus of a state of strain in a universal medium (equation J). More than this, the entire theory unfolds from the central concept of the motion of a connected mechanical system—that is, everything is related to the initial vision of a continuous set of processes filling the field. These "processes" are connected, solenoidal displacements, and they exist throughout the field.

The equations are consistently referred to observation-points in the field. Nowhere, however, does Maxwell mean to assert that a motion or displacement *exists at the point of sampling*; the representation remains, as Faraday would have it, liberated from specific hypotheses, and thus open to further investigation and speculation.[27]

We have, in one sense, seen nothing at all of the great connected system of matter in motion to which Maxwell has devoted this sequence of

27. Similarly the bell-ropes, most recently recalled on page 288 above. Measurements that are made at each rope do not reveal *properties that exist in the rope itself*; rather they manifest properties that are distributed throughout the whole system in a way we do not know; attempts to articulate the details of that distribution would constitute "specific hypotheses."

chapters in the *Treatise*—we have no definite idea of where any of it is, though the force of analogy has often led us to speak as if the "displacement" were that of a fluid in the conductor. Yet in another sense we now know a great deal about the electromagnetic field as a dynamical system, in terms of its generalized properties and its measurable behavior. We have spoken throughout the dynamical theory with a certain ambivalence, which made it possible to delineate the field by filling space with the vectors **A**, **B**, **D**, and $\mathscr{E}$ without asserting that the moments, displacements, or forces were spatially located at the points at which their measures were placed. What results is in a sense a ghostly structure, for it is a field of measures, not a field of the matter about which the theory is written. Beginning with an *a priori* mathematical idea of the motion which Maxwell is convinced must exist, we move in the direction of physical reality; but this progress consists only in giving to the mathematical theory a shape that corresponds to phenomena observed in the laboratory. In this spirit, Maxwell has conformed the mathematical characterization of the field to Faraday's method of the moving wire. But when all is done, the result is a body of mathematical relations which reflect, but do not touch, the physical world.

The dynamical theory of electromagnetism remains a generalized representation of the physics of electromagnetism; strictly, the dynamical field is what Maxwell elsewhere called a *diagram*:

> The configuration of mechanical systems may be represented in models, plans, or diagrams. The model or diagram is supposed to resemble the material system only in form, not necessarily in any other respect.[28]

Such is, I think, the final significance of the fact that Maxwell built his electromagnetic theory upon the Lagrangian equations of motion. He has presented a system of relations which he believes are true, and which express the form, without at any time identifying the substance, of the electromagnetic universe as Faraday had conceived it.

Maxwell is not content to stop at this point, which is no more than a secure first stage in the advance of natural philosophy. He will go beyond the dynamical theory, both drawing consequences from it (such as the electromagnetic theory of light) and advancing hypotheses that aim to bridge the gap between abstract dynamic theory and a physical theory. To follow him in that progress, however, would be matter for another study.

28. Maxwell, *Matter and Motion*, p. 3.

Conclusion

This study has been, for the writer, an experiment in reading one of the great and yet problematic works of our scientific literature. It has been a critical study, not so much in the sense of passing judgment on Maxwell's production—for such sentence has been passed, justly and adequately, by such critics as Duhem—but rather in the sense in which we speak of criticism of other forms of literature. Here the aim of the critic is to distinguish among possible meanings, to discover purposes which lie beneath the immediate surface, and to recognize the relation which the form and manner of the work bear to the author's intent.

In this sense, the question at stake has been how to read a work of scientific literature. As an exercise in addressing this problem, I think the study has been fruitful, probably because Maxwell wrote the *Treatise* with the counterpart question in mind, how to *write* a work of scientific literature. For Maxwell, the central literary challenge was to invest the symbols of mathematical physics with meaning, by shaping the equations in such a way as to suggest those fundamental ideas which he believed the course of scientific search and research was awakening in the human intellect. In particular, Maxwell's book is bathed in the curious light of Faraday's thought; its express task is to reveal the essential clarity of Faraday's insight, by giving it the form any insight into nature finally deserves, that of equations rightly articulated.

The reader is drawn into this motion of thought as well, for Maxwell writes as a teacher and knows that equations, however well formed, will be of no use unless the student has been prepared for them. Writing as he always does in two phases, first of illustration and then of rigor, Maxwell hopes to initiate the student through the use of example and analogy, and thus to suggest in advance the conepts which the mathematical symbols are to bear. The symbols, he fears, will not speak for themselves unaided. The very history of the science must be woven into the process by which the equations are introduced if they are to be meaningful for the reader, as they have become for Maxwell. This process of interpretation is so vivid in the *Treatise*, and the result achieved so evidently tentative, that one often feels that the "reader" is, to a degree, Maxwell himself, for whom each successive formulation is still only exploratory.

If the *Treatise* were no more than an effort to articulate what had already been understood fully by Maxwell, it would be a great text. But especially in the dynamical theory, we have the sense of a work that is deliberately designed to break through its own bounds, and to open the way to forms of thought which are not yet known. Faraday had constantly disciplined himself to do this, and had become an artist in a rhetoric which would remain open to new possibilities at every crucial point. In the dynamical theory, I have suggested, Maxwell found a mathematical instrument to achieve this same freedom of thought.

The dynamical theory yields a generalized account of the electromagnetic field—an account in which the spatial fields of the electromagnetic quantities are no more than diagrams of the state of a system which we cannot localize, and yet which we know, Maxwell believes, with a high degree of clarity and deserved intellectual satisfaction. We know it as a connected mechanical system; we know its configuration and its most essential properties, though we cannot specify its detail.

The same theory is "generalized" in a deeper sense, one probably closer to Maxwell's ultimate purposes. It opens the very terms of discourse to enlargement, so that it becomes the vehicle for a search for new meanings at the foundation of all explanation of nature, in dynamics itself. In this sense, Maxwell's *Matter and Motion* is a scholium to the dynamical theory of the *Treatise*, for it is an essay in which terms such as force, momentum, mass and energy are re-examined. They are terms which Faraday had used with a blessed naïveté, and which Maxwell wishes to approach with renewed innocence. The same is true also, I suspect, of the notions of space and time themselves: Maxwell appears to be searching for a way to admit a thoroughgoing relativity into our account of nature, a larger view both of the concepts, and of the coordinate framework. If this is indeed Maxwell's effort, we see in the *Treatise* not a statement of a final position, but a motion of mind temporarily arrested; Maxwell has indeed incorporated a sense of this suspended intellectual progression in the text itself.

Not every view of the world would be compatible with an effort of this kind. The *Treatise* could not have taken the form in which it has come to us unless its author believed deeply in the Cartesian project of seeking intelligibility in nature through the clarity and distinctness of mathematical form. But a subjective principle is woven into the work as well: the *Treatise* is as much an invocation of mind, as it is an account of an objective reality. Maxwell is constantly seeking to arouse the mind to do new work, to bring a clarity of its own to bear upon the physical world, and thus to transmute the first phase of empirical inquiry into a second phase of *a priori*, mathematical knowledge. This can be a progressive effort, as Whewell had proposed, because the mind is only gradually—over the centuries, or in the course of reading or writing a book—awakened to the exercise of its full powers. It is a view in which learning has a special role, since the foundations of knowledge are themselves conceived as in flux, only gradually opening to our grasp. It is a view singularly sensitive to the relation of teacher and student, since the subjective contribution of the mind of each must be so great.

The *Treatise* then, though it is a work of mathematical physics, is carefully shaped to meet the demands of a very special philosophy of the world and its relation to mind. This viewpoint, which I have traced in general to Hamilton and Whewell, so much influences the plan and style of the *Treatise* that it would hardly be excessive to claim that though it is a text in physics, it implies a whole philosophy. It is, then, a work of philosophy at the same time it is a work of physics—which is only to say that it succeeds in being a work of natural philosophy. Perhaps Maxwell's *Treatise on Electricity and Magnetism* is best understood as one of the last of the great works of natural philosophy.

Bibliography

(Note: For short-form notation used in references to certain works frequently referred to in the text, see page xxiii.)

I. PRIMARY SOURCES
(writings of Ampère, Faraday, William Thomson, and Maxwell)

1. ANDRÉ MARIE AMPÈRE

Ampère's work has been incidental to the present study, but it has played a greater role than I would originally have expected. I have depended mainly on Ampère's principal work on electrodynamics, in which he himself drew together the results of many earlier papers into one connected statement:

Théorie mathématique des phénomènes électro-dynamiques uniquement déduite de l'expérience (Paris: 1826). In this study I have referred to the following edition: *Nouveau tirage* (Paris: Librarie Scientifique Albert Blanchard, 1958).

Strangely, Ampère's works have apparently never been collected. There has been, however, a most intelligent anthology on early electromagnetism, including the *Théorie mathématique* as well as other writings of Ampère's, along with Oersted, Biot and Savart, Davy, Faraday, etc., with valuable notes:

Jules Joubert, ed., *Collection du mémoirs relatifs à la physique*, 5 vols. (Societé Française de Physique: 1864–91). Volume I contains Coulomb's memoirs; Volumes II and III contain the writings on electromagnetism.

Early memoirs of Ampère's have been collected in:

Mémoirs sur l'électromagnetisme et l'électrodynamique (Paris: Gauthier-Villars, 1921)

and in:

R. A. R. Tricker, *Early Electrodynamics* (Oxford: Pergamon Press, 1965), which contains an extensive account of the work of Ampère and others, and translations from Oersted and Biot and Savart as well.

William Thomson refers to an 1827 collection of six of Ampère's memoirs (EM:410n). I believe this is the same as the *Théorie mathématique* listed above and cited by Maxwell (*Tr* ii/141). The 1826 edition of the *Théorie* bore a subtitle listing six memoirs which were "*réunis*" in its pages; the dates nearly correspond to those Thomson gives; the differences are probably slips.

A serious study of Ampère's scientific thought mast take into account his:

Essai sur la philosophie des sciences (Paris: 1834).

In the present study, I have been concerned with Ampère's significance for Maxwell, thus I have not attempted to give an account of Ampère's actual views.

Ampère's correspondence has been collected in a very valuable edition:

L. de Launay, ed., *Correspondence du Grand Ampère* (3 vols.; Paris: Gauthier-Villars, 1936–43).

2. MICHAEL FARADAY

There is an abundance of material in print to aid the study of Faraday. On the whole, I have found it sufficient for purposes of this study to make use of the following works:

Faraday, *Experimental Researches in Electricity*. (3 vols.; London: 1839, 1844, 1855); a facsimile reprint edition of all three volumes is available (Santa Fe: Green Lion Press, 2000). Note that the *Experimental Researches* comprises reprints of materials that had already appeared elsewhere from 1821 on; it therefore constitutes a partial collection of Faraday's published writings on electricity.

Thomas Martin, ed., *Faraday's Diary* (7 vols. and Index vol.; London: G. Bell & Sons, 1932–36).

H. Bence-Jones, *The Life and Letters of Faraday* (2 vols.; Philadelphia: 1870). I have used the Philadelphia edition because of its availability; I believe it to be identical with the London edition of the same year.

L. Pearce Williams, *Michael Faraday* (New York: Basic Books, 1965).

An important selection of Faraday's correspondence is reprinted in the above-mentioned biographies. A student of Maxwell should note that Faraday's paper "Conservation of Force," which Maxwell discusses in his letter to Faraday of November 13, 1857 (Campbell, *Life*, pp. 202–04), is reprinted in:

Faraday, *Experimental Researches in Chemistry and Physics* (London: 1859). Reprinted London: Taylor and Francis (1991). The essay on Conservation of Force appears at pp. 443 ff.

It is also important to note that Faraday's "Historical Sketch of Electromagnetism," which was the occasion of his beginning research on this subject, is not included in the *Experimental Researches in Electricity*, but was printed in the *Annals of Philosophy*, *2* (1821), pp. 195 ff., 274 ff., and *3* (1822), pp. 107 ff.

Finally, a comprehensive bibliography of Faraday's writings is:

Alan E. Jeffrey, *Michael Faraday: A List of his Lectures and Published Writings* (London: Chapman & Hall, 1960).

3. WILLIAM THOMSON (LORD KELVIN)

For the purposes of the present study, three works have sufficed as sources of Thomson's writings:

Mathematical and Physical Papers (6 vols.; Cambridge: Cambridge University Press, 1882–1911).

Reprint of Papers on Electrostatics and Magnetism (London: Macmillan & Co., 1872).
My references in this study are to the second edition:

_______ Second edition (London: Macmillan & Co., 1884). Note Maxwell's review of this work at (*SP* ii/301); concerning the "Mathematical Theory of Electricity in Equilibrium" and the "Mathematical Theory of Magnetism" in this collection, see page 93nn of the present study.

Silvanus P. Thompson, *The Life of William Thomson* (2 vols.; London: Macmillan & Company, 1910) I have drawn upon this for Thomson's letters as well as for the biographical account; it contains a bibliography that is of great help in sorting out Thomson's writings.

4. JAMES CLERK MAXWELL

Maxwell's output was huge. I list here primarily the writings to which I have referred in this study of the *Treatise*.

A. BOOKS
in chronological order of publication

On the Stability of the Motion of Saturn's Rings (Cambridge: 1859). This was Maxwell's Adams Prize essay of 1856, which was apparently published separately as a book. I have seen this only as an entry in the British Museum Catalogue and do not know whether it differs from the *Scientific Papers* (*SP* i/288 ff.); cf. H.107.

Theory of Heat (London: Longmans, Green & Co., 1871); in the series "Textbooks of Science, Adapted for the Use of Artisans," edited by Thomas N. Goodeve.

The work was popular, and it went through many editions. Goodeve was delighted by it, regarding it as a model for the series, and promptly asked Maxwell, through Longman, to contribute the *Preliminary Dissertation* to the series (letter of William Longman to Maxwell, June 20, 1871; ULC Add.MSS.7655 II/48). This request might account for some of the manuscript fragments on "Methods of the Physical Sciences" (see page 333, below), and may have been the beginning of *Matter and Motion*. Apparently the principal revision of the *Theory of Heat* occurred with the edition of 1891:

Theory of Heat, with corrections and additions by Lord Rayleigh (London: Longmans, Green & Co., 1891).

A Treatise on Electricity and Magnetism (2 vols.; Oxford: The Clarendon Press, 1881).

_______ Second edition: edited by William David Niven (2 vols.; Oxford: The Clarendon Press, 1881).

_______ Third edition: edited by Joseph John Thomson (2 vols.; Oxford; The Clarendon Press, 1892).

The third edition was reprinted, preserving the original pagination and editorial material (Stanford, California: Academic Reprint, 1953), and the same version is currently available from Dover Publications, New York. References in the present study are to the third edition, except where otherwise noted.

German and French editions of the *Treatise* appeared within the decade after Maxwell's death:

_______ Trans. Weinstein (Berlin: Springer Verlag, 1883).

________ Trans. O. Seligmann-Lui, with notes by the translator (2 vols.; Paris: Gauthier-Villars, 1885–87).

I note the above information concerning editions, but a review of foreign editions, and in general the reception of Maxwell's *Treatise*, though it would be of the greatest interest, has been no part of the present study.

Maxwell undertook the *Treatise* in 1867 as the "Senate-House Treatise on Electricity"—that is, the text on electricity which would make it possible to give Tripos examinations in that subject at Cambridge University (letter from P. G. Tait to Maxwell, No. 2, November 27, 1867, ULC Add.MSS.7655 I, a/3). The Clarendon Press accepted it for publication, and Maxwell was at work on Part ii, in August of 1868 (Maxwell to Tait, August 3, 1868; H.299). By December 1869 he had completed the section of Part iv to which the present study is devoted (Maxwell to Tait, December 10, 1869, H.333). In 1870 he writes about quaternion operations, evidently in relation to the *Treatise* (H.346, 348); printer's proof sheets in the manuscript collection bear dates from September, 1871, through January, 1873. The Preface is dated February 1, 1873.

Maxwell had been engaged in preparing a second edition of the *Treatise* at the time of his death in November, 1879. This was a project which apparently paralleled the writing of the *Elementary Treatise*, discussed below. William D. Niven explained in his preface to the second edition, which he produced two years after Maxwell's death, the extent of the changes Maxwell had made. The first nine chapters were significantly altered, but the remainder of the second edition is simply a reprint of the first, with certain comments by Niven added in footnotes identified by square brackets. Since Maxwell's revisions do not extend beyond topics of electrostatics, the later editions of the *Treatise* are identical with the first insofar as the dynamical theory of electromagnetism is concerned. The difference between the first end later editions has not, therefore, been an important consideration in the present study, and I have felt free to refer to the readily available third edition in references throughout the text. In connection with the *Elementary Treatise*, however, a comparison of the first and second editions of the early chapters of the *Treatise* might constitute a valuable study.

The Cambridge University manuscript collection contains a considerable number of manuscript pages and printer's proof sheets of the *Treatise*, but they appear to correspond to only scattered parts of the work and, regrettably, to pertain primarily to the first edition. Only a few sheets might constitute revisions which did not appear in Niven's edition, and I have not studied these to determine whether this is indeed the case.

J. J. Thomson's comments as editor of the third edition of the *Treatise* became so extensive, as he explains, that they required publication in a separate volume, entitled *Recent Researches in Electricity and Magnetism* (Oxford: 1893). This work belongs to the category of extensions of Maxwell's theories, however, and has not contributed to the present study.

Matter and Motion (London: Society for the Promotion of Christian Knowledge, 1876); in the series: "Manuals of Elementary Science." Also (New York: Pott, Young, & Co., 1876).

Reprinted with notes and appendices by Sir Joseph Larmor (London: Society for the Promotion of Christian Knowledge, 1920). This edition, reprinted by Dover

Publications [1952], is currently available. Note that the Preface marked "1877" in the Dover reprint actually appeared in the 1876 edition cited above.

I have remarked on the importance of this little volume in the course of Chapter 5 of the present study. In it, I believe, Maxwell attempted to state, without the use of analytical mathematics, the principles of a new, more generalized understanding of the foundations of mechanics. If I am right, it has the status of a scholium to Chapter V of Part iv of the *Treatise*, and I have therefore treated it as an integral part of the "dynamical theory" with which the present study is concerned. The discipline of restating the results of a mathematical investigation in terms which are devoid of mathematical symbols was very close to Maxwell's heart; the production of an elementary treatise was therefore for him not a passing effort, but perhaps the most profound phase of scientific thought. *Matter and Motion* was perceptively reviewed by Tait in *Nature, 16* (1877), pp. 119 ff.

The Electrical Researches of the Honorable Henry Cavendish, F.R.S., Written between 1771 and 1781; edited from the original manuscripts by James Clerk Maxwell (Cambridge: Cambridge University Press, 1879). A second edition, revised by Joseph Larmor, was issued as:

The Scientific Papers of the Honorable Henry Cavendish, F.R.S.: Volume I. The Electrical Researches (Cambridge: Cambridge University Press, 1921).

Work on this book, which constituted the first publication of many of Cavendish's important electrical researches, occupied a great—one would think, disproportionate—part of Maxwell's time during the first years of his directorship of the Cavendish Laboratory. Maxwell repeated certain of Cavendish's experiments and carried out long calculations concerning them. Maxwell's long notes to the edition are of great interest to a student of Maxwell's electrical work; for the most part, however, they do not pertain to electromagnetism, and thus have not entered into the present study. Although Maxwell's work on the Cavendish writings was no doubt in part a burden imposed upon him and a misfortune for science, there can at the same time be no doubt of its great interest to Maxwell, a fact that reflects his fundamental concern for the history and methodology of science.

An Elementary Treatise on Electricity and Magnetism, ed. William Garnett (Oxford: The Clarendon Press, 1881).

________ Second edition: (Oxford: The Clarendon Press, 1888).

The 1888 edition has been reprinted with notes and an introduction by Peter Pesic (New York: Dover Publications, Inc., 2005)

The *Elementary Treatise* was promptly translated and published abroad:

________ Trans. L. Grätz (Braunschweig: Vieweg, 1883).

________ Trans. Gustave Richard (Paris: 1884).

Like the second edition of the *Treatise*, the *Elementary Treatise* was left unfinished at Maxwell's death. William Garnett, Maxwell's demonstrator at the Cavendish Laboratory, and co-author with Lewis Campbell of the biography of Maxwell, undertook to bring the work into print. The status in which he found it, and the arrangements for publication, are described in his preface. Whereas the second edition of the *Treatise* does not get beyond electrostatics, Maxwell had completed electrostatics and written a good deal on currents for the *Elementary Treatise*. Garnett rounded out the topic of currents with material from the larger *Treatise*; these sections are identified with asterisks in the text. The *Elementary Treatise* does not, therefore, deal directly with the topic of the present study; but it has nonetheless great interest as an indication of the direction Maxwell's thought was taking.

Maxwell had written a "Preface" outlining rather fully what was evidently intended to become a text in optics (H.491). Campbell says, "The MS. of a considerable part of this book is still extant" (*Life*, p. 147n), but I have not recognized it among the Cambridge manuscripts. Niven remarks on the project as well (*SP* i/xxv). The "Preface" I have mentioned, which breaks off abruptly, is in the form of an address to the Cambridge Philosophical Society and relates to an item in the *Scientific Papers* (*SP* ii/591).

B. SCIENTIFIC PAPERS

Most of Maxwell's published papers were collected by Niven in 1890:

W. D. Niven, ed., *The Scientific Papers of James Clerk Maxwell* (2 vols.; Cambridge: Cambridge University Press, 1890).

The same edition has been reprinted twice with original pagination:

________ (2 vols.; Paris: Hermann, 1927), and

________ (2 vols. in one; New York: Dover Publications, [1952]).

Although nearly complete insofar as Maxwell's published papers are concerned, this edition is sadly lacking in bibliographical information and omits a number of valuable papers which are to be found intact in the Cambridge manuscript collection (ULC Add.MSS.7655), as well as certain very interesting fragments. The recent work by Peter Harman represents a dramatic improvement in these and other respects:

P. M. Harman, ed., *The Scientific Letters and Papers of James Clerk Maxwell* (3 vols.; Cambridge: Cambridge University Press 1990, 1995, 2002).

This monumental work was not available to me during the composition of this study in the late 1960's. For the convenience of today's reader I have, where applicable, provided references to the Harman edition.

Published articles cited in the present study, listed in approximate chronological order:

"Experiments on Colour as Perceived by the Eye," *Transactions of the Royal Society of Edinburgh, 21* (1857), pp. 275–98. (*SP* i/126–154); H.59 is an abstract.

"On Faraday's Lines of Force" (Part I: 1855; Part II: 1856), *Proceedings of the Cambridge Philosophical Society, 10* (1864), pp. 27–83 (*SP* i/155–229); cf. H.84, 85, 87. The German edition, by Ludwig Boltzmann, contains valuable notes: *Ostwalds Klassiker der exakten Wissenschaften* Nr. 69 (Leipzig: 1895).

"On Physical Lines of Force," published in four parts in the *Philosophical Magazine*: Part I, "The Theory of Molecular Vortices Applied to Magnetic

Phenomena," *21* (1861), pp. 161–175; Part II, "...Electric Currents" *21* (1861), pp. 281–91, 338–48; Part III, "...Statical Electricity," *23* (1862), pp. 12–24; Part IV: "...the Action of Magnetism on Polarized Light," *23* (1862), pp. 85–95 (*SP* i/451–513). In this case also, the German edition, by Ludwig Boltzmann, includes extensive notes: *Ostwalds Klassiker der exakten Wissenschaften* Nr. 102 (Leipzig: 1898).

"A Dynamical Theory of the Electromagnetic Field" (1864), *Philosophical Transactions, 155* (1865), pp. 459–512. (*SP* i/526–597) Abstracted in *Proceedings of the Royal Society, 13* (1864), pp. 531–36; and in *Philosophical Magazine, 29* (1865), pp. 152–57. (In reference to this, see Cranefield, Paul F., "Clerk Maxwell's Corrections to the Page Proofs of 'A Dynamical Theory...,'" *Annals of Science, 10* (1954), pp. 359 ff.) See also H.231, 238, 239.

This paper is reprinted as:

James Clerk Maxwell. *A Dynamical Theory of the Electromagnetic Field*, edited and introduced by Thomas F. Torrance (Eugene, Oregon: Wipf and Stock Publishers, 1996)

and also in:

Thomas K. Simpson. *Maxwell on the Electromagnetic Field* (Rutgers University Press, 1997).

"Note on the Electromagnetic Theory of Light," appended to the paper "On a Method of Making a Direct Comparison of Electrostatic with Electromagnetic Force" (1868), *Philosophical Transactions, 158* (1868), pp. 643–58 (*SP* ii/125–43); cf. *Proceedings of the Royal Society, 16* (1868), pp. 449–50; and *Philosophical Magazine, 36* (1868), pp. 316–17; also H.289, an abstract.

"Address to the Mathematical and Physical Sections of the British Association," *British Association Reports, 40* (1870), pp. 1–9 (*SP* ii/215 ff.). *Nature, 2* (1870), pp. 419–22. See also H.344, a draft.

"On Colour-Vision at Different Points of the Retina," *British Association Reports, 40* (1870), pp. 40–41 (*SP* ii/230 ff.).

"Introductory Lecture on Experimental Physics" (October 25, 1871). Separately printed: London and Cambridge, 1871. (British Museum Catalogue entry) (*SP* ii/241 ff.).

"On Colour Vision," *Proceedings of the Royal Institution, 6* (1872), pp. 260–71 (*SP* i/267 ff.).

"On the Proof of the Equations of Motion of a Connected System" [1873?], *Proceedings of the Cambridge Philosophical Society, 2* (1876), pp. 292–94 (*SP* ii/308 ff.). Compare the manuscript "On the Interpretation of Lagrange's and Hamilton's Equations of Motion," discussed under The Cambridge Manuscript Collection, page 333 below.

"On Action at a Distance," *Proceedings of the Royal Institution, 7* (1873), pp. 44–54; *Nature, 7* (1873), pp.323–25, 341-43 (*SP* ii/311 ff.). Compare the manuscript "On Faraday's Lines of Force," the James Watt anniversary lecture, discussed under The Cambridge Manuscript Collection, page 333 below.

"Elements of Natural Philosophy," *Nature, 7* (1873), pp. 399–400 (*SP* ii/324 ff.); a review of Thomson and Tait's *Elements of Natural Philosophy*.

"On the Dynamical Evidence of the Molecular Constitution of Bodies" (February 18, 1875), *Journal of the Chemical Society, London, 13* (1875), pp. 493–508; *Nature, 11* (1875), pp. 357–59, 374–77; *Gazzetta Chimica Italiana, 5* (1875), pp. 190–208 (*SP* ii/418 ff.). Compare R. C. Nichols, *Nature, 11* (1875), pp. 486–87, an extensive note on this article in a letter to the editor.

"Whewell's Writings and Correspondence," *Nature, 14* (1876), pp. 206–08 (*SP* ii/528 ff.): a review of Isaac Todhunter's *William Whewell*.

"Capillary Action," *Encyclopædia Britannica*. Ed. 9, Vol. 5 (1876), (*SP* ii/541 ff.).

"Constitution of Bodies," *Encyclopædia Britannica*. Ed. 9, Vol. 6 (1877) (*SP* ii/626 ff.).

"Ether," *Encyclopædia Britannica*. Ed. 9, Vol. 8 (1878) (*SP* ii/763 ff.).

"Thomson and Tait's Natural Philosophy," *Nature, 20* (1879), pp. 213–16 (*SP* ii/776 ff.): a review of the second edition of the *Treatise on Natural Philosophy*.

C. CORRESPONDENCE

The most complete collection of Maxwell's side of his scientific correspondence is P. M. Harman's:

P. M. Harman, ed., *The Scientific Letters and Papers of James Clerk Maxwell* (3 vols.; Cambridge: Cambridge University Press 1990, 1995, 2002).

Note, however, that Harman publishes only letters written *by* Maxwell; letters written *to* Maxwell are listed, but not reproduced, in appendices. Correspondence of Maxwell's has been published in several other places; I list here those biographies and memoirs in which letters of Maxwell's appear.

General correspondence

The original biography of Maxwell is a rich source of his correspondence, verses, and essays:

Lewis Campbell and William Garnett, *Life of James Clerk Maxwell with a Selection from his Correspondence and Occasional Writings, and a Sketch of his Contributions to Science* (London: Macmillan & Co., 1882).

Lewis Campbell, who had been a friend of Maxwell's since their student years at the Edinburgh Academy, wrote the first half, which is the genuinely biographical part of this work. Campbell went to Oxford to become a classical scholar under Jowett; one gets the impression that his contact with Maxwell over the years thereafter may have become rather occasional. He apparently knew nothing of science. The second half of the *Life* was supplied by William Garnett, Maxwell's demonstrator at the Cavendish Laboratory, in an effort to compensate for Campbell's total inability to deal with Maxwell's scientific work. But Garnett tells us little more (and a good deal less) than can be learned from a perusal of Maxwell's published papers. The result is not really a biography at all, but a book the first half of which is highly valuable as a primary source of Maxwell's incidental writings and correspondence, together with the recollections of a friend. In a revised edition, some material was *added* to the first half (despite the fact that the edition is "abridged"); material of no real value was removed from the second part. I have preferred to refer to this second edition:

________, new edition, abridged and revised (London: Macmillan & Co., 1884).

Correspondence with Stokes

Joseph Larmor, ed., *Memoir and Scientific Correspondence of Sir George Gabriel Stokes* (2 vols.; Cambridge: Cambridge University Press, 1907).

Correspondence with Maxwell is in the second volume, as well as referee's reports of Maxwell's, mentioned earlier. Larmor reproduces only letters of Maxwell's to Stokes; letters from Stokes are to be found in the Cambridge manuscript collection, discussed below. Aside from the valuable correspondence, this work tells us little about the relation between Maxwell and Stokes.

Correspondence with Tait

The correspondence between Maxwell and his intimate scientific friend Peter Guthrie Tait is of the greatest interest to students of Maxwell; it is extensive and direct, and at the same time chronicles the interest of the moment and trends of thought. It is witty, cryptic, and sometimes penetrating. On the other band, it does not go to the more reflective side of Maxwell's thought, which shows up better in correspondence with others, in letters reproduced in the *Life*. Some of the Tait-Maxwell correspondence was published by Knott:

Cargill Gilston Knott, *Life and Scientific Work of Peter Guthrie Tait* (Cambridge: Cambridge University Press, 1911).

Knott complains that while Tait kept Maxwell's letters, Maxwell failed to preserve Tait's (*op. cit.*, p. vi); apparently Knott was unfamiliar with the Tait-Maxwell letters now in the Cambridge University manuscript collection. Knott has published only a fraction of the whole set.

Correspondence with Faraday

So far as I know, the correspondence between Maxwell and Faraday was quite limited; with the exception of one minor item it is published in the 1884 edition of Campbell's *Life*. Williams' claim, that the letter he prints at the close of his life of Faraday is published for the first time, is a slip. I have not, however, consulted Faraday manuscript materials.

Correspondence with William Thomson

An extremely valuable collection of the scientific correspondence between William Thomson and Maxwell was published by Larmor:

Joseph Larmor, ed., *Origins of Clerk Maxwell's Electric Ideas* (Cambridge: Cambridge University Press, 1937).

A few items that are not included in this collection may be found among the Cambridge manuscripts.

Correspondence with Strutt

Correspondence with John Strutt (Lord Rayleigh) is printed in:

Robert John Strutt, *Life of John William Strutt, Third Baron Rayleigh* (London: E. Arnold & Co., 1924)

Additional Maxwell correspondence is to be found at a number of centers. Correspondence with Lord Rayleigh, consisting of eight letters of the period 1871–73, will be part of the Rayleigh Archives at the AFCRL Center, Bedford, Massachusetts (communication from Dr. John N. Howard). There are letters from Maxwell to Henry A. Rowland at the Johns Hopkins University. I understand that Maxwell-Gibbs correspondence is preserved at Yale University.

ii. THE CAMBRIDGE MANUSCRIPT COLLECTION

There is a large, extremely valuable collection of Maxwell materials in the Library of Cambridge University (ULC Add.MSS. 7655). These have been arranged to a certain extent by categories, and the correspondence has been placed in alphabetical order and numbered. The material was microfilmed and very kindly made available on loan to the Johns Hopkins University, where I was privileged to work with it in 1967 and 1968. Proper study of these documents would, of course, be a project in itself; and I was able to draw upon them only irregularly for the present study.

The Cambridge manuscript holdings in connection with the Cavendish Laboratory have been described in general by Derek Price:

Price, Derek J., "The Cavendish Laboratory Archives," *Notes and Records of the Royal Society, 10* (1953), pp. 139 ff.

In relation to Maxwell, the collection contains several complete unpublished papers or whole sections of intended works, as well as fragments which are often extremely interesting. There are notebooks and scraps of calculation; the notebook entries are largely in the nature of jottings—there is no diary or journal. It is striking that on the whole the material used by Campbell in writing the *Life* and reproduced there is missing from this collection; this includes correspondence with Campbell himself, and with a number of friends with whom Maxwell entered into serious discussion of his thought, as well as the intriguing "Apostle's [s'?] Club" and "Eramus Club" essays, which are reproduced by Campbell only in fragments. It is also disappointing that the collection apparently no longer includes the extensive notes of Maxwell's lectures of 1878–79, which were said to have been deposited with the Cavendish Laboratory by Andrew Fleming (lectures of the October term [Thermodynamics] and Lent and May terms [Electricity], 1878–79: Ambrose Fleming, in *James Clerk Maxwell: a Commemoration Volume, 1831–1931* (New York: Macmillan & Co., 1931), p. 118.

With reference to the *Treatise* itself, the collection contains a considerable number of manuscript fragments, scattered among other materials. Only a fraction of the *Treatise* is represented, however, and as I mentioned earlier, only a few items might prove to belong to the projected second edition.

Other Maxwell manuscript materials from Edinburgh end Aberdeen have been copied in the Cambridge collection, as well as letters from acquaintances and members of the family, relating to attempts to locate materials.

Manuscripts in the Cambridge University collection referred to in the text.

"Introductory Lecture at King's College, London, 1860" (H.183). Complete manuscript .

"Introductory Lecture at the University of Aberdeen, 1856" (H.132). Complete manuscript.

"Mathematical Theory of Polar Forces." Fragment (cf. H.36)

"Methods of Physical Science" (H.512). A section of a work referred to as "this treatise"—evidently intended for the work which appeared as *Matter and Motion*; note the reference to "Article II" "to define the limits of a material system." The first edition of *Matter and Motion* identified the articles with Roman numerals; Article II concerned the "Definition of a Material System."

"On an Experiment to Determine whether the Motion of the Earth Influences the Refraction of Light." Marked "Received April 26, 1864" (H.227). Complete manuscript. Evidently submitted for publication in the *Philosophical Transactions* and rejected; see page 265n, above.

"On Faraday's Lines of Force" (cf. H.84, 87). Complete manuscript: possibly the form in which Part I of "On Faraday's Lines of Force" was read to the Cambridge Philosophical Society (*SP* i/155 ff.).

"On Faraday's Lines of Force" (H.437). A card photographed with this says "Draft of paper read at Glasgow in memory of James Watt"; the text confirms the occasion as a memorial to Watt. After the opening remarks, the substance is very close to "Action at a Distance" (*SP* ii/311 ff.).

"On the Application of the Ideas of the Calculus of Quaternions to Electromagnetic Phenomena" (H.347). Complete manuscript; compare the *Treatise*, 1st ed., pp. 8 ff.

"On the Interpretation of Lagrange's and Hamilton's Equations of Motion"(H.419). A card attached to this manuscript in the Cambridge collections says "(about 1876) published," but I have found no record of it as a published paper.

There are reported to be some notebooks of Maxwell's, as well as early notes for the *Treatise*, at King's College, London (Sir John Randall, "Aspects of Maxwell's Life and Work," in C. Domb, ed., *Clerk Maxwell and Modern Science*, p. 2).

II. GENERAL BIBLIOGRAPHY

The general literature on Maxwell and the development of electromagnetic theory has been reviewed before; it is sufficient to refer to a very useful bibliography in the American Institute of Physics series of "Resource Letters":

William T. Scott, "Resource Letter FC–1 on the Evolution of the Electromagnetic Field Concept," *American Journal of Physics, 31* (1963), pp. 819 ff.

An excellent study of Maxwell's electrical theory is:

Bromberg, Joan. *Maxwell's Concept of Electric Displacement* (Ph.D. thesis, University of Wisconsin, 1967: University Microfilms 67–475).

It includes a valuable discussion of the critical literature on Maxwell's electromagnetism. I very much regret that this thesis did not become available to me in time to draw upon it for the present study.

Another important work, which I unfortunately discovered too late to take properly into account in this study, is:

Davie, George Elder. *The Democratic Intellect: Scotland and her Universities in the Nineteenth Century* (Edinburgh: Edinburgh University Press, 1961).

I believe this work lends confirmation to the view I have taken, that the influence of Hamilton and Scottish education is of considerable importance to Maxwell's work, and hence to British science of the nineteenth century.

I should like to call attention to another work of significant interest, which is not always noted in bibliographies of Maxwell:

Boltzmann, Ludwig. *Vorlesungen über Maxwells Theorie der Elektrizität und des Lichtes* (Leipzig: 1893).

This work contains an extensive bibliography, reflecting the reception of Maxwell's theory on the Continent.

An important study of Maxwell's theory of the method of science is:

Turner, Joseph. *The Methodology of James Clerk Maxwell* (Ph.D. thesis, Columbia University, 1933: University Microfilms 6725).

WORKS CITED IN THE PRESENT STUDY

I list below the works which have been referred to either in footnotes or in the body of this text. Within the entry for each author, works are listed in approximately chronological order. I have not, however, listed separately the many papers of Thomson, Faraday, and Ampère to which reference has been made throughout the text. In these cases, references have been to readily available standard collections and, as I have discussed under Primary Sources, above, bibliographic information is easily accessible to the interested student.

Ampère, André-Marie. *Théorie mathématique des phenomènes électrodynamiques uniquement déduite de l'éxperience* (see Primary Sources, above).

________ and Savary, F. (Notes on a passage in Faraday's *Experimental Researches* [*XR* ii/127]) *Annales de chimie, 18* [1827], p. 127).

Adams, George. *An Essay on Electricity*. 2nd ed. London: 1875.

Anderson, P. J. "David Thomson," *Dictionary of National Biography*.

Aristotle. *Poetics*.

________. *Posterior Analytics*.

________. *Rhetoric*.

Bacon, Francis. *The Works of Francis Bacon*. 14 vols. Ed. James Spedding, Robert Ellis, and Douglas Heath. London: 1858–74. Reprinted Stuttgart: Friedrich Fromann Verlag, 1963.

Ball, Walter W. Rouse. *A History of the Study of Mathematics at Cambridge*. Cambridge: 1889.

Barnett, S. J. "Magnetism by Rotation," *Physical Review, 6* (1915), pp. 239 ff.

________. "A New Electron-inertia Effect and the Determination of m/e for the Free Electrons in Cu," *Philosophical Magazine, 12* (1931), pp. 349 ff.

Becquerel, Antoine. *Traité de l'électricité.* 7 vols. Paris: 1834–40.

Bianchi, N. *Carlo Matteucci et l'Italia del suo tempo.* Rome: 1874.

Biot, J. B. and Savart. "Sur l'aimanation imprimée aux métaux par l'électricité en mouvement," *Annales de chimie et physique, 12* (1820), pp. 222–23.

Bork, Alfred. "Maxwell, Displacement, Current, and Symmetry," *American Journal of Physics, 31* (1963), pp. 857 ff.

Boscovich, Roger Joseph. *A Theory of Natural Philosophy* (1763). Cambridge, Massachusetts: M.I.T. Press, 1966.

Bromberg, Joan. "Maxwell's Displacement Current and his Theory of Light," *Archive for History of Exact Sciences, 4* (1967), pp. 219 ff.

________. "Maxwell's Concept of Electric Displacement." Ph.D. Thesis, University of Wisconsin, 1967. University Microfilms 67–475.

Brush, Stephen G. *Kinetic Theory.* Vol. 1, *Nature of Gases and of Heat.* Oxford: Pergamon Press, 1965.

Campbell, Louis and Garnett, William. *The Life of James Clerk Maxwell* (see Primary Sources, above).

Carnot, Sadi. *Reflections on the Motive Power of Heat* (1824). Trans. R. H. Thurston (1890). Ed. B. Mendoza. New York: Dover Publications, 1960.

Clausius, Rudolf. "Ueber die von Gauss angeregte neue Aufassung der elektrodynamischen Erscheinungen" (1868), translated in *Philosophical Magazine, 37* (1869), pp. 445 ff.

Crowther, J. G. *British Scientists of the Nineteenth Century.* London: Kegan, Paul, and Co., 1935.

Davie, George E. *The Democratic Intellect: Scotland and her Universities in the Nineteenth Century.* Edinburgh: Aldine Press, 1961.

Davy, Humphry. "On a New Phenomenon of Electromagnetism," *Philosophical Transactions, 113* (1823), pp. 153 ff.

Dalton, John. *A New System of Chemical Philosophy* (1808). New York: The Citadel Press, 1964.

de la Rive, Auguste. *A Treatise on Electricity.* Translated by Charles V. Walker. 3 vols. London: 1853–58.

Descartes, René. *Oeuvres de Descartes.* 12 vols. Ed. Charles Adams and Paul Tannery. Paris: Leopold Cerf, 1904.

Ducasse, Curt J. "William Whewell's Philosophy of Scientific Discovery," Chapter IX of *Theories of Scientific Method: the Renaissance through the Nineteenth Century.* Ed. E. H. Madden, R. M. Blake, and Curt J. Ducasse. Seattle: University of Washington Press, 1960.

Duhem, Pierre. *The Aim and Structure of Physical Theory* (1906). Translated by Philip Wiener. New York: Atheneum, 1962.

________. *Les théories électriques de J. Clerk Maxwell.* Paris: Hermann, 1902.

Ettingshausen, Andreas von. "Ueber Ampère's elektrodynamische Fundamentalversuche," *Königliche Akademie der Wissenschaften, Wien: Mathematische-Naturwissenschaftliche Klasse, Sitzungsberichte, 11* (1878), p. 109.

Euler, Leonhard. *Letters of Euler on Different Subjects in Physics and Philosophy Addressed to a German Princess* (1770–72). Translated by Henry Hunter. 7 vols. London: 1802.

Faraday, Michael. (For works by Faraday, see Primary Sources, above. References to the *Experimental Researches in Electricity* are made in the text by a short-form notation which is explained on page xxiii of the Preface. Since the places of original publication are identified in the *Experimental Researches*, and since thorough bibliographical information on Faraday's writings is readily available in works discussed under Primary Sources, individual articles are not listed again here.)

Felici, Riccardo. *Annales de chimie, 34* (1852), pp. 64 ff.

Forbes, James David. *Dissertation Sixth ... of the Progress of Mathematical and Physical Science.* Edinburgh: [1856?].

Fourier, Joseph. *Théorie analytique de la chaleur* (1822). Translated by Alexander Freeman (1878). Reprinted; New York: Dover Publications, 1955.

Gauss, Karl Friedrich. "General Theory of Terrestrial Magnetism" (1838), translated in *Scientific Memoirs*, ed. Richard Taylor, *2* (1841), pp. 184 ff.

________. "Intensitas Via Magneticae Terrestris ad Mensuram Absolutam Revocata," *Werke.* 12 vols. Göttingen: 1863–1935.

Gilbert, William. *On the Loadstone and Magnetic Bodies* (1600). Translated by P. F. Mottelay (1892). Ann Arbor: Edwards Brothers, 1938.

Grassman, Hermann. "Neue Theorie der Elektrodynamik," *Annalen der Physik, 64* (1845), pp. 1 ff.

Hamilton, William. *Lectures on Metaphysics and Logic.* Vol. I: *Metaphysics.* Ed. H. L. Mansel and J. Veitch. Boston: 1859.

Helmholtz, Hermann. "Ueber die Bewegungsleichungen der Elektrodynamik" (1870), *Wissenschaftliche Abhandlungen, 1.* 3 vols. Leipzig: 1882–95, pp. 545 ff.

Hertz, Heinrich. "Versuche zur Feststellung einer oberen Grenze für die kinetische Energie der elektromotorischen Strömung," *Annalen der Physik, 10* (1880), pp. 414 ff.

Hesse, Mary. *Forces and Fields.* London: Nelson and Sons, 1961.

Jones, H. Bence. *The Life and Letters of Faraday.* 2 vols. Philadelphia: 1870. (Identical to the London edition of the same date.)

Joos, Georg. *Theoretical Physics*, 3rd ed. New York, Hafner Publishing Company (n.d.)

Joubert, J[ules] (ed.). *Collection de mémoires relatifs à la physique*, Vols. 2 and 5: *Mémoires sur l'électrodynamique.* Paris:1885, 1887.

Knott, C. G. *Life and Scientific Work of Peter Guthrie Tait.* Cambridge: Cambridge University Press, 1911.

Lagrange, Joseph-Louis. *Méchanique analytique* (1811). 2 vols. Paris: Albert Blanchard, 1965. (Reprint of 4th ed. Paris: 1888–89.)

Lanczos, Cornelius. The *Variational Principles of Mechanics*. Toronto: University of Toronto Press, 1949.

Laplace, Pierre Simon. *Oeuvres de Laplace*. 7 vols. Vols. 1–7: *Traité de méchanique céleste* (1823). Paris: Gauthier-Villars, 1843–47. Translated by Nathaniel Bowditch (reprinted in *Celestial Mechanics*, 4 vols., Bronx, New York: Chelsea Publishing Co., 1966).

Larmor, Joseph. *Memoir and Correspondence of Sir George Gabriel Stokes*. 2 vols. Cambridge: Cambridge University Press, 1907.

________. (ed.). *Origins of Clerk Maxwell's Electric Ideas*. Cambridge: Cambridge University Press, 1937.

Launay, Louis de. *Correspondence du grand Ampère*. 3 vols. Paris: Gauthier-Villars, 1936–43.

________. *Le grand Ampère: d'après des documents inédits*. Paris: 1925.

Lindsay, Robert and Margenau, Henry. *Foundations of Physics*. 2nd ed. New York: Dover Publications, 1957.

Long, Robert N. *Mechanics of Solids and Fluids*. Englewood Cliffs, New Jersey: Prentice-Hall, 1961.

Lowe, Victor. *Understanding Whitehead*. Baltimore: The Johns Hopkins Press, 1962.

Lucretius, Carus. *De Rerum Natura*. Trans. W.H.D. Rouse. Cambridge, Massachusetts: Harvard University Press, 1947.

Mansel, H. L. *The Philosophy of the Conditioned*. London: 1866.

Martin, Thomas (ed.). *Faraday's Diary*. 7 vols, and index vol. London: G. Bell and Sons, 1932–36.

Matteucci, Carlo. "Mémoire sur la propagation de l'électricité dans les corps solides isolants," *Annales de chimie*, *27* (1849), pp. 133 ff.

________. "Mémoire sur la propagation de l'électricité dans les corps isolants, solides et gazeux," *Annales de chimie*, *28* (1850), pp. 385 ff.

________. "On the Laws of Magnetism and Diamagnetism, in a Letter to Dr. Faraday," *British Association Reports* (1852), Part 2, pp. 6–10.

________. *Cours spécial sur l'induction*. Paris: 1854.

Maxwell, James Clerk. (See under Primary Sources, above)

Merz, John Theodore. *A History of European Thought in the Nineteenth Century*. 1st ed. 1896–1914. 2nd. ed. (1904–12) reprinted: 4 vols. New York: Dover Publications, 1965.

Mill, John Stuart. *An Examination of Sir William Hamilton's Philosophy*. London: 1865.

Moon, Parry and Spencer, D. E. "Electromagnetism without Magnetism: an Historical Sketch," *American Journal of Physics*, *22* (1954), pp. 120 ff.

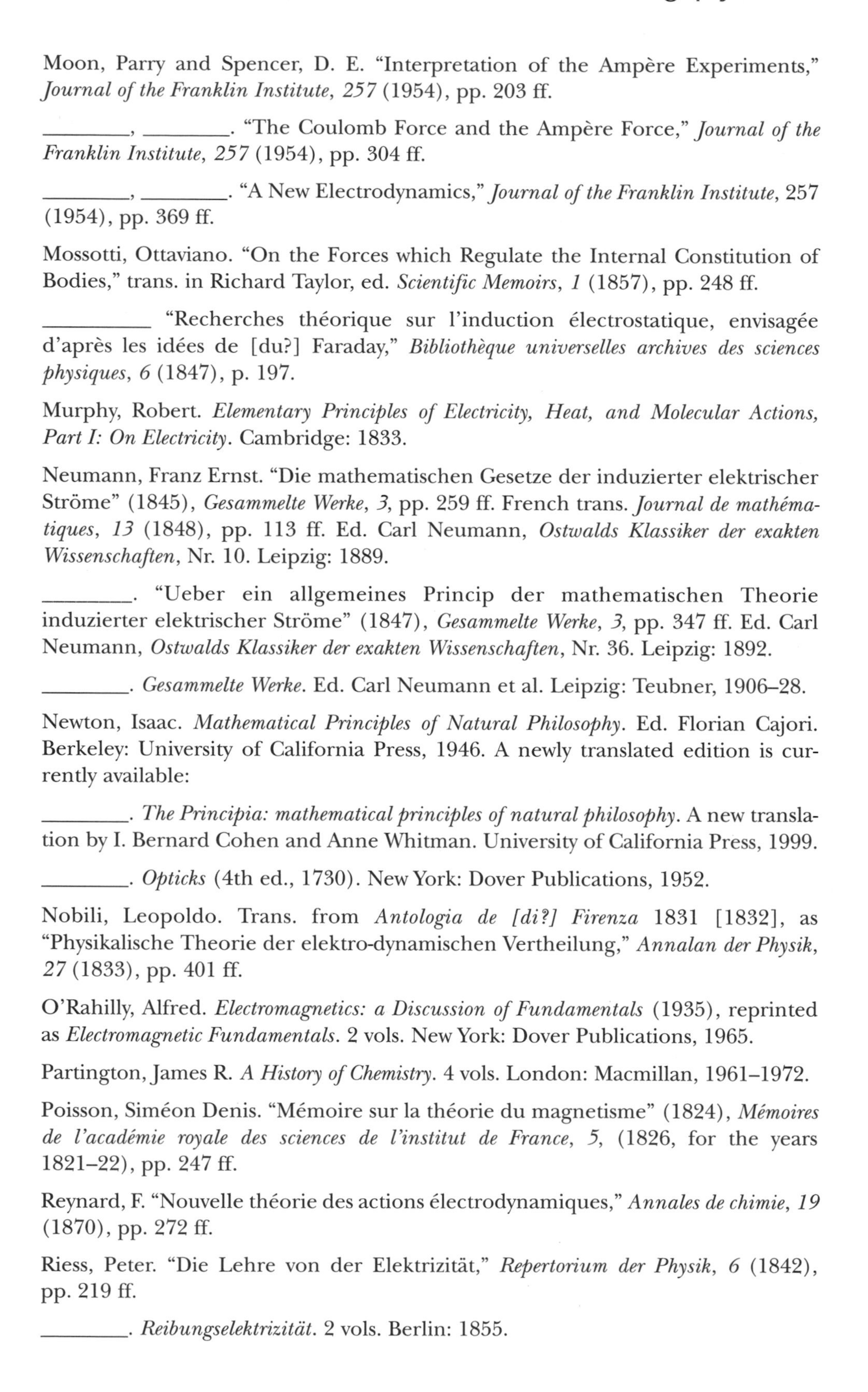

Moon, Parry and Spencer, D. E. "Interpretation of the Ampère Experiments," *Journal of the Franklin Institute, 257* (1954), pp. 203 ff.

________, ________. "The Coulomb Force and the Ampère Force," *Journal of the Franklin Institute, 257* (1954), pp. 304 ff.

________, ________. "A New Electrodynamics," *Journal of the Franklin Institute,* 257 (1954), pp. 369 ff.

Mossotti, Ottaviano. "On the Forces which Regulate the Internal Constitution of Bodies," trans. in Richard Taylor, ed. *Scientific Memoirs, 1* (1857), pp. 248 ff.

__________ "Recherches théorique sur l'induction électrostatique, envisagée d'après les idées de [du?] Faraday," *Bibliothèque universelles archives des sciences physiques, 6* (1847), p. 197.

Murphy, Robert. *Elementary Principles of Electricity, Heat, and Molecular Actions, Part I: On Electricity*. Cambridge: 1833.

Neumann, Franz Ernst. "Die mathematischen Gesetze der induzierter elektrischer Ströme" (1845), *Gesammelte Werke, 3,* pp. 259 ff. French trans. *Journal de mathématiques, 13* (1848), pp. 113 ff. Ed. Carl Neumann, *Ostwalds Klassiker der exakten Wissenschaften*, Nr. 10. Leipzig: 1889.

________. "Ueber ein allgemeines Princip der mathematischen Theorie induzierter elektrischer Ströme" (1847), *Gesammelte Werke, 3,* pp. 347 ff. Ed. Carl Neumann, *Ostwalds Klassiker der exakten Wissenschaften*, Nr. 36. Leipzig: 1892.

________. *Gesammelte Werke*. Ed. Carl Neumann et al. Leipzig: Teubner, 1906–28.

Newton, Isaac. *Mathematical Principles of Natural Philosophy*. Ed. Florian Cajori. Berkeley: University of California Press, 1946. A newly translated edition is currently available:

________. *The Principia: mathematical principles of natural philosophy*. A new translation by I. Bernard Cohen and Anne Whitman. University of California Press, 1999.

________. *Opticks* (4th ed., 1730). New York: Dover Publications, 1952.

Nobili, Leopoldo. Trans. from *Antologia de [di?] Firenza* 1831 [1832], as "Physikalische Theorie der elektro-dynamischen Vertheilung," *Annalan der Physik, 27* (1833), pp. 401 ff.

O'Rahilly, Alfred. *Electromagnetics: a Discussion of Fundamentals* (1935), reprinted as *Electromagnetic Fundamentals*. 2 vols. New York: Dover Publications, 1965.

Partington, James R. *A History of Chemistry*. 4 vols. London: Macmillan, 1961–1972.

Poisson, Siméon Denis. "Mémoire sur la théorie du magnetisme" (1824), *Mémoires de l'académie royale des sciences de l'institut de France, 5,* (1826, for the years 1821–22), pp. 247 ff.

Reynard, F. "Nouvelle théorie des actions électrodynamiques," *Annales de chimie, 19* (1870), pp. 272 ff.

Riess, Peter. "Die Lehre von der Elektrizität," *Repertorium der Physik, 6* (1842), pp. 219 ff.

________. *Reibungselektrizität*. 2 vols. Berlin: 1855.

Riess, Peter. "On the Action of Non-conducting Bodies in Electrical Conduction" (1854), *Philosophical Magazine, 9* (1855), pp. 401 ff.

Richardson, O. W. "A Mechanical Effect Accompanying Magnetism," *Physical Review, 26* (1908), pp. 248 ff.

Robison, John. *A System of Mechanical Philosophy*. 4 vols. Edinburgh: 1822.

Roget, Peter. *Treatise on Electricity, Galvanism, Magnetism, and Electromagnetism*. London: Society for the Diffusion of Useful Knowledge, 1832.

Routh, Edward J. *Elementary Treatise on the Dynamics of a System of Rigid Bodies* (1860). 2nd ed. Cambridge: 1868

Shairp, John Campbell, Tait, Peter Guthrie, and Adams-Reilly, A. *Life and Letters of James David Forbes*. London: 1875.

Simpson, Thomas K. "Maxwell and the Direct Experimental Test of his Electromagnetic Theory," *Isis, 57* (1966), pp. 423 ff.[1]

Smith-Rose, R. L. *James Clerk Maxwell*. London: Longmans, Green, and Co., 1948.

Spencer, J. Brookes. "Boscovich's Theory and its Relation to Faraday's Researches: an Analytical Approach," *Archive for History of Exact Sciences, 4* (1967), pp. 184 ff.

Stokes, George Gabriel. *Mathematical and Physical Papers*. 5 vols. Cambridge: 1880–85.

Tait, Peter Guthrie. "Quaternion Investigations Connected with Electrodynamics and Magnetism," *Quarterly Journal of Pure and Applied Mathematics, 3* (1860), pp. 331 ff.

________. *Sketch of Thermodynamics*. Edinburgh: 1868. There apparently existed an earlier, pamphlet version of this work: see page 230n, above.

Thompson, Silvanus P. *The Life of William Thomson, Baron Kelvin of Largs*. 2 vols. London: MacMillan and Co., 1910.

Thomson, Joseph John. "Report on Electrical Theories," *British Association Reports* (1885), pp. 97 ff.

________. "James Clerk Maxwell," in *James Clerk Maxwell: a Commemoration Volume*. Ed. J. J. Thomson et al. Cambridge: Cambridge University Press, 1931.

Thomson, William, Baron Kelvin. "Magnetism, Dynamical Relations of," in *Nichol's Cyclopedia of the Physical Sciences*. 2nd ed. London: 1860. (Compare page 204n, above.)

________ and Tait, Peter Guthrie. *A Treatise on Natural Philosophy*: *Vol. I*. Oxford: 1867. No further volumes ever appeared, but the first volume was enlarged in a second edition, in two parts, in 1878. The final edition, to which I have referred in this study except as otherwise noted, has been reprinted as: *Principles of Mechanics and*

1. Other writings by Thomas K. Simpson include: *Maxwell on the Electromagnetic Field* (Rutgers University Press, 1997), *Figures of Thought: A Literary Appreciation of Maxwell's* Treatise on Electricity and Magnetism (Green Lion Press, 2005), and *Newton, Maxwell, Marx: Reflections on the Idea of Science* (Green Lion Press, forthcoming).

Dynamics. 2 vols. New York: Dover Publications, 1962. Note the review of the second edition by Maxwell, page 330 above.

Thomson, William and Tait, Peter Guthrie. *Elements of Natural Philosophy*. Oxford: Clarendon Press, 1873. Note the review of this work by Maxwell, page 330 above.

(For other works by Thomson, see Primary Sources, above. References to the *Reprint of Papers on Electrostatics and Magnetism* are made in the text by a short-form notation which is explained at page xxiii of the Preface. Since thorough bibliographical information on Thomson's writings is readily available in the biography by S. P. Thompson, listed above, individual articles referred to in the text have not been listed again here.)

Todhunter, Isaac. *William Whewell: an Account of his Writings*. 2 vols. London: 1876. Note the review of this work by Maxwell, listed above.

Tolman, Richard C. and Stewart, T. D. "The Electromotive Force Produced by the Acceleration of Metals," *Physical Review, 8* (1916), pp. 97 ff.

________, Karrer, S., and Guernsey, E. W. "The Mass of the Electric Carrier in Copper, Silver, and Aluminium," *Physical Review, 9* (1917), pp. 164 ff.

________ and Mott-Smith, L. M. "Further Experiments on the Mass of the Electric Carrier in Metals," *Physical Review, 21* (1923), pp. 525 ff.

________, ________. "A Further Study of the Inertia of the Electric Carrier in Copper," *Physical Review, 28* (1926), pp. 794 ff.

Truesdell, Clifford. *The Kinematics of Vorticity*. Bloomington, Indiana: Indiana University Press, 1954.

________. "Rational Fluid Mechanics, 1687–1765," in: *Leonhardi Euleri Opera Omnia*. Ser. 2, Vol. 12. Lausanne: 1954, pp. ix ff.

Turner, Joseph. "The Methodology of James Clerk Maxwell." Ph.D. Thesis, Columbia University, 1953. University Microfilms No. 6725.

________. "Maxwell on the Method of Physical Analogy," *British Journal for the Philosophy of Science, 6* (1955–56), pp. 226ff.

Tyndall, John. *Researches on Diamagnetism and Magnecrystallic Action*. London: 1870.

________. *Faraday as a Discoverer*. New York: 1873.

Van Rees, Richard. "Ueber die Faradaische Theorie der Magnetischen Kraftlinein," *Annalen der Physik, 90* (1853), pp. 415 ff.

Watts, Isaac. *The Improvement of the Mind*. Washington, DC: 1813.

Weber, Wilhelm. "Elektrodynamische Maasbestimmungen über ein allgemeines Grundgesetz der elektrischen Wirkung" (1846–48), *Wilhelm Weber's Werke, 3*, pp. 25 ff. A partial English translation was given under the title "On the Measurement of Electrodynamic Forces," in: R. Taylor, ed. *Scientific Memoirs, 5* (1852), pp. 489 ff.

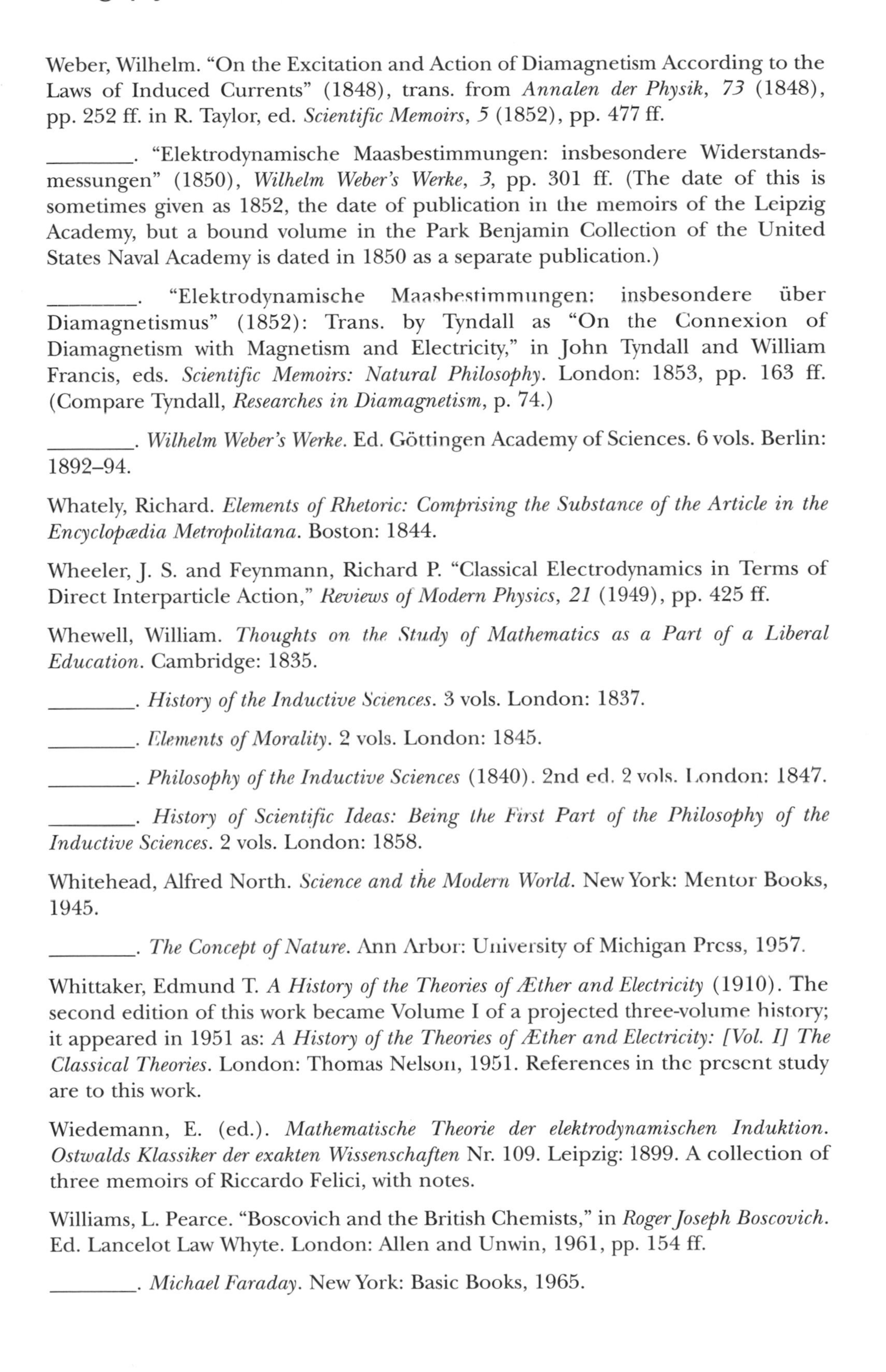

Weber, Wilhelm. "On the Excitation and Action of Diamagnetism According to the Laws of Induced Currents" (1848), trans. from *Annalen der Physik, 73* (1848), pp. 252 ff. in R. Taylor, ed. *Scientific Memoirs, 5* (1852), pp. 477 ff.

________. "Elektrodynamische Maasbestimmungen: insbesondere Widerstandsmessungen" (1850), *Wilhelm Weber's Werke, 3,* pp. 301 ff. (The date of this is sometimes given as 1852, the date of publication in the memoirs of the Leipzig Academy, but a bound volume in the Park Benjamin Collection of the United States Naval Academy is dated in 1850 as a separate publication.)

________. "Elektrodynamische Maasbestimmungen: insbesondere über Diamagnetismus" (1852): Trans. by Tyndall as "On the Connexion of Diamagnetism with Magnetism and Electricity," in John Tyndall and William Francis, eds. *Scientific Memoirs: Natural Philosophy.* London: 1853, pp. 163 ff. (Compare Tyndall, *Researches in Diamagnetism,* p. 74.)

________. *Wilhelm Weber's Werke.* Ed. Göttingen Academy of Sciences. 6 vols. Berlin: 1892–94.

Whately, Richard. *Elements of Rhetoric: Comprising the Substance of the Article in the Encyclopædia Metropolitana.* Boston: 1844.

Wheeler, J. S. and Feynmann, Richard P. "Classical Electrodynamics in Terms of Direct Interparticle Action," *Reviews of Modern Physics, 21* (1949), pp. 425 ff.

Whewell, William. *Thoughts on the Study of Mathematics as a Part of a Liberal Education.* Cambridge: 1835.

________. *History of the Inductive Sciences.* 3 vols. London: 1837.

________. *Elements of Morality.* 2 vols. London: 1845.

________. *Philosophy of the Inductive Sciences* (1840). 2nd ed. 2 vols. London: 1847.

________. *History of Scientific Ideas: Being the First Part of the Philosophy of the Inductive Sciences.* 2 vols. London: 1858.

Whitehead, Alfred North. *Science and the Modern World.* New York: Mentor Books, 1945.

________. *The Concept of Nature.* Ann Arbor: University of Michigan Press, 1957.

Whittaker, Edmund T. *A History of the Theories of Æther and Electricity* (1910). The second edition of this work became Volume I of a projected three-volume history; it appeared in 1951 as: *A History of the Theories of Æther and Electricity: [Vol. I] The Classical Theories.* London: Thomas Nelson, 1951. References in the present study are to this work.

Wiedemann, E. (ed.). *Mathematische Theorie der elektrodynamischen Induktion. Ostwalds Klassiker der exakten Wissenschaften* Nr. 109. Leipzig: 1899. A collection of three memoirs of Riccardo Felici, with notes.

Williams, L. Pearce. "Boscovich and the British Chemists," in *Roger Joseph Boscovich.* Ed. Lancelot Law Whyte. London: Allen and Unwin, 1961, pp. 154 ff.

________. *Michael Faraday.* New York: Basic Books, 1965.

Index

D

E

H

I

K

L

M

PHOTO: ERIC SIMPSON

ABOUT THE AUTHOR

Thomas King Simpson is Tutor Emeritus at St. John's College in Annapolis, Maryland and Santa Fe, New Mexico. He has taught at the American University at Cairo and is a co-founder of The Key School in Annapolis, Maryland. He was educated at Rensselaer Polytechnic Institute, St. John's College, Wesleyan University, and the Johns Hopkins University, where in 1968 he received his doctorate in the history of science. His background also includes engineering and the classics.

Simpson has written extensively on both Michael Faraday and James Clerk Maxwell. His wide range of interests, suggestive perhaps of those of Maxwell himself, extends also to schools, museums, and in general towards broadening the role of what George Elder Davie has called "the democratic intellect."

Other books by Simpson include *Maxwell on the Electromagnetic Field* (Rutgers University Press, 1997) and *Figures of Thought: A Literary Appreciation of Maxwell's* Treatise on Electricity and Magnetism (Green Lion Press, 2005). He has also written numerous articles for Encyclopædia Britannica's *The Great Ideas Today*; three of those articles are being reissued by Green Lion Press as *Newton, Maxwell, Marx: Reflections on the Idea of Science.*